Back on Track

HAGLEY LIBRARY STUDIES IN BUSINESS, TECHNOLOGY, AND POLITICS

Richard R. John and Erik Rau, *Series Editors*

Back on Track

American Railroad Accidents and Safety, 1965–2015

MARK ALDRICH

Johns Hopkins University Press Baltimore

Printed in the United States of America on acid-free paper
9 8 7 6 5 4 3 2 1

Johns Hopkins University Press
2715 North Charles Street
Baltimore, Maryland 21218-4363
www.press.jhu.edu

Library of Congress Cataloging-in-Publication Data

Names: Aldrich, Mark, author.
Title: Back on track : American railroad accidents and safety, 1965–2015 / Mark Aldrich.
Description: Baltimore, Maryland : Johns Hopkins University Press, 2018. | Series: Hagley library studies in business, technology, and politics | Includes bibliographical references and index.
Identifiers: LCCN 2017012954 | ISBN 9781421424156 (hardcover : alk. paper) | ISBN 1421424150 (hardcover : alk. paper) | ISBN 9781421424163 (electronic) | ISBN 1421424169 (electronic)
Subjects: LCSH: Railroad accidents—United States—History. | Railroad accidents—United States—Prevention. | Railroads—Safety regulations—United States. | Railroads—United States—Safety measures.
Classification: LCC HE1780 .A75 2018 | DDC 363.12/2097309045—dc23
LC record available at https://lccn.loc.gov/2017012954

A catalog record for this book is available from the British Library.

Special discounts are available for bulk purchases of this book. For more information, please contact Special Sales at 410-516-6936 or specialsales@press.jhu.edu.

Johns Hopkins University Press uses environmentally friendly book materials, including recycled text paper that is composed of at least 30 percent post-consumer waste, whenever possible.

For Michele Aldrich, 1942–2016, once again, for everything

My heart is warm with friends I make,
And better friends I'll not be knowing;
Yet there isn't a train I wouldn't take,
No matter where it's going.

Edna St. Vincent Millay, Second April

Contents

Figures

Tables

Preface

This is a book about the safety of modern American railroads. There is certainly no shortage of books on railroads. Indeed, the library shelves creak under their weight. But while the literature on American railroad history is vast, it also reveals some curious gaps. There is little writing on the modern period in comparison with the ocean of work on earlier eras, and none of these modern works deals with safety in any detail. The general literature on health and safety is also vast, but most of it focuses on the environment, or drugs, or other consumer products, or the Occupational Safety and Health Administration (OSHA). There are extensive discussions of railroad economic regulation and deregulation, but the corresponding literature discussing their impact on safety is far more modest. In short, as I discovered when I researched *Death Rode the Rails* (2006), except for a few scattered articles, there was not much historical literature on any aspect of railroad safety and that remains true to this writing.

Some of the motivation to write this book has been my continued fascination with railroads which, like all addictions, is difficult to explain to someone who doesn't share it. Like lots of small boys and girls in times past, I liked to watch trains. As a teenager I joined the ranks of trespassers, walking the tracks of the Boston & Maine, with a dog and a shotgun, hunting pheasant, jumping off just before the train arrived and waving at the engine driver, who eventually got used to me. As an adult, a historian, and an economist, I continue to be intrigued by railroads because of their historical significance, their complexity, and their brute efficiency. A mile-long freight of 100-ton cars traveling 70 mph crewed by only 2 people, backed by an immense, complicated technical and organizational system, is an impressive human achievement. My interest in safety and its regulation originated with a brief stint working for OSHA about 1980. A natural extension has been to inquire into railroad safety—an interest reinforced each time I ride a commuter line.

In *Death Rode the Rails* I told the story down to the end of Interstate Commerce Commission (ICC) safety regulation in 1965. That seemed like a good place to stop, for with the formation of the Department of Transportation (DOT) and the Federal Railroad Administration (FRA) about that time, the story changed radically. Yet in other ways, stopping in 1965 is misleading, for the carriers' safety was then beginning to crumble, while today, the railroads' record is in most ways exemplary. If we are shocked when a train of North Dakota crude oil blows up, as happened at Lac-Mégantic in 2013, it is because we have forgotten how common such events were during the 1970s. Thus, the past 50 years, in which railroad safety worsened for the first time in decades and then once again began a dramatic improvement, constitutes an important and mostly ignored story that is well worth remembering and recounting.

While the focus of this book is railroad safety from 1965 to 2015, I have not been compulsive about either date. For comparative purposes I often delve into the years between World War II and 1965—and indeed the book's first chapter provides background well into the nineteenth century. Similarly, if relevant events have occurred after 2015, I have sometimes incorporated them into the analysis too.

The railroads that are the subject here are those regulated by the FRA. These include the freight railroads, Amtrak, the Alaska Railroad, and commuter lines connected to the network, but not subways and light rail in stand-alone systems, although much of the analysis is germane to safety on those lines as well. I also omit the tourist railroads and because of inadequate information there is less on the smaller (Class II and Class III) carriers than I would like.*

This is a work of economics as well as history, for despite occasional protestations to the contrary, safety and accidents always involve economic choices. Thus, one set of questions that carries through the book is how well the private economy does in providing safety—when it fails and why, and what a powerful engine for better safety it can sometimes be. However, this is a work of history too, and if critics sometimes parody economists as individuals who believe the world is perfectly rational and businesses are models of efficiency, the reader will find little of that here. Like all managers, those of modern railroads operate under a fog of uncertainty. They make mistakes as do their underlings, and efficiency is an eternal quest, not a state of being. I have tried to be faithful to these realities.

I employ such economic ideas as safety incentives, risk trade-offs, transaction costs, externalities, information availability, and principal-agent problems. These are important, useful concepts, but they and the others herein are in fact quite simple and should scare no one away. There are also tables and graphs embedded in the narrative, and there is some statistical analysis, but I have locked it up in footnotes and an appendix. I

* As of 2007, Class I carriers are those with revenues of $250 million or more; Class II carriers are those with revenues less than $250 million but more than $20 million; while Class III carriers have revenues less than $20 million. See 49 CFR Part 1201, General Instructions 1-1.

have also left out most jargon and while economic and business history is sometimes derided as being wholly without people, I have tried not to lose sight of the real men and women who are the subject of the story.

If economics has powerfully shaped railroad safety, so have public policies and technological and organizational changes. Their collective interplay constitutes a theme in what I hope is an interdisciplinary narrative. Any discussion of technology requires an understanding of the nuts and bolts—why and how things work—and a context that shapes its workings and I have tried to provide both. Railroad enthusiasts, many of whom have worked for the carriers, are famous for their technological expertise and while I have endeavored to ensure that my descriptions of complexities are correct, for the sake of the narrative I have often simplified matters considerably. I have similarly abbreviated descriptions of accidents, the causation of which can sometimes be mind-bendingly complicated. While I expect critics to call out errors, I hope they will forgive my simplifications.

Historians insist that technology is socially constructed; economics is one of those shaping forces. Expectations of rising accident costs, for example, have induced innovation to develop safer maintenance of way equipment. Organizational innovations also respond to economic incentives and railroad work reveals many experiments with new management ideas and methods to reduce employee accidents. Regulatory policies and their evolution have been central to modern railroad safety. Thus, if one set of questions revolves around how well private markets supply safety, another focuses on how public policies have shaped safety outcomes. Inevitably, therefore, this study is not only interdisciplinary but a venture in political economy as well.

In recent years company safety departments have metamorphosed into departments of health, safety, and the environment, but for several reasons I have resisted the urge to broaden this narrative in similar fashion. Except for noise and perhaps exposure to diesel exhaust, the health problems associated with modern railroad work seem comparatively modest. In contrast, environmental difficulties are potentially vast—every old coal yard and creosote works might be a potential Superfund site—and different in kind from safety problems. Hence, I leave both health and the environmental issues to other scholars. I also ignore matters of railroad security for it is an entirely separate topic. There is as well no discussion of matters such as what to do about Amtrak or high-speed rail, for they too are tangential to the topic of the book. I have also ignored OSHA's effects on rail worker safety.

Two points of focus and style need discussion. Statistics, as historians remind us, are also a social construct. From the earliest days, the carriers, their unions, and regulators have fought over what counts as accidents or injuries and how to count them. Inflation, for example, can confound the counting of train accidents. In addition, while statistics can sometimes depoliticize issues, they can inflame debate as well. For years, the ICC and the FRA employed the question-begging term "negligence" as a cause of train accidents. Far from neutralizing disagreements, such terminology polarized debate. Moreover, the rail unions have recently claimed that company pressures lead to large-scale injury underreporting, calling into question the entire process of safety measurement, while changes in injury definitions can result in dramatic changes in the numbers. Even data on such apparently straightforward injuries as fractures and amputations sometimes exhibit strange statistical behavior, while a change in definition reduced recorded passenger fatalities in 1958. Still, because fatalities are less subject to these foibles I have largely relied on such data to chart safety trends.

As a matter of style, in previous work I employed masculine terminology to discuss workers, for railroading was then overwhelmingly male-dominated. This is less true now and it seems appropriate that writing should reflect the new realities. Thus, I eschew terms such as "man hours" and "engineman" (unless the individual discussed is a man) and employ instead "employee hours" and "engine driver." The term "engineer" is reserved for an individual technically trained in those disciplines.

No one writes a book without a lot of help and my debts are many. Smith College is more generous than many institutions in providing emeritus faculty office space, technical support, and finance for research trips, and I am deeply appreciative of its assistance. Smith's interlibrary loan librarians—Susan Daily and Chris Ryan—tirelessly ordered materials they must have thought no sane person could want, while reference librarians—especially Sika Berger and Bruce Sajdak—helped with many bibliographical issues. Smith's technical support people, most notably Pat Billingsley, Liane Hartman, Bill Krieger, and Steve Stander, kept my computer running. At Cornell, I have benefited enormously from the help of the librarians at the Catherwood Library and the Kheel Center. The response of Cammie Wyckoff and her staff at the Cornell Library Annex to my many requests was above and beyond the call of duty and John Howard deserves special thanks for his help. Constance Carter speeded and improved my research efforts at the Library of Congress, while Tab Lewis and David Pfeiffer provided equally valuable help at National Archives.

I also want to thank those individuals and institutions who generously allowed me to use their materials. They include John Previsich, president of the Transportation Division of the Sheet Metal, Air, Rail and Transportation Workers Union; William Vantuono, editor of *Railway Age*; Kaitlyn Wilson, manager, Corporate Social Responsibility at CSX Transportation; Dennis Pierce, national president of the Brotherhood of Locomotive Engineers and Trainmen; Clark Ballew, associate editor of the *Journal* of the Brotherhood of Maintenance of Way Employees, Division of the International Brotherhood of Teamsters. Thanks too to David Tyrell, Technical Advisor for Train Crashworthiness and Safety at the Volpe Center, and Luis Maal, program manager at the Transportation Technology Center. I greatly appreciate the efforts of Amy Vedra of the Indiana Historical Society and Lauren Baker of the Indiana State Police Museum for helping track down images of the Dunreith, Indiana, hazmat disaster. Officer Joe Pitcher, retired from the Knightstown, Indiana Police Department, and Officers John C. Linegar and Mike Smith, retired from the Indiana State Police, were kind enough to share with me their experiences as first responders at Dunreith. To Eric Cox, editor of the Knightstown, Indiana *Banner*, my deepest appreciation for his willingness to share the newspaper's images of that tragedy. Thanks too to the Bancroft Library, Hagley Library and King Features, Syndicate. If I have forgotten anyone or any institution I hope they will forgive me.

George Bibel patiently answered some of my questions on mechanical engineering. My thanks to John K. Brown, who read and greatly improved Chapter 3, and to an anonymous critic who provided the same valuable services for the entire book. Bob Buchele assisted with statistics. None of them is responsible for any remaining errors.

In researching this book I have benefited from the work of those railroad historians who discuss post–World War II developments. These include a number of Roger Grant's valuable corporate histories. Maury Klein's superb biography of the Union Pacific traces that company's history into the twenty-first century. Don Hofsommer's *Southern Pacific* ends in 1985, while his *Grand Trunk Corporation* tells that company's story from the 1970s to the 1990s. Robert Gallamore and John Meyer's *American Railroads* is almost entirely on the post–World War II years. James Ely's *Railroads and American Law* covers the recent past, and Richard Saunders's works on rail mergers discuss these events well into modern times. For much of the broader history of railroads, I have relied on the work of these authors as well as others too numerous to mention.

Railroad safety is an aspect of business and economic history and a number of writers in these fields have influenced my thinking. These include Joseph Schumpeter, Alfred Chandler, and several modern scholars, including Naomi Lamoreaux, Richard Langlois, Daniel Raff, Peter Temin, and others who have tried to extend and modernize Chandler's work. Louis Galambos's revision of his organizational synthesis also provides insights that are helpful in placing railroad safety in a wider context. My understanding of safety issues also draws on the risk and risk perception literature by Jeanne and Roger Kasperson, Daniel Kahneman, Paul Slovic, Cass Sunstein, Amos Tversky, Kip Viscusi, and many others. The economic analysis of modern railroad safety by Ian Savage has been helpful while his econometric work and that of several other researchers helps shore up the argument in a number of places. Finally, my greatest debt is to my wife, Michele, who died this past year. She was a distinguished historian of science, an archivist, and a superb proofreader and critic of my work and that of many others as well. By long tradition, errors of fact or interpretation are entirely on my head.

Back on Track

Introduction

About 2:39 in the morning, on January 6, 2005, an improperly lined switch diverted Norfolk Southern freight Train 192 onto an industry track near Graniteville, South Carolina, where it collided with a parked train and derailed. The wreck ruptured a tank car of chlorine, which resulted in the death of nine people and caused the evacuation of everyone within a 1-mile radius. A bit over three years later, at 4:22 in the afternoon of September 12, 2008, Southern California Regional Rail Authority Metrolink Train 111 collided head-on with a Union Pacific freight near Chatsworth, California. The collision caused the locomotive of Train 111 to telescope 52 feet of the lead passenger coach, killing 25 people and injuring another 102. Both train wrecks made national headlines. By contrast, an accident the day before Graniteville, on January 5, 2005, in which a Union Pacific freight struck and killed a maintenance of way employee near Laramie, Wyoming, excited little interest. Nor was it particularly newsworthy when a Norfolk Southern freight hit a pickup on January 4, 2009, at a crossing marked only by crossbucks near Jackson, Georgia, killing its driver.[1]

The wrecks of Trains 192 and 111 made big news in part because they were so rare. The carriers once killed passengers by the hundreds each year and hardly a month went by without disasters like Metrolink being splashed over the front page of newspapers. Freight wrecks are also spectacular—especially when they involve hazardous materials (hazmat)—and they too are rare these days. One might suppose that this decline in train accidents simply reflects the general decline in railroading, for it is true that railroads are no longer central to the lives of Americans as they were two or three generations ago. While some of us ride commuter lines like Metrolink to work, many more of us drive, and for long-distance travel we are far more likely to consult Southwest's website than that of Amtrak. Yet railroads have not declined; they have merely grown more slowly than other forms of transport; in fact, freight traffic has reached unprecedented levels, while passenger volume has grown steadily since the 1980s.

Train accidents have always caught the public's attention and they have often propelled public policy. Accordingly, I provide lists of important accidents in Appendices 1 and 3. But what I have termed "little accidents," such as the death of the Union Pacific employee, have never been newsworthy. Their decline too, although largely unreported, has been spectacular. Railroads were once the largest source of industrial work—employing about two million men and women, or nearly 5 percent of the labor force at the time of World War I. Today, fewer than 200,000 make their living working on the railroad—a fraction of those employed at Walmart. That the carriers no longer kill thousands of workers each year is partly a result of this reduction in railroad employment. Nevertheless, risks to remaining employees have fallen dramatically as well. Even the deadly grade crossing has become more benign. There are fewer crossings and more and better guards. Accordingly, accidents at unguarded crossings like that in Jackson, Georgia, have receded even as rail and road traffic has risen.

Modern Railroad Safety in Historical Context

The decline of the risks of railroading is part of a much broader set of events that have steadily improved our life expectancy and well-being. All forms of travel are far safer than they were a generation ago. Death on the job is shocking now, and it once was not. Infectious diseases have receded—although they may be waiting to stage a comeback—and even cancer rates are falling. We are safer and healthier now than the men and women of a century ago, and even since World War II the gains have been dramatic. As the political scientist Aaron Wildavsky has summarized these events: "richer is safer." Yet historians remind us that safety and health are contingent; there is a connection between dollars and death rates but it is by no means straightforward. Thus, railroads did not become as safe as they are today in straight-line fashion. Rather, there have been two long periods when railroad safety improved, separated by a wilderness in which the carriers and their safety fell apart. The broad themes that shape the carriers' destiny throughout these three periods all involve in various measure the roles of incentives, institutional innovations, technological changes, resources, and regulations.[2]

It is useful to begin that story in the middle, with the carriers in the wilderness, and then discuss the first period of safety gains, for it provides the necessary background to modern times (and I treat it in more detail in Chapter 1). In 1958, James Symes, president of the Pennsylvania Railroad, testified before the Senate Committee on Commerce. "We are falling apart," he bluntly informed the committee. It was a stunning admission. Fifty years before, the Pennsylvania had called itself the "Standard Railroad of the World," and that was not simply advertising puffery. It was the largest railroad in the world. It had the best track, well signaled and maintained to high standards; it had led the railroads into steel passenger cars after 1900 and into electrification in the 1930s. Its named trains were famous. But by 1958, it was running on fumes and when Symes testified, he could have been speaking for all the eastern carriers. The Erie, the Lackawanna, the Lehigh Valley, the New York Central, the Boston &

Maine, the Reading, the Jersey Central, and, of course, the New Haven all faced crumbling finances and deteriorating physical plants. Even the southern and western carriers were feeling the pinch. However, if the railroads were falling apart when Symes spoke to Congress, their safety record had not yet raised any red flags. Passenger and worker safety had been improving for decades and even grade crossing accidents were declining. All that was about to change, however, for if richer was safer, poorer was about to become a lot more dangerous.[3]

A half century before Symes testified, about 1905, the railroads had been rich, arrogant, and powerful. They had dominated transportation. The economist Joseph Schumpeter described a kind of disruptive competition that resulted from innovations of new products and processes that swept away older goods and ways of doing things. The railroads were the classic example of a Schumpeterian innovation that was both enormously creative and simultaneously destructive. As one writer put it, "everywhere the railroad went it created something, as if by magic." Yet railroads also destroyed canals and stagecoaches, and for about 80 years they faced no significant land competition. Everything Americans bought arrived by railroad. Honeymooners took the train to Niagara Falls and wealthy New Yorkers traveled by Pullman, on business or to summer in the Adirondacks or the Berkshires. Railroads captured the imagination as well, and they changed our language—get the red carpet treatment; don't get sidetracked; put it on the fast track; she was all steamed up—and many more metaphors.[4]

However, the American love affair with railroads was a stormy one because in popular telling the railroad became the octopus, ruining farmers and small businesses, looting shippers and travelers, corrupting politicians. Railroad leaders—E. H. Harriman, the Vanderbilts, J. P. Morgan—symbolized the robber barons as grasping plutocrats. Railroad accident metaphors also entered the language—don't get derailed; he was asleep at the switch—and they were the source of popular songs such as "Casey Jones" and "The Wreck of Old '97." Railroads were also responsible for many an epitaph: "off the track the engine rushed—some were drowned, and I was crushed."[5]

As train accidents jumped with a traffic boom after 1900, popular outcry rose accordingly, fanned by sensationalist journalism. Safety advocates usually need both victims and villains—a virtuous "us" and an evil "them." The victims, of course, were those injured in accidents and the villains were the railroad leaders. In popular telling, train accidents were yet another result of the robber barons' greed. Indeed, in many a headline, accidents were simply "railroad murders." By inference, profits were extracted at the expense of safety—a position that remains popular to this day. It is true that companies can trade off safety and profitability in the short run—they can postpone a safety investment in favor of a higher dividend—but over the longer run, as is argued throughout this book, they typically come as a package.[6]

Wealth, power, and unpopularity in a democracy make an unstable mix. Congress responded to public concerns in 1906, giving the Interstate Commerce Commission (ICC) power over railroad rates. Federal safety regulation had begun even earlier: in response to popular outcry, Congress mandated air brakes and semi-automatic couplers for freight trains in the 1890s. Federal hours of service regulations, requirements for accident reporting and investigation, and locomotive inspection all arrived by 1910, and more was threatened. To turn away popular wrath, the carriers responded with a concerted Safety First campaign. It worked: accidents and injuries began to decline and after about 1910 there was little more federal safety regulation. This arrangement was what I have termed "voluntarism." It involved an implicit contract: safety would be largely left to the railroads, remaining the domain of private enterprise and market forces, but only as long as the carriers continued to deliver the goods. The government's role was to allow the carriers high enough profits to provide the funds and incentives for safe railroading. Voluntarism was well suited to an era of small government, and for a time it worked, as is briefly depicted in Chapter 1. Worker and passenger safety steadily improved down to the 1950s, and with the aid of state and modest federal funding, even grade crossing accidents declined.

As Schumpeter understood, however, innovations and creative destruction are the essence of capitalism, and a regulated industry unable to adapt and compete with the new ways will eventually get into trouble. In the years after 1920, federal regulation had spread to include exit, entry, and much else, and it hamstrung the carriers while new competition from automobiles, trucks, barges, pipelines, and airlines increasingly stole their markets and eroded their profits, thus diminishing the incentives and resources available for safety. Because safety and profitability typically come as a package, as profits disappeared, companies' physical plants crumbled, morale deteriorated, and safety worsened. The package, in short, fell apart. In 1958, when Symes spoke to Congress, railroad safety had been improving for a half century. In the subsequent two decades, derailments would skyrocket and worker safety would deteriorate. Voluntarism would be scrapped by 1970, to be followed by economic deregulation in 1980 and a new period of safety progress.

A Change in Regimes

As the carriers' prominence and prosperity gradually declined after 1920, they also faded from the headlines and from public consciousness. There is not much railroad poetry anymore, and while my parents' generation listened to Hank Snow's "big eight wheeler coming down the track," I switched the station to Chuck Berry's "Maybellene." In disaster songs

"Casey Jones" got sidetracked by "Tell Laura I Love Her." What brought the carriers back on center stage again was the upsurge in train accidents that began in the late 1950s.[7]

"You can never exaggerate the impact of a railroad disaster in the courts of government policy and public opinion," noted an editor of *Trains*. The railroad wrecks that excite public interest have typically been passenger disasters like Metrolink, but passenger safety was not the reason railroad accidents reemerged as a public concern. Rather, it was a series of spectacular freight wrecks such as that at Graniteville that spilled a stew of hazmat, resulting in fires, explosions, and mass evacuations. Such accidents excited public concern out of proportion to their lethality, while blaring news headlines and television stories amplified public worries. Although hazmat risks had existed since the early days of railroading, their political importance in the 1960s was new. A major theme of this book is that more than any other single issue they have shaped and continue to shape modern concerns with railroad safety.[8]

In response to these disasters, between 1965 and 1970 the federal government constructed a new regime in railroad safety, with nearly every aspect of railroading now a public responsibility. Robert Higgs has proposed that crises are the major cause of government expansion, and Higgs's model surely fits railroad safety regulation, for a new regime began to materialize when Congress created the Federal Railroad Administration (FRA) in 1966 and transferred safety regulation to it from the ICC. That same year saw the birth of the National Transportation Safety Board (NTSB) as well and both were important institutional innovations. Neither organization immediately set out to scrap voluntarism, but they provided far more searching accident reports than had the ICC. The NTSB especially publicized its findings and called for change.[9]

As many observers have noted, the 1960s marked a major change in public attitudes toward the role of government. Indeed, so great have been the changes that one text describes them as a new policymaking system. Higgs points out that the earlier crisis of the 1930s prepared the way for these events as it knocked down the ideological barriers to expansion of government. The burden of proof had once been on those who wished to increase the functions of the federal government. Increasingly, however, for the men and women who experienced the 1930s or grew up in their shadow, the shoe came to be on the other foot. While these days robber baron imagery of heartless capitalists sacrificing helpless victims on the altar of mammon is largely confined to labor journals—in 2006, one publication claimed railroads "gamble with lives of workers to fatten bankrolls"—the notion that public safety cannot be trusted to private market incentives has become ubiquitous. Both the FRA and the NTSB regularly pointed out that the rising tide of wrecks resulted from failures over which the government had no control. Congress responded, and the

Federal Railroad Safety Act of 1970 changed all that, providing the FRA broad powers over all aspects of railroad safety similar to those that the Federal Aviation Administration (FAA) and its predecessors had been enforcing on airlines since the 1920s. The act thus marked the transition from voluntarism to modern times.[10]

A second regime change—this one in economic regulation—also transformed railroading during these years. As noted, voluntarism collapsed in part because the railroads could no longer deliver better safety. Ironically, therefore, government economic regulation of railroads was a major cause generating the conditions justifying an expansion of government safety regulation of railroads. Yet, as Congress discovered in the 1970s, stricter safety regulation alone could not solve the accident problem, for some aspects of railroad safety continued to deteriorate throughout much of that decade. Thus, while Higgs proposed that crisis expands government, crisis was also what finally motivated economic deregulation of the railroads.

It is tempting to date the return to safety progress that began about 1980 with the Staggers Rail Act, which led to partial economic deregulation and transformed railroading in many ways. It encouraged end-to-end mergers, led to contracting out (vertical disintegration), and yielded much institutional innovation. However, the beginnings of better safety lie in an earlier period. They resulted from the dawning realization by some railroad managers that in an increasingly competitive world, their salvation lay in better marketing and service. For a host of reasons, better service required better safety—fewer wrecks and fewer work accidents. These struggles by a group of carriers to transform organizations in order to compete within the regulatory straitjacket began the modern safety turnaround. The liability revolution has given safety a push as well, for it has made injuries to workers, passengers, and anyone else harmed by a railroad extraordinarily expensive. Economic deregulation as codified by the Railroad Revitalization and Regulatory Reform Act ("Four-R Act") of 1976 and the Staggers Rail Act of 1980 provided the flexibility to compete and the incentives and profitability to ensure the eventual spread and success of these beginnings. Thus, another theme of this book is that the turnaround in railroad safety resulted less from federal regulation than from market incentives, and that while economic deregulation was central to the safety gains, it did not originate them.

If the beginnings of the safety transformation of the 1970s and its subsequent driving force lay in the private sector, public policies have shaped these developments in many ways—including for the first time since the post–Civil War years the provision of resources for renovation. Federal money rebuilt Conrail, much of the Northeast Corridor, and some other lines as well, while governments routinely subsidize Amtrak and commuter lines. A large infusion of federal funds for grade crossing safety has generated major reductions in accidents and casualties. After 1970, federal support for research also began to reshape railroading. Until then, railroad research came largely from equipment suppliers or involved small-scale projects initiated by the carriers and the Association of American Railroads (AAR). Federal efforts began with the Federal Railroad Safety Act of 1970, and while the sums were modest by defense-industry standards, they were large in comparison to previous private-sector railroad research. Federal efforts also midwifed important new transportation research intuitions. At about the same time, the AAR sharply increased research funding—in response to worsening safety problems and perhaps as well to ensure that federal efforts did not dominate the future shape of technology. Drawing on research and technology generated elsewhere as well, these efforts have paid off in myriad ways.

Company and federal research built on the economy-wide digital and communications revolution to yield a rapid rate of technological innovation that has sharply improved both productivity and safety. Information, communication, and control have always been central to railroad safety. The consequences of their increasing sophistication and drastic cheapening constitute another important theme. The new methods have also yielded continuing learning by doing, resulting in a host of incremental improvements while there has also been much complementary investment in enhanced worker training. The direction of research and technological change has also reflected the new economics of deregulated railroading. Pressures for lower costs and improved service have induced the innovation and diffusion of a host of remote sensors and testing devices, employing ultrasound, infrared, laser, and other technologies that have been integrated into computer networks, allowing instantaneous access to safety information.

Sharper competition has shaped the direction of technology in other ways as well. The railroads have increasingly blended industrial engineering and economic research in ways that enhanced the productivity of safety investments—for example, by determining the proper spacing of remote sensors, the use of premium rail, and many other engineering-economic issues. Moreover, as railroads shed marginal track, the increase in traffic density required faster, more mechanized maintenance equipment—an imperative reinforced by rising wages. Rising traffic density has made track tie-ups from accidents increasingly expensive, encouraging safer machine design and improved worker training as well as studies of fatigue and other human factors that shape safety.

Federal safety regulations have come to encompass nearly every aspect of railroading, but the following chapters reveal few instances where they have had dramatic effects. The most likely reason is that train wrecks, grade crossing accidents, and worker injuries are expensive, and their high cost has been a powerful stimulant to better safety. Regulations, therefore, usually reinforce what the dictates of self-interest encourage the carriers to do anyway. Good safety practice requires redundancy, however, and modern regulations provide a kind of institutional backup, with regulators looking over the shoulders of management. This does not mean that regulation has been irrelevant, but neither has it been the primary driver in the safety gains since 1970.

A striking feature of FRA safety regulation has been how little controversy it has stirred until recently—at least compared to Occupational Safety and Health Administration (OSHA) health and safety regulations or Environmental Protection Agency (EPA) environmental rules—which may explain why it has been less well studied. From the beginning, FRA regulation has been more attuned to consensus and cooperation than that of other agencies—more "European," as Stephen Kelman and David Vogel describe Sweden and Great Britain—and rules that reinforce self-interest are not likely to be controversial. In addition, ex-railroaders initially staffed the agency, and it proved a good listener, willing to tweak regulations that pinched too sharply. In addition, critics—especially the carriers' unions—have charged that enforcement has been lax. Certainly trade journals reveal no chorus of complaints complete with rhetoric over mindless, rigid rule enforcement as resulted from early OSHA activities. Finally, the ability of FRA to dispense funds for railroad research has surely helped mute the carriers' criticism of that agency. Yet if most regulation has reinforced self-interest, recent congressional mandates for positive train control (PTC) as well as regulations mandating electro-pneumatic brakes and locomotive crew size are exceptions. The recent round of crude oil derailments motivated these new rules; they are contentious and the carriers have claimed they are poor safety investments even as they add to paperwork burdens.[11]

Another theme of the following chapters stresses the role that public policies other than regulations have played in improving safety. In addition to its support of research, the FRA also performs a number of what—following Mancur Olson—I term "market-augmenting activities" that facilitate safety work of the private sector by spreading information, improving trust, and reducing transaction costs. The agency carries out detailed safety inspections of carriers that amount to an overall performance review. In various ways it has tried to break down the wall of mistrust that separated the carriers and their unions and often impeded joint action to improve safety. The FRA has developed joint labor-management advisory committees to study particular types of accidents and draft new rules, and it has innovated a Confidential Close Call Reporting System (C3RS) to encourage reporting and learning from near-accidents without disciplinary threats.[12]

The improvement of railroad safety also reflects innovations in the understanding of accident and injury causation in the past half century—not only on the railroads but in other industries and in academic discourse as well. On the carriers, safety work had initially been largely the province of safety departments, which emphasized unsafe acts and exhorted "negligent" employees to work more safely, employing contests, posters, and discipline. Such an approach did not encourage a wider focus on equipment design and it concentrated on worker mistakes without inquiring too deeply into the causes of those mistakes. Symptomatic of a newer approach was the decision by the FRA in 1973 to substitute the term "human factors" for "negligence" as a cause classification for train accidents, which encouraged a deeper probing into the human causes of accidents. Gradually, behaviorism, which emphasizes positive reinforcement rather than discipline, has begun to influence railroad approaches to safety.

Studies of human behavior have also appeared that focus on fatigue, alcohol and drug use, corporate safety culture, and much else. Students of safety also began to see injuries as the outcome of complex causation. The idea of system safety—a concept imported from aerospace and nuclear work—began to enter railroading in the form of fault tree analysis and management tools such as PERT (Program Evaluation and Review Technique). In the 1980s, all the large carriers embraced Total Quality Management (TQM), and while the motive was to improve service, the new concepts such as continuous improvement, zero tolerance (for defects or injuries), root cause analysis, and intense employee involvement spread into safety work, gradually reducing both train and little accidents. The early railroad Safety First Movement had been long on exhortation, contests, posters, and discipline. Twenty-first-century safety programs still include some of these elements, but compared to earlier days, there is a much larger dose of work programming, job study, worker training, and improved equipment, as well as employee and union involvement.

There have been policy failings as well. Until recently, neither the commuter lines nor Amtrak nor the regulatory bodies have spent much time worrying about injuries and fatalities to passengers that result from events other than train accidents. In addition, these groups have been slow to address the safety of railroad passenger cars. Work by Hugh De Haven and others in the 1940s and 1950s that focused on crash energy management (CEM) resulted in seatbelts and padded dashboards in automobiles by the 1960s. Various agencies within the Department of Transportation (DOT) had pushed both devices, yet curiously the technology failed to transfer within the DOT because neither the FRA nor the Federal Transit Administration (FTA) showed any interest in designing safer railroad passenger cars during these years. Not until the 1990s would safer equipment become a regulatory priority. Many passengers, perhaps including some of those on Metrolink Train 111, paid the price for this tardiness.

"There has been a rail revolution, but few people know about it," Richard Saunders concluded in *Main Lines*. Saunders was referring to the revolution in production and productivity that has characterized the freight railroads since 1980, but he could just as well be discussing their accident and injury record. Railroad safety has had a colorful, checkered, and important history. In modern times, like the railroads themselves, it has fallen apart, and then has come storming back from disaster. Public safety and economic policies toward railroads have been almost entirely reinvented, while multiple revolutions in technology and organization have reshaped the carriers and the lives of their workers. The results have dramatically diminished the risks that railroads present to workers, passengers, and the public to levels undreamed of in the 1960s, and the process is by no means complete. That is the story this book tells.[13]

1

The Long View

American Railroad Safety, 1828–1955

What is life? 'tis but a vision,
here I died by a collision.
Twenty more died by the same,
Verdict—"nobody to blame."

Unknown author, 1856

Life's a railroad. Hurry on!
Always keep a-going.
Never stop to look at flowers
by the roadside growing.
Never mind what's on the track,
On—though headlong—faster!
If the engine Progress stops
that's the great disaster.

Culma Croly, "Life's a Railroad," 1854

Railroads and America grew up together and the relationship has not always been an easy one. The railroads were a spectacular Schumpeterian innovation, and they rearranged the economic and cultural landscape as perhaps nothing had before. From the very beginning, Americans have loved their trains. They changed the language as they came to symbolize the engine of progress through science, man's journey through life, and even American destiny, and their metaphors invaded religion, politics, and nearly everything else. Many men and women of a certain age can still recall the romance of the named trains—the *Twentieth Century Limited*, the *Super Chief*—and a train whistle at night, even from a diesel, still stirs the soul. But if trains meant romance, Americans also quickly developed a healthy respect for the new technology. Progress, it seemed, was Janus-faced; for a long time, railroads were the largest moving things on land, and they brought a new set of risks to an emerging industrial society. By about 1900, railroads were the greatest single source of violent death. Yet these dangers gradually receded throughout the nineteenth and into the middle of the twentieth centuries. How and why that happened is the subject of this chapter.[1]

Dangerous Beginnings

Train accidents—collisions, derailments, boiler explosions in the age of steam, and a miscellany of others—have always fascinated, and they began early. The first commercial railroad, the Baltimore & Ohio, broke ground in 1828. Three years later, in 1831, a locomotive boiler exploded on the South Carolina Railroad, killing its fireman; it was probably the first of what became a rapidly increasing list of fatal train accidents (see Table A1.1). Such disasters focused public attention but they by no means exhausted the dangers that came with the new technology. Well into the twentieth century, most of the injuries and fatalities resulted from little accidents that did not involve train wrecks. Like sniper fire, they picked off men, women, and children usually one or two at a time in a nearly infinite number of minor tragedies. These were the accidents that befell the passenger who stuck her head out the window of the train—and was decapitated. Or the man in a hurry who jumped from a moving passenger car and fell under the wheels. Or the workman, crushed while coupling cars. Or the children playing on the tracks, or the man who didn't see the train at the crossing.

All of these accidents multiplied as the railroads multiplied but—much the way a patient develops antibodies to fight infection—the carriers developed a technological community that discovered and diffused the information and innovations needed to improve and tame the iron horse. Accident costs and passenger injuries proved powerful safety incentives, and they were assisted by state and later federal interventions as aspects of railroad safety came increasingly to be seen as a public responsibility. The risks associated with railroading can be assessed in two ways. Fatalities per passenger or per worker capture dangers to individuals. Fragmentary state data suggest these declined unevenly from the 1840s. A broader measure looks at the social costs of railroading—fatalities per unit of output—and at least for Massachusetts, these too fell from the earliest days.[2]

The Emergence of Train Accidents

Early train accidents reflected the primitive state of technology and business organization. As Joanne Yates has stressed, safety is a matter of information, communication, and control; yet everything about the railroads was new, largely unknown, and therefore hard to control. Boiler explosions were mysterious and metallurgy largely unstudied. Early trains had no way to stop save the use of hand brakes. Signaling was equally primitive. The properties of track, rail, wheels, springs, cylinders, and nearly everything else in a train were largely unstudied and nearly everything could cause a wreck. Similarly, by the 1850s, railroads had become the first very large businesses. Running them required novel organizational structures and rules, all of which had to be learned the hard way.

By the 1850s, it was also clear that American railroads were far more lethal than their European counterparts. Many European travelers in America remarked on the differences and on what appeared to be Americans' blithe indifference to safety. They found the utter absence of fencing and crossing guards astounding. Writing of his 1839 trip, the Englishman

Thomas Grattan told of a violent jolt when his train traversed a crossing. A conductor laconically explained that the train "was going over a chaise and horse." In 1853, Charles Weld, another English traveler on the B&O, discovered after his train wrecked that the rails "were worn in many places to a mere ribbon." The conductor admitted to him that they had many derailments but, he claimed, they "rarely killed people."[3]

These dangers, as perceptive travelers realized, were the outcome of a distinctive set of economic conditions that shaped the American system of railroading well into the twentieth century. Antebellum America was rich in resources—especially wood—and poor in capital and labor, and its rickety railroads reflected that fact. Writing in 1839, the English traveler Frederick Marryat explained that accidents resulted because the railroads "are not so well made," and he pointed out that "nothing is made in America but to last a certain time," which reflected the scarcity of capital. Railroads were therefore built cheaply and would be rebuilt out of future profits. He gave as an example the complete lack of fencing in America, which resulted in numerous accidents and "many cows cut into atoms." The French economist Michel Chevalier pointed out that sharp curves and steep grades also saved on grading, while the pony truck allowed American locomotives to navigate where European equipment would derail. "The Americans are unequalled in the art of constructing wooden bridges," which were "remarkable for their cheapness," he pointed out. They were perhaps equally remarkable for their propensity to burn, rot, and collapse under heavy loads. Similarly, American rails were originally wooden with but a strap of iron; when these proved inadequate the carriers shifted to a self-supporting T-rail that required less labor to install than British designs. American locomotives were designed for ease of repair, while the purpose of the eponymous cowcatcher was self-evident.[4]

American labor markets also differed from those in Europe. Chevalier observed the extreme scarcity of civil engineers in America. One result was that men almost entirely innocent of engineering theory sometimes designed bridges. In addition, the scarcity of all kinds of labor in America undercut threats to dismiss workers, as jobs were nearly always available. Here again safety reflected problems of information and control. Trainmen—enginemen, firemen, and conductors—worked largely without supervision. And without the threat of dismissal, companies had no way to enforce fidelity to rules and regulations. Thus, sometimes enginemen simply ignored timetables or train orders, and sometimes low-level supervisors told them to do so in order to get the train over the road. When a train made an unscheduled stop, the flagman was supposed to go back to warn following trains. But there was no one to see and so on dark and cold nights he might not go far enough or perhaps not go at all. In short, shirking was common because monitoring was difficult and incentives inadequate. Economists refer to all of these instances that involved a potential divergence between the aims of an employer and a lower-level manager or employee as "principal-agent problems" or simply "agency problems," and they plagued American railroads from the start.[5]

Labor scarcity combined with the unreliability of early technology also contributed to a work culture that shaped safety in peculiar ways. Managers and workers developed a "get the train over the road" culture that encouraged risk-taking. In addition, trainmen were largely free from supervision; they faced enormous responsibilities and took great risks. The result was an ethos of bravery and independence that sometimes saved lives, but also encouraged the men to decide for themselves which rules to obey and when to do so. Rule violations were therefore a regular source of disaster and such behavior was protected by tight labor markets.

For these reasons, therefore, American lines were more dangerous than their European counterparts. The first recorded derailment—on the Camden & Amboy—occurred in 1833. An early collision on the Pennsylvania Railroad at Mount Union, Pennsylvania, in 1853 reflected both employee nonfeasance and the primitive methods of train control (Table A1.1). Three years later, similar failings led to a head-on collision on the North Pennsylvania at Camp Hill, Pennsylvania, that killed 66 (Table A1.1). The wrecks appearing in Table A1.1 come from a long list. These and the many others were what Nathan Rosenberg has called "focusing devices": they generated strong incentives to improve safety and, properly interpreted, the information to do so.[6]

The pressures for improvement came partly from public opinion. In the early days, an occasional accident had seemed a small price to pay for the benefits of faster travel. But by the 1850s, much more modern attitudes had evolved and newspapers routinely sensationalized train accidents.[7] Thus, the New Orleans *Times-Picayune* referred to the Norwalk bridge disaster of 1853 (Table A1.1) as "railroad slaughter." The Albany *Evening Herald* favored a different metaphor, referring to that wreck as "The Railroad Murder." The *New York Weekly Herald* also referred to a recent wreck as murder and blamed it on the cupidity and stupidity of management. Thus, it was less the "machine in the garden" that concerned Americans of that day than the businessman in the garden. Lurid descriptions of accidents abounded. After an 1859 accident that resulted from the collapse of a culvert on the Michigan Southern, one eyewitness reported that "the ground was strewn with heads, arms, legs and dead bodies."[8]

Poets and cartoonists found rich pickings in the idea that accidents simply reflected capitalist indifference and cupidity.

> A stands for Accident, frequent alas!
> B for the Bungling that brings them to pass.
> C is for Cheapness, the sole end and aim
> D of Directors who're "free from all blame."[9]

While pundits seemed to think that railroad managers were indifferent to such tragedies, in a democracy public worries ensured that killing travelers would be expensive. Courts quickly made the carriers absolutely liable for passenger injuries from train accidents, while the newspapers' harsh attitudes were often reflected in jury awards. The *New York Weekly Herald* pointed out that the Norwalk wreck would cost the New Haven more than $250,000, which—with skilled workers' wages at about $1.50 a day—was a considerable sum. Moreover, train wrecks might drive away business and they ruined equipment, tore up roadbed, and disrupted schedules. Companies therefore had strong motives to prevent them.[10]

As noted, accidents also provided information, for failure can be instructive; boiler accidents like that on the *Best Friend*—as well as the epidemic of steamboat explosions—led to scientific investigations of the problem. The Norwalk disaster resulted in state intervention, while the many collisions as at Mount Union and Camp Hill speeded the introduction of the telegraph and train order system. Ashtabula (Table A1.1) and many other bridge accidents spurred engineering understanding, state inspections, and improved technology. Such hard-won changes did not come by magic; they were in part the fruits of the emerging railroad technological community.[11]

The Rise of a Railroad Technological Community

Safety and efficiency are both good things, but they can conflict: on early railroads, for example, speed was both productive and dangerous. Such relationships can be depicted by the lower trade-off curve AA in Figure 1.1. The curve represents a frontier; at any given time the state of technology and organization make impossible all combinations of safety and output that are outside it. Point X represents an attainable combination on curve AA. In fact, most firms will be not be perfectly efficient and may reside inside the curve (point Z), but competition will ensure that they try to move toward the curve. Competition will also shift the trade-off curve out (BB) as innovations improve technology and organization. Over time, therefore, safety and efficiency can both improve (from X to Y).

The quest for safer and more efficient railroads resulted in the development of a technological community. It was a loose network composed of scientists, engineers and engineering societies, the carriers themselves, and their suppliers, glued together by engineering publications that discussed and disseminated the new information. Because railroads are themselves a network technology—freight cars and tracks interact as do intercompany technological choices—an important function of the technological community was not only to improve technology but also to ensure commonality.[12]

The first of the important publications was the *American Railroad Journal*, which dated from 1832. Under Henry Varnum Poor (1812–1905),

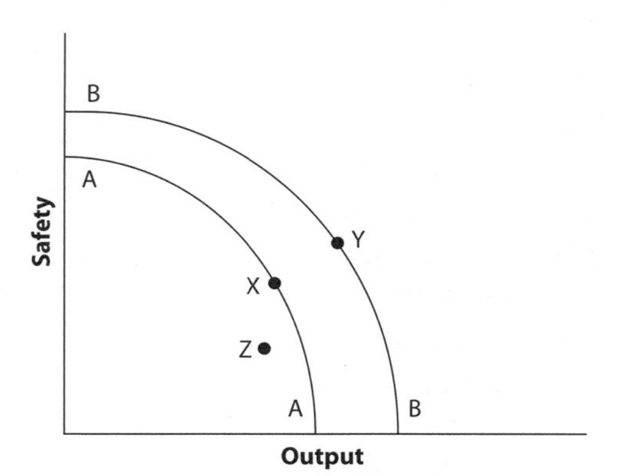

Figure 1.1. The Changing Trade-off Between Safety and Output

it reported accidents, criticized bad management, and publicized better technology. It was followed by the *American Railway Times* (1849), *Railway Age* (1876), and *Railway Review* (1879). The most influential of these technical publications was the *Railroad Gazette* (1870). From 1870 to 1883 it was edited by Mathias Forney (1835–1908). During these years, the *Gazette* published the first nationwide statistics on train accidents. It helped disseminate improved methods and was a forceful voice for safer railroading. Railroad problems were a major fixture in general engineering publications during these years. Railroad technology initially dominated *Engineering News* (1875). Edited from 1887 to 1895 by the distinguished engineer Arthur Mellen Wellington (1847–1895), the *News,* like the *Gazette*, shined the light of publicity on bad practice, crusading especially for better track, roadbed, and bridges. Railroad technical men formed the Master Car Builders (1867) and the Master Mechanics (1867), which were devoted to improving and standardizing equipment. The American Railway Association (ARA), which loosely coordinated all the others, came in 1872. Many other more specialized groups (the Airbrake Association, the American Association of Railroad Superintendents) performed similar functions. There were also regional railroad "clubs" that were devoted to organizational and technical matters.

Competitive striving, of course—among railroads, among their suppliers, and among individual employees—was the driving force motivating these improvements. Aided and abetted by public opinion and regulatory threats, these institutions brought rapid improvements in railroad technology. These were the individuals and societies that published and discussed the lessons of accident statistics, discovered metal fatigue, developed the standard code of train rules, employed the telegraph to transmit train orders, debated the proper shape and weight of rail, standardized car wheels and axles, evaluated car couplers, introduced steel rails, invented block signals, and adopted the air brake.

They also pioneered in bridge safety. Bridge accidents had plagued American railroads from the beginning. Some, like Ashtabula, reflected poor design by amateur engineers, but the more basic problem was that economics led the carriers to choose flimsy designs. Thus, wooden bridges and trestles rotted and burned and yielded a stream of wrecks. Derailments on bridges without floors also resulted in disasters, while American pin-connected iron bridges would fold like a hinge if a single member failed. In the 1880s, led by Wellington of the *Engineering News*, critics pressed for improvements, and by 1900 better maintenance and technology brought the problem to heel.[13]

As these and literally thousands of other technological and organizational innovations improved and diffused, railroad travel gradually became more reliable and safer. In the era before federal regulations, all innovations had to promise returns that would exceed their cost of capital. Hence, few were adopted simply because they improved safety; rather, whether the product was air brakes or the telegraph or better axles, most promised returns in the form of both enhanced productivity and better safety. The logic of profitability also ensured that the carriers would introduce these innovations where they did the most good. So, for example, steel rails appeared first on high-density lines, where safety and economic payoff were greatest.

Profit considerations also implied that safety improvements often generated offsetting behavior—sometimes referred to as risk compensation. Air brakes, for example, increased train speed, thereby partly reducing their safety impact. Figure 1.1 depicts this form of a trade-off. Suppose firms with old-fashioned brakes run passenger trains at 20 mph. This is point X in Figure 1.1, resulting in a certain number of passenger miles and accidents. Faster trains mean greater output but less safety. The invention of air brakes shifts the line out to BB. The carriers can hold speed (and thus output) constant and take the whole gain from air brakes in terms of safety (moving straight up), but they will likely increase speed too, moving to point Y and thus taking some of the gains from better brakes as higher speed.[14]

Most other innovations had these same characteristics. The telegraph and train order system originated on the Erie in the 1850s and it bettered both safety and efficiency. In the 1860s, Ezra Miller invented and patented a combination trussed platform and hook coupler for passenger cars that companies quickly adopted; it improved the ride and helped absorb shock during collisions. Steel rails also dated from the 1860s; they were safer and allowed heavier trains. Block signals, which separate trains by a space interval, are inherently safer than separation by a time interval as occurs under the Timetable and Train Order system. They also allow greater traffic density, and the carriers began to install them in the 1870s.[15]

Many more modest innovations also improved safety. While some of the new technology was patented, there was also much collective invention and learning by using. The use of steel and better designs for early locomotives, for example, were endlessly debated by the Master Mechanics and gradually adopted. Rails, wheels, axles, and nearly every other piece of equipment and track were continuously upgraded as understanding of metallurgy improved and systematic testing spread. Economic incentives ensured that, like steel rails, most such innovations would go first to the high-density lines, which, of course, was also where they had the biggest safety payoff.

Yet the railroads as a network technology had to exchange freight cars and this proved a major impediment to both safety and efficiency. Because compensation for cars traveling over foreign roads was on a mileage (later per diem) rate, the benefits of maintenance and improvements tended to accrue to others—they were external benefits, or "externalities" in the language of economics. Thus, a carrier had little incentive to improve a car that traveled on others' tracks—and similarly, it experienced few risks from poor maintenance. Most spectacularly, freight car interchange impeded the development and diffusion of better couplers and brakes. The carriers endlessly experimented with a substitute for the old, dangerous link-and-pin freight car coupler, which required men to go between moving cars, and with some form of better brakes for freight cars, but little was accomplished until state and later federal pressures began to build (see below). Freight car interchange has continued to impede the development and diffusion of improvements to rolling stock down to modern times.

While some innovations such as standard codes for signaling and train rules required organizational changes, most new technologies involved new investment and almost from the day they were completed the carriers engaged in a continuing massive upgrading of both track and equipment. The *Railroad Gazette* began collecting data on train accidents in 1873. Its statistics captured most such accidents that resulted in fatalities. These data expressed relative to train miles suggest that the carriers' investments in better technology showed up most clearly in the decline of derailments down to 1890. The increasing collision rate despite the introduction of the telegraph and block signals reflects both the enormous rise in traffic density, especially on eastern lines, and the carriers' continuing difficulties in enforcing safer behavior on the part of workers and mid-level managers.

Little Accidents

As noted above, while train accidents made most of the headlines, little accidents—to workers, to passengers, to trespassers, and at grade crossings—did most of the killing. Such accidents reflected some of the same economic conditions that separated American from European railroading. In America, passenger stations were unfenced, while American cars with a vestibule encouraged passengers to run for a departing train or sometimes leap off early—but not always successfully. In Europe, companies stationed a watchman at rail-highway crossings, but men were expensive in the United States and so most crossings were unguarded except for a sign reading "Look Out for the Train." In many cities, tracks went down the streets and children picked coal or played on the tracks. With rail fares perhaps 3 cents a mile, train travel was beyond the means of poor or working-class families throughout much of the nineteenth century, and so they walked. European railroads developed in settled country with comparatively decent roads. In the United States, tracks were typically straighter and cleaner than winding, dirt roads—which is one reason they make good rail trails today—and so the poor walked the tracks, and many died in the process.

In addition, while train accidents that killed passengers were universally seen to be the fault of the carriers and were accordingly expensive at law, accidents to passengers in stations, to workers, or to trespassers or individuals at crossings seemed less due to the fault of the railroads than to the carelessness of the injured. Such attitudes, reflected in court judgments, reduced the carriers' incentives to prevent these kinds of tragedies.

The data in Table 1.1 are among the earliest nationwide figures available; they demonstrate that out of 2,451 rail workers killed in action during 1890, 541 died in train accidents. Thus, even though most of the casualties in train accidents were employees, most employee casualties resulted from little accidents. Train accidents also killed 113 passengers, but far more died from a miscellany of causes including station accidents and hopping on or off moving trains. Other fatalities were mostly individuals at rail-highway crossings or trespassers. Massachusetts's data allow a longer-term perspective on these accidents to others. From 1852 on, fatalities to "others" declined relative to train miles, as Massachusetts modestly improved crossing safety and station design, but rose relative to population. Since Fatalities/Population = (Fatalities/Train Mile) $*$ (Train Mile/Population), this implies that even as fatalities per train mile declined fatalities per capita rose because of increased traffic density. For this reason, public concerns rose even as safety was improving.

Because little accidents were widely seen to be the fault of the injured, liability costs played a modest role in inducing safety, which usually arose

Table 1.1. American Railroad Fatalities, 1890[a]

Cause	Passengers	Workers	Others	All Fatalities
Train Accidents	113	531	346	990
Crossings and Stations	45	120	741	906
All Other Causes	128	1,800	2,511	3,728
Coupling Cars	—	369	—	369
Riding Freights[b]	—	650	—	650
All Causes	286	2,451	3,598	6,626

Source: ICC, *Statistics of Railways*, 1890.
[a]Year ending June 30, 1890.
[b]Includes falling from cars and trains plus overhead obstructions.

as a byproduct of other improvements. Trainmen's accidents from locomotives slowly declined as the Master Mechanics gradually increased the efficiency and reliability of the iron horse. Miller's combination coupler and platform was intended to reduce the likelihood of telescoping (when one car overrides and crushes another car in a train accident), but it also reduced the risk that a passenger would fall off the train when trying to go between cars.

A fault-based view of the world also ensured that few of the carriers' critics worried over trespasser fatalities, while such attitudes also impeded public action to reduce rail-highway crossing accidents. However, by the 1880s, more modern ideas began to intrude and crossing accidents started to attract public concern. Initially, the focus was on crossing elimination, but elimination was expensive and towns and cities had little interest in spending public funds to protect the unwary. They thought it should be a railroad responsibility, while the carriers wanted governments to pay because their citizens benefited. Since municipalities could require that companies guard crossings with a watchman, they could effectively raise the railroads' payoff to elimination, but this seems to have accomplished little. By the 1880s, both Connecticut and Massachusetts were encouraging cities to pursue crossing elimination by apportioning most of the costs on the railroad. By the turn of the century, railroads began massive programs of track elevation in big cities such as Chicago, Philadelphia, and Newark. Crossing accidents continued to grow, however, as train and pedestrian traffic increased.

Regulatory Beginnings

While there has never been a time that railroads were unregulated—they were, after all, creatures of the state, cloaked with the power of eminent domain—before the 1890s, public policies played a modest role in shaping safety. In the 1830s, Massachusetts began to require the carriers to place warnings at highway crossings and the engineman to ring the bell. It began to mandate fencing in 1846 and the reporting of casualties in 1847. Soon, New Hampshire, New York, and other states followed with similar laws. In the 1870s and 1880s, fiery wrecks such as the bridge collapse at Ashtabula and the head-on collision at Republic (Table A1.1) generated public campaigns to replace car stoves—the usual origins of the fire—with safer stoves or, better yet, steam heat. By about 1890, New York and the New England states had banned the "deadly car stove" and steam heat was making its debut.

An alternative to these direct regulations evolved out of Massachusetts's experience. In the 1840s, after a number of accidents on the Western Railroad, the state legislature investigated its practices, which—in the face of this veiled threat—quickly improved. Investigation and publicity with a regulatory "gun behind the door" was a kind of "soft regulation" that Thomas McCraw has termed "voluntarism." In 1869, Massachusetts established a state railroad commission modeled after British practice. While it had no regulatory power, under the guidance of Charles Francis Adams it practiced voluntarism—employing "sunshine" (publicity and investigation) as a potent force for better railroading. Adams used the Eastern Railroad collision near Revere, Massachusetts, on August 16, 1871 (Table A1.1), to force Massachusetts's carriers to adopt better and more consistent train rules and signals that were widely copied elsewhere.[16]

Railroad commissions gradually spread to other states and they performed a variety of safety functions. In Illinois, Massachusetts, New York, and some other states they inspected bridges. And in the 1880s, Connecticut, Massachusetts, New York, and other states began to require use of automatic couplers. While freight car interchange dulled company incentives to install the new equipment, collectively the carriers welcomed such improvements, for the link-and-pin coupler was inefficient as well as dangerous and trains with air brakes could not use it. But they did not welcome state intervention, worrying—with good reason—that laws might mandate differing couplers or simply choose the wrong one.

The Spread of Safety Regulation

Between about 1890 and the end of World War I, American railroads took on the characteristics that would dominate the system down to the 1960s. On main lines, massive investments in plant and equipment rebuilt the early flimsy track with much more substantial rails, bridges, and roadbed. Technological change continued at a rapid pace as old innovations like block signals continued to spread, while the carriers and suppliers improved rail design, the air brake, and nearly everything else. Locomotives and rolling stock continued their steady growth in size. System building also proceeded, and by the 1890s the Pennsylvania, the New York Central, the Union Pacific, and other large carriers had emerged. Size raised the payoff to knowledge, and the Pennsylvania began the first corporate laboratory in the country in 1876, followed soon after by a number of other lines.

Federal Beginnings

Size, along with autocratic and monopolistic behavior, had long made the carriers a popular target, and by the late 1880s they were the most hated industry in the country. The critics' indictment—which I call the "robber baron narrative"—focused on monopolistic behavior. Railroad managers, it was alleged, skinned small shippers via price discrimination in the form of rebates—thereby contributing to the rise of other monopolies such as Standard Oil. In addition, they bribed legislatures with free passes and routinely swindled investors. It was no great stretch to cast them as

HOW TO INSURE AGAINST RAILWAY ACCIDENTS.
Tie a couple of Directors upon every Engine that starts with a Train.

The robber baron view of railroad safety: accidents reflected managerial greed (*top*) or indifference (*bottom*). (Top image, *New York World*, June 14, 1913; bottom image, *Harper's Monthly*, July 1853)

villains in the safety drama. In this role, their cupidity led them to skimp on safety, resulting in the slaughter of passengers and workers. As *The Boston Globe* concluded in 1887, there was "no excuse for dreadful railway holocausts. Corporation greed is responsible." It was a view that has never entirely died out.[17]

These public perceptions resulted in the Interstate Commerce Commission (ICC), which opened for business in 1887. The railroads did not oppose legislation; they hoped it might help them reduce competition and they preferred unified federal rules to the inconsistent and sometimes conflicting enactments of individual states. Although the ICC initially had little power over rates or safety, in 1889 it inaugurated the first detailed nationwide collection of data on railroad casualties. This was an important development, for the numbers shocked. In 1890, they demonstrated that 1,019 railroad men were killed coupling cars or falling from freights as they worked hand brakes (Table 1.1). Knowledgeable observers also understood that this was not simply a matter of carelessness, for under the pressure to produce men were encouraged to take chances—trading safety for output. Even work accidents, therefore, were coming to be seen as a social rather than simply an individual matter, and with the information public, their control also became a public concern.[18]

By this time, the carriers and their suppliers had finally developed workable semi-automatic couplers and air brakes for freight trains. Combined with the ICC statistics on trainmen's mortality, the availability of the new technology created a moral imperative to use it. In 1893, without much carrier opposition, Congress passed the first federal railroad safety regulation—the Safety Appliance Act, which mandated air brakes on freight trains and a coupler that did not require men to go between cars. After many delays, the law finally became operational in 1900. Regulations, like efficiency improvements, can also generate offsetting behavior, and carriers noticed that with the advent of semi-automatic couplers, switching became much rougher while air brakes encouraged longer, heavier, and faster freights. But the new equipment steadily reduced the carnage and proved profitable as well. To reformers it demonstrated that technological fixes were efficacious and that they knew more about railroading than did the carriers.[19]

Effective economic regulation waited another decade. In 1903, Congress passed the Elkins Act—which the railroads did not oppose—giving the ICC authority to prevent rebates. In addition, in 1906 the Hepburn Act finally gave the commission power to set maximum freight rates and outlawed free passes. The years between 1896 and 1920 were marked by almost continuous inflation (a result first of gold discoveries and then World War I) and in 1910, with rising costs, the railroads applied for a general rate increase. In a celebrated decision that came to be called the Eastern Rate Case, the ICC turned them down. Although later appeals resulted in some relief, the damage had been done. Adjusted for inflation, railroad earnings declined as did the carriers' appeal on the stock market, retarding investments and improvements. By 1917, they were unable to cope with the flood of wartime traffic and were temporarily taken over by the federal government and run by the US Railroad Administration—which promptly raised rates.[20]

This expansion in federal economic regulation occurred at a time of heightened worries over railroad safety. While passenger safety had been improving for a long time, a business expansion that began in 1896 yielded a boom in railroad traffic, which in turn brought with it a sharp rise in passenger fatalities from train accidents that peaked in 1907. A head-on collision between two passenger trains on the Southern near Hodges,

Tennessee, that killed 63 on September 24, 1904 (Table A1.1), was typical of those that increased public pressures. The many such wrecks led Congress to pass an Accidents Reports Act in 1901, which inaugurated modern injury and accident reporting. Statistical definitions are invariably contentious and contested, and the ICC definitions were a compromise between railroad and labor interests. The new rules required reports of train accidents to the ICC that either resulted in a casualty or met a monetary threshold of $150—which remained unchanged until 1948—and so the continuing rise in prices artificially inflated accidents.

Both reported and inflation-adjusted collisions rose sharply in the years up to 1907 as the old Timetable and Train Order system buckled under the increasing onslaught of traffic. Derailments, a problem that had seemed to be in remission in the 1880s, also rose as longer trains and the new 50-ton cars put unanticipated stresses on couplers, rails, and wheels. The rise in train accidents heightened public worries, which led to the Hours of Service Act in 1903, which limited trainmen's time on duty. The rail unions also got a law requiring self-cleaning ash pans that year and the federal Locomotive Inspection Act in 1910. Public fears also led to a second Accident Reports Act in 1910 that empowered the ICC to investigate train accidents. While the commission had little power, like the Massachusetts Railroad Commission under Charles Francis Adams, its early investigations would bring sunshine into the dark corners of railroad accidents. As with the act of 1901, the consequences were far-reaching, for making the information public allowed the carriers to learn from each other's accidents. The rail unions began to campaign for train limit laws that would have set a maximum car length for freight trains as well as "full crew" laws requiring a third man in the cab. While labor claimed that such laws were safety measures, the railroad press argued persuasively that these were make-work projects that had little to do with safety. At the federal level, nothing came of either proposal. The full crew laws, however, proved popular among the states. While the carriers fought them relentlessly, 22 states had such requirements by 1917. They provide a good example of how advocates can sometimes employ safety as a cover to advance other goals.[21] After about 1907, although reported collisions and especially derailments soared, this was entirely a result of inflation: adjusted for price increases, such accidents receded as the carriers responded with a host of safety investments, new technologies, and Safety First (see below). Yet the reported numbers made the headlines. Ironically, therefore, public worries led to a public policy—reporting of train accidents—that in turn magnified public worries. Moreover, the rise in reported accidents occurred in a context that greatly magnified their importance, for these years were also the heyday of the urban newspaper. While the press had a long history of denouncing railroad managers as murderers, as passenger wrecks rose after 1896 the outpouring of criticism was truly staggering.

Research on risk perception emphasizes that individuals sometimes overestimate the dangers from minor risks and they find some risks especially frightening. Train wrecks have characteristics that can make them seem particularly horrifying. They are largely involuntary, and they are uncontrollable: no citizen can influence the likelihood of an accident. Modern writing also reveals that risk perceptions are only loosely related to actual risks. In particular, the media can amplify perceived risks simply by the volume of writing on the topic (an "information cascade") and by the use of signal words ("railway slaughters") and dramatic illustrations. The robber baron narrative was also widely employed to explain safety lapses, as writers routinely implied that companies were recklessly indifferent to safety. In 1904, Francis Lynde in "Death and the Drumming Wheels" described corporate greed as the "quicksand at the bottom of the goodly river of disaster." This he blamed on the need to pay dividends on watered stock.[22]

No company felt the public's wrath more than the New Haven; its president was Charles Mellen, a representative of the Morgan interests, which in turn symbolized Wall Street greed. The New Haven was also a New York commuter line and carried readers and writers of that city's newspapers and opinion journals to and from work. Its accident record was, therefore, of particular concern to the scribbling classes. Like all the carriers, agency problems plagued the New Haven—flagmen who did not go out far enough or engine drivers who ran signals or otherwise violated rules. In 1912, one engine driver took a curve too fast and derailed the train, resulting in a gruesome wreck at Westport, Connecticut, in which the old wooden cars burst into flames and immolated seven passengers. The ICC, newly empowered by the Accident Reports Act of 1910, investigated the wreck. As Westport was almost identical to an earlier accident after which the company had ignored the ICC's recommendations, the commission placed the blame squarely on the company's management. A *New York World* editorial promptly termed the New Haven a "Morgan-Rockefeller road" and suggested that accidents would cease if management became as interested in passengers as they were in profits. Later, a group of frightened New Haven commuters led by Representative Lynn Wilson (D Connecticut) of Bridgeport blamed Westport and other wrecks on "watered stock which caused a cutting down of all running expenses, which resulted in the accidents." The New Haven, they concluded, was "a cancer of imperialism . . . [and] a financial despot." Cartoonists made the same point, routinely drawing the stock ticker tape of death. Thus, just as greedy managers justified the need for federal supervision of railroad economics, so they also required expanded federal controls over safety, or so the popular press claimed. After several more accidents that might well have simply been bad luck, the press, urged on by the ICC, finally drove Mellen from office.[23]

When the merry old New Haven opened its Summer slaughterings at Stamford, Conn., there was brought to mind a remark of Mr. Hearst that the motto of the road was "Your money or your life," to which a certain New England wag in the Senate replied, "No, your money and your life"

The New Haven was a Morgan-dominated road. A string of accidents in 1911–1913 confirmed to critics that it sacrificed safety for profits. (*Hearst's Magazine*, August 1913)

The carriers' Progressive critics also had important safety legislation in the works. The ICC began to urge laws that would empower it to mandate block signals and it advocated experimentation with automatic train control—a radical new technology to prevent collisions that would stop the train if the engine driver ran a stop signal. Several wrecks such as that at Westport, where old wooden cars were crushed and burned, contrasted with other accidents such as one at Tyrone, Pennsylvania, on July 30, 1913 (Table A1.1), where steel passenger cars apparently prevented many fatalities, leading to public clamor for their introduction. By about 1912, in response to the New Haven's wrecks, Congress began hearings on both steel cars and block signals, while the ICC was suggesting that to improve safety it might need additional authority to control railroad operations.

The years after 1900 also brought federal regulation of the risks from transporting hazardous substances. Unlike most safety laws, however, this one came at the request of the carriers. Transport of explosives had caused problems for decades. Most railroads had developed procedures to ensure safe transport, but because of competition private company rules were inadequate to force proper packaging of hazardous substances—a situation made worse by freight car interchange. Finally, in 1903, an explosion of dynamite in a freight car in the Pennsylvania Railroad's Crestline, Ohio, yard blew a hole 40 feet deep and galvanized that company to press for federal rules. The Pennsylvania was the largest carrier in the nation and it mobilized the ARA and lobbied important shippers such as DuPont to support regulation. Congress promptly responded, passing P.L. 174 in 1908, giving the ICC power to regulate transport of explosives. Just before this, the ARA had organized its own Bureau for the Safe Transportation of

Two wrecks in 1913 demonstrated the value of steel cars. On June 30, a collision on the New Haven near Stamford, Connecticut, telescoped the wooden cars, killing six passengers (*bottom*). A month later, a collision on the Pennsylvania near Tyrone, Pennsylvania, involving steel cars resulted in no passenger fatalities (*top*). (Top image, courtesy of Hagley Museum and Library; bottom image, Library of Congress)

Explosives and Other Dangerous Articles, which developed and enforced a set of industry-wide rules to improve safety. As was no doubt intended, the ICC, which had no expertise of its own, adopted the bureau's rules, thereby giving them the force of law. The bureau moved aggressively, developing standards for packaging and shipping first of explosives and then a widening array of other hazards. By World War I, the bureau was working with the American Petroleum Institute and a tank car committee of the Master Mechanics to improve tank car safety. Until modern times, the bureau's annual reports provide the only statistics on what we now call "hazmat accidents," and they gradually declined. This rather strange arrangement whereby industry self-regulation was given the rule of law persisted largely unchanged into the late 1960s (Chapters 2 and 3).[24]

Safety First

The New Haven's experience provided railroad managers a glimpse of the future. "Nothing is more unprofitable than to ignore or defy public sentiment" was the message *Railway Age* took from that company's problems. These regulatory threats galvanized the railroads into action, just as Charles Francis Adams had predicted. A number of major carriers began campaigns to improve their public image. Edward Harriman had emphasized safety as long as he had been associated with the Union Pacific, and about 1907 the Harriman lines (that included the Southern Pacific and Illinois Central as well) opened their accident investigations to the public. The Pennsylvania Railroad hired Ivy Lee as one of the first corporate publicity agents to improve its reputation, and he too opened the company's accident investigations. Lee also worked for the Baltimore & Ohio and the Union Pacific and later went on to burnish the public image of John D. Rockefeller. These and other lines began to publish news releases and craft the public utterances of top officials. Moreover, thoughtful executives also realized that the industry as a whole needed better public relations. About 1910, the carriers established the Bureau of Railway Economics and the Special Committee on the Relation of Railway Operation to Legislation to advance their cause. By 1912, the ARA was beginning to realize the role accidents played in generating regulatory threats to industry autonomy.[25]

In 1910, Ralph Richards, a claim agent on the Chicago & North Western, had persuaded management to back a safety campaign. In modern business terminology, Richards was trying to reorient the North Western's culture away from the old "get the train over the road" attitudes. Richards's approach emphasized safer work practices to reduce employee casualties. But he understood that safety involved a kind of gift-exchange relationship and that for workers to cooperate the company would have to prove its sincerity by spending money on the safety of plant and equipment. To management he stressed that the focus on worker safety was in part a means to motivate a reduction in all kinds of accidents, and only by harnessing workers' self-interest could such a broader safety campaign be successful. He also realized that mid-management agency problems could sink his efforts, and so his first move was to inform all intermediate-level managers that he had the backing of the top brass. His efforts quickly began to show results. While safety work on the North Western and other lines was in part a price-induced organizational innovation—liability costs for passengers and workers were both rising—its main motive was to improve the carriers' public image and help fend off unwanted and intrusive legislation. In 1912, mindful of its public relations value, the ARA urged all carriers to follow the North Western example. Railroad safetymen formed a section of the newly created National Safety Council (NSC) and they adopted Richards's motto: Safety First.[26]

The carriers backed their public relations offensive with deeds. As late as 1909, half of all new passenger car orders were for wood; by the first six months of 1913, that number had dropped to zero as they shifted to steel cars. After a few false starts, the builders greatly strengthened early steel cars, and in accidents such as that at Tyrone they proved their worth. About this time, several car builders also introduced a different design where a collapsible vestibule would absorb some of the collision energy. Yet the carriers ultimately chose to go with stronger superstructures instead, and this early example of what is now called crash energy management (CEM) seems to have been largely forgotten until the 1990s (Chapter 5). The companies were also rapidly expanding the block signal system, and like steel rails a generation earlier, the profit motive directed both steel cars and block signals to their highest payoff areas, which were where traffic and therefore risks were also highest. By 1920, while block signals protected only about 40 percent of road miles, they governed at least 70 percent of passenger miles. After numerous hearings, Congress decided to do nothing. Cocking the regulatory "gun behind the door" had been all that was necessary.

Coping with Little Accidents

After 1907, passenger and worker fatality rates began what proved to be a long-term decline (Figure 1.2). While passengers gained from the spread of block signaling and steel cars, most of their safety improvements came from declines in little accidents. In the 1890s, Massachusetts began to press for safer stations. Companies responded, designing urban stations with an eye to passenger safety. The shift in passenger traffic from rural to urban areas with better facilities also improved safety. So did the move to longer journeys, which reduced station exposure and entraining and detraining relative to passenger miles. Companies also instituted gates on cars, which—while intended to catch scofflaws—generated safety benefits too, as they reduced the incentive to run for a train or jump off early.

For workers, the safety gains were an outcome of myriad improvements in track and equipment as well as the impact of the Safety Appliance Act. The act reflected a spreading belief that work accidents were neither the inevitable result of carelessness nor simply an individual problem. Safety First partly mirrored these views, for while it emphasized safer work habits and employee fidelity to rules, by implication these were a company responsibility.

Even trespasser fatalities declined sharply after 1910. While this was mostly a result of the rise in auto-mobility, attitudes were changing as well. A report from the New York Central in 1913 emphasized that the dead were not always hobos but many times "well-to do and respected citizens." They were, in short, solid members of the community and their protection was therefore a community responsibility. While such claims

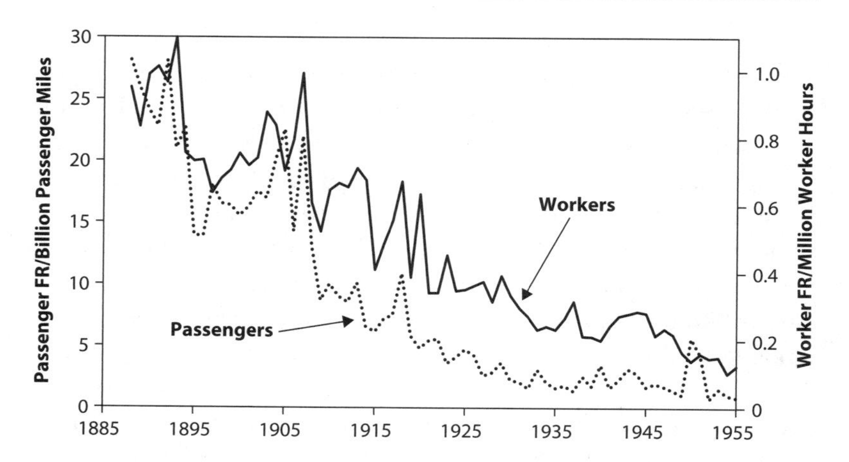

Figure 1.2. Passenger and Worker Fatality Rates (FR), 1888–1955. (ICC, *Accident Bulletin*, and author's calculations)

implied that trespassing accidents were no longer simply an individual matter, they also conveniently absolved the railroads of responsibility. However, if auto-mobility reduced trespassing fatalities, it increased those at grade crossings. While the carriers continued to elevate tracks and remove crossings from a growing number of cities, such efforts were swamped by the expansion of the road network and driving. The railroads initially supported the Good Roads Movement that began in the 1890s. By the twentieth century, however, its successes led to a boom in highway construction that created dangerous crossings faster than they could be eliminated or guarded. Problems of grade crossings would eventually come to dominate railroad safety concerns as other forms of accident receded (Chapter 6).[27]

With minor exceptions, by World War I the basic institutions were in place that would shape railroad safety for a generation. Economic matters such as rates, service, and mergers were subject to hard regulations—rule makings that had the force of law. Voluntarism, or "soft regulation," largely governed safety, however, with a "gun behind the door" that was only occasionally brandished. The carriers' technological network that developed and diffused better methods of railroading, along with a newfound commitment to Safety First, would reduce accidents largely without formal government intervention. This was a double bargain, as the *Railway Age* pointed out several times: the carriers would deliver the safety goods and the government would provide a regulatory environment that yielded the funds to do so. For about 40 years, it worked.

Wartime pressures briefly interrupted these events. Worker and passenger fatality rates rose sharply from 1914 to 1918 (Figure 1.2). The immense press of traffic also yielded a number of spectacular train accidents. In one of the worst wrecks in American history, a misread train order led to a head-on collision on the Nashville, Chattanooga & St. Louis on July 9, 1918, that killed 90 (Table A1.1). The year 1919 brought another spate of collisions. While such disasters were an outcome of the wartime traffic boom, to the ICC they seemed to portend a return to the bad old days and it urged all the carriers to experiment with a new technology—automatic train control.

Safety from War to War

With the Transportation Act of 1920, the federal government returned the carriers to private control in 1920 and for about a decade they thrived. With the economy booming, prices stable, and competition from automobiles, trucks and airlines not yet serious, profits rose. Rail stocks even participated modestly in the great bubble of the late 1920s. Such profitability supported safety-enhancing investments along with a host of new technologies, some of which originated in the prewar years. Labor market changes contributed to these improvements by reducing agency problems. With the carriers' labor demands no longer expanding after 1920, turnover declined; accordingly, companies could invest more in training. Pensions and other policies raised the cost of job loss, while safety organizations partially substituted for monitoring. As a result, even during the darkest years of the Depression, safety steadily improved and the carriers weathered the disruptions of World War II far better than they had those of World War I.

Sources of Safety—and Danger

Yet the hard side of safety regulation was not quite dead. The sensational collisions during World War I had resulted from employee mistakes. Responding to such accidents, the ICC had campaigned for authority to institute automatic train control, which it obtained in the Transportation Act of 1920. It was, essentially, the last gasp of Progressive-era railroad safety reform. In 1922, the ICC mandated that large carriers install some form of automatic train control on some passenger divisions.[28]

The order proved contentious. Previously, most safety-enhancing innovations such as air brakes and improved couplers had more than paid for themselves by increasing productivity. Automatic train control promised no such gains, and *Railway Age* wondered who would pay for it—a conflict that would become more serious in future decades. Moreover, the carriers argued that they could make better use of the money elsewhere, and they were probably correct. Indeed, to the extent that the order transferred resources from investments such as block signals that would have had a higher safety payoff it might have worsened safety.

While automatic train control contributed little to improving safety, the railroad technological community continued to produce and diffuse a steady stream of applied research. Much of this involved the transfer to railroads of technologies developed elsewhere and all of which reduced accidents and raised productivity. Research at Corning Glass resulted in signal lenses with improved color and transmissibility and employed early versions of Pyrex to prevent breakage, while a better-designed and manufactured wheel that sharply reduced failures emerged from investigations by wheel makers. Electromechanical car retarders appeared in the 1920s, reducing the need for yardmen to ride freight cars. Managers obtained new sources of low-cost information about risks as new methods of monitoring and nondestructive testing arrived. Manufacturers of artillery first employed magnetic particle crack detection (Magnaflux) to test gun barrels; the carriers imported it to search for defects in car wheels. Ultrasound, X-rays, and electromagnetic induction proved similarly useful. Centralized traffic control, which first appeared in a small way in the 1920s, was also an information technology. Finally, roller bearings for rolling stock and continuously welded rail (CWR) dated from the interwar years as well.[29]

The new technologies contributed to what *Railway Age* called in 1927 a "record breaking . . . program of track betterment." Use of block signaling continued to expand, protecting about 80 percent of all passenger miles on the eve of World War II. The carriers improved ballast and installed heavier rail on main lines; they replaced old bridges and began to shift to preserved ties. Wooden passenger cars all but disappeared except on branch lines. In addition, steel cars replaced wooden freight cars, while the arch-bar truck, which had been a fruitful source of derailments, generated such problems under the heavier cars that the ARA finally banned it on new equipment in 1928 in favor of cast steel. Again, virtually all of these improvements promised to return at least their cost of capital because they enhanced productivity as well as safety.[30]

Despite generally improving safety, the gradual spread of larger equipment continued to generate system problems, as it had in earlier decades. One of the ICC's first investigations under the Accident Reports Act of 1910 had been of a broken rail that caused a derailment on the Lehigh Valley near Manchester, New York, on August 25, 1911, that killed 28 (Table A1.1). It laid the blame on a transverse fissure in the rail, a new and mysterious form of failure that it attributed to the high wheel loads. Such accidents would continue to bedevil the carriers for decades. In the 1920s, studies also emerged of what modern researchers call track-train dynamics that pointed to interactions between roadbed and the new, 70-ton freight cars that could cause derailments. Other investigations emerged of rail and wheel stress that led to improved track, roadbed, and rolling stock.[31]

The driving force for larger equipment within both individual carriers and the ARA was the Master Mechanics, and it performed like a political interest group that pressed for better, larger equipment and technology. Yet railroads were an interacting technological system: changing any one part would affect all the rest. Larger freight cars were bound to have far-reaching ramifications for rails and roadbed. Yet neither the carriers nor the ARA ever developed a mechanism that coordinated the Master Mechanics' equipment decisions with those of the civil engineers whose domain was track and roadbed. Their professional organizations were correspondingly compartmentalized too. While engineers had been aware of these kinds of technological interactions for decades, the failure to develop the organizations needed to control interaction problems would return to haunt them in the 1960s as equipment continued to grow in size (Chapter 2).[32]

ICC oversight shaped these developments in several ways. In the early 1920s, following in the footsteps of Charles Francis Adams, it began hectoring the carriers to improve freight train brakes to offset the risks from heavier, faster trains. After about a decade of research and testing, Westinghouse developed the type AB brake that was easier to maintain, faster, and capable of an emergency stop after a service application. The ARA required it on all new equipment after 1933, but the Depression and freight car interchange slowed its diffusion. The ICC investigations of wrecks from transverse fissures also kept the pressure on both the carriers and the steel companies to solve that problem. In the early 1920s, Elmer Sperry devised an electromagnetic rail flaw detector that could discover such fissures and marked the first significant use of nondestructive testing. In the early 1930s, research at the University of Illinois supported by both railroads and steel makers finally demonstrated that transverse fissures resulted from shearing stress combined with tiny shatter cracks in the rail. The cracks resulted from hydrogen coming out of solution as the rails cooled, and so controlled cooling could prevent their formation.

The carriers' Safety First Movement, which had been born out of the Progressive-era threats to managerial freedom, prospered during the 1920s, in part no doubt because the cost of work accidents to employers continued to increase. Before World War I, the movement had met with much skepticism from the rail unions, which saw it as yet another publicity stunt intended to blame accidents on workers. Yet the safety movement had thrived under the US Railroad Administration during World War I. When the railroads returned to private control, the ARA took over safety work and managed to persuade unions of its good faith and to support the program. The ARA Safety Section meetings helped develop and disseminate information and expertise from disparate sources and provide it widespread publicity. The Safety Section also reflected the growing centralization of railroad institutions: in 1919, the Master Mechanics and Master Car Builders became the ARA Mechanical Section; and in 1934, the ARA and several other groups merged to form the Association of American Railroads (AAR). It was not accidental that the carriers formed the latter organization at a time when government was rapidly expanding, for the AAR was primarily concerned with lobbying.

One modern writer has described the traditional approach to work safety as the "Three E's": engineering, education, and enforcement. However, while Ralph Richards had emphasized the need to improve the safety of plant and equipment as a testimony of management's good faith, exhortation, not education, characterized railroad safety work during the interwar years. In 1913, the widow of Edward H. Harriman had established the Harriman Award for the railroad with the best safety record. It was a fitting tribute, for Harriman, until his death in 1909, had been as safety-conscious as any railroader of his day. Along with the Harriman competition, there were endless contests and jingles and posters intended to encourage workers to work safely. The goal, as noted, was to reduce work injuries, but also, by fostering safer work habits to generate a kind of organizational commitment that would reduce all forms of accidents.[33]

In Lewis Carroll's *Through the Looking-Glass, and What Alice Found There*, the Red Queen notes that "it takes all the running you can do, to keep in the same place." Safety work, the companies discovered, suffered from a kind of "Red Queen effect," for without constant shaking up it might get stale with both managers and workers simply going through the motions. Companies employed games, contests, and prize competitions, while employee publications such as *Frisco Man*, *C&O Rail*, and *B&O Employees' Magazine* all exhorted workers to work more safely and efficiently. Increasingly, however, enforcement reinforced exhortation with new rules, improved monitoring, and better employee selection and discipline. As noted, changes in the labor market after the war reinforced company efforts because with employment no longer expanding, dismissal became a more serious threat.

The bias in safety work toward what would later be termed "human factors" as a source of accidents reduced its effectiveness, for safety men rarely pressed for equipment improvements. Yet there were many areas where small, inexpensive changes might have yielded benefits. For example, safety departments were silent on the value of better water glasses and blowout plugs for steam locomotives, while labor unions were the driving force behind power reverse gears and better hand brakes for freight cars.

Interwar Safety Gains

Every area of railroad safety improved during the interwar years. Collisions and derailments fell sharply, and passenger fatality rates from such accidents, which had averaged about 1.8 per billion passenger miles in 1912–1914, dropped about 73 percent to 0.5 from 1937 to 1939. Not only did collisions and derailments become rarer, they became less lethal too. In the age of wooden cars (1902–1910), fatalities per passenger involved in such accidents averaged 2 to 3 per 1,000, but with the spread of steel cars they fell to less than 1 on the eve of World War II. Little accidents to passengers also fell, reflecting in part a shift in traffic to better-designed

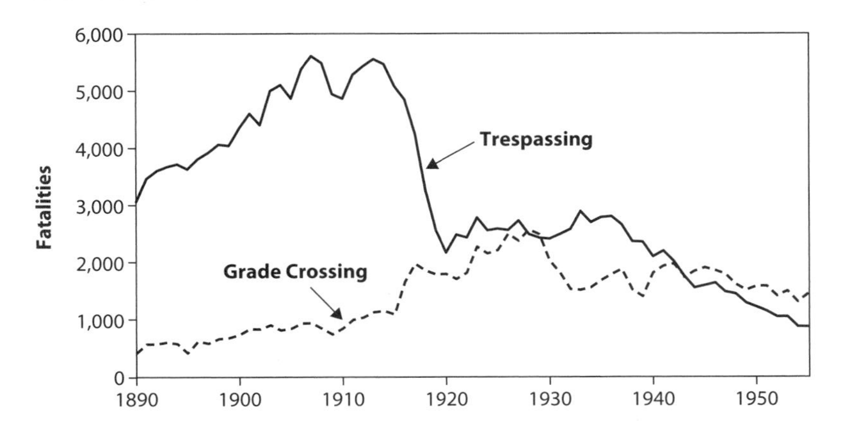

Figure 1.3. Fatalities of Trespassers and at Grade Crossings, 1890–1955. (ICC, *Statistics of Railways/ Accident Bulletin*, and author's calculations)

urban stations. In addition, as cars and buses nibbled away at the carriers' short-haul traffic, average length of journey rose, thereby reducing accidents from entraining and detraining relative to passenger miles.

Worker fatality risks fell about 62 percent—from 0.71 per million employee hours in 1912–1914 to 0.27 in 1937–1939 (Figure 1.2). For train crews the gains were even greater: fatality rates fell about 73 percent—from 4.69 to 1.26 per 1,000 employed during the same period. The breadth of these gains suggests that they were not due to any single dramatic change in technology. Instead, they reflected a wide range of investments and improvements in addition to organizational innovations such as Safety First and improved rules and discipline.

The decline in casualties to trespassers that began about 1910 continued during the 1920s as rising incomes and the spread of better roads and automobiles reduced incentives to walk the tracks and ride the rails. Fatalities jumped again during the Great Depression as a whole generation of young people took to the rails, and the long-term downward trend only returned with the advent of wartime prosperity (Figure 1.3).

The Grade Crossing Problem

Even grade crossing accidents finally began a long-term decline after 1929 (Figure 1.3). While the automobile reduced trespassing casualties, for a long time it had ballooned accidents at crossings. In the nineteenth century, crossing accidents were often lethal to the horse and riders in the buggy, but rarely dangerous to the train. Nor had liability costs been a major concern, and accordingly, railroads paid such accidents little attention. The automobile and especially the truck changed that equation because crossing accidents might derail the train and, if the truck contained flammables, yield a potentially expensive disaster. An early example occurred on the Soo on August 18, 1922, when a passenger train plowed into an oil tank truck at a crossing guarded only by crossbucks, killing 10, including 6 passengers (Table A1.1). A second threat to the carriers' bottom

"The Grade Crossing Monster." The railroads inveighed against this view that crossing accidents were the fault of the carriers. (Library of Congress)

line involved regulation. In 1925—the first year for which figures are available—there were 205,595 public crossings, or about 1 for every mile of track. Should states or cities require the railroads to fund either grade separation or improved guarding, the costs would be immense. Accordingly, in 1922 the Safety Section of the ARA instituted a "Cross Crossings Carefully" campaign similar to today's Operation Lifesaver, while individual carriers joined up too. The carriers' goals of these campaigns were not only to reduce accidents, but also to implant in the public's consciousness that they were the responsibility of drivers, not the railroads. Hence the public—not the railroads—should shoulder the expenses of improving crossing safety.

The early years of the twentieth century had seen a growing role for engineers and other experts in policymaking, and as Bruce Seely has shown, nowhere was their role more important than in road building. Inevitably, they began to think about grade crossing problems. While the railroads officially portrayed crossing accidents as the result of driver carelessness, in the 1920s their engineers and those in the new state highway departments tended to look for physical causes, such as bad visibility. They also understood that as the value of motorists' time increased with rising incomes, they might become impatient waiting for work trains. Thus, just as the public health movement had begun to "medicalize" childhood injuries instead of attributing them to parental failings, so a new emphasis on engineering and social causes began to replace moral failings such as carelessness as an explanation for crossing as well as worker accidents.[34]

Yet the new views proved equally serviceable to the carriers, for emphasis on the complexity of accident causes made them—like trespassing—a community responsibility, shifting the focus of crossing safety away from railroads and opening the door to public funding. As states—and especially the federal government in the 1930s—began to assume financial responsibility for crossing safety, a pressing question immediately arose: with more than 200,000 crossings, many of which saw little traffic, which should be guarded and which removed? This was a textbook resource allocation problem; in the 1930s, engineers developed the first mathematical models to assess hazards and allocate protection to the most dangerous crossings. Partly as a result, crossing accidents peaked in 1928 and then slowly declined thereafter, despite a 172 percent rise in vehicle miles traveled between 1925 and 1941. The gains reflected railroad safety programs that heightened awareness of risks, along with the closing and protection of a comparatively small fraction of crossings, and the decline in train miles that resulted from the Depression and the carriers' shift to longer trains.

Carrying the Arsenal of Democracy to War

World War II not only ended the Depression for the carriers, it also brought the greatest expansion of traffic in any comparable period in their history as the railroads carried the arsenal of democracy to war. Between 1939 and 1943, freight train mileage rose from about 452 million to a peak of 701 million, while with longer trains and heavier cars, revenue ton-miles rose even more—from about 333 billion to a peak of more than 737 billion between 1939 and 1944. In addition, with gasoline and tires rationed, people flocked back to the trains: passengers per train leaped from 57 in 1939 to 199 in 1944, while total passenger miles jumped from about 22 billion to 96 billion over the same period. The contrast with World War I, when gridlock had resulted in government seizure, could not have been sharper.

As traffic expanded, collisions and derailments exploded. The increase also reflected the effects of inflation on accident reporting, for most such wrecks were minor and as usual most involved freights. However, a number of spectacular train collisions and derailments reflected the usual sorts of errors and equipment failures. Thus, on September 6, 1943, a hotbox burned off the axle journal on the Pennsylvania's *Congressional Limited* near Philadelphia and the ensuing wreck killed 79. Not quite 2 years later, on August 9, 1945, a rear collision on the Great Northern near Michigan, North Dakota, killed 34 (Table A1.1). One reason such accidents had so many casualties was because, as noted above, the war increased the number of passengers per train, which increased fatalities per collision or derailment.

Despite such occasional disasters, railroad safety held up remarkably well throughout the war (Figure 1.2). Passenger fatality rates rose, but still averaged lower every year of the conflict than they had been in 1940. And while World War I had increased worker fatality rates, despite the explosion in traffic and dilution of employee experience, the carriers' safety institutions largely rose to the challenge. Worker fatality rates rose only to 0.28 in 1942 from their average of 0.27 in 1937–1939, and declined thereafter.

Postwar Progress

In 1942, the AAR had set up a Committee for the Study of Transportation, to chart the carriers' postwar future. Ultimately, the committee issued some 20 reports with recommendations on a range of issues from pipeline competition to taxes. Julius Parmalee, director of the Bureau of Railway Economics, authored the report on future traffic trends, and he was cautiously optimistic, predicting that both freight tonnage and passenger traffic in 1950 would exceed 1940 levels. There was a report on service, belying claims by later critics that the carriers were indifferent to customers' needs. Another covered public relations—a major concern in a regulated industry—and safety was an integral part of the carriers' public face. As Samuel Dunn, editor of *Railway Age*, pointed out, Ralph Richards's original safety campaign had been the "best . . . public relations work ever done." Finally, another outcome of the committee's work was the establishment of a new research department in 1945, which underscored the carriers' understanding of the need to improve technology. These reports collectively suggest that the railroads were hoping for a replay of the 1920s. In particular, they expected that their safety institutions, which had withstood the test of World War II, would continue to deliver ever fewer accidents, overseen, of course, by a benign regulatory environment. In fact, for about a decade that is what occurred.[35]

The Course of Train Accidents

Reported train accidents declined unevenly for about a decade after World War II. Some of this decline resulted from sharp increases in the ICC's dollar threshold for accident reporting, which rose from $150 in 1946 to $375 in 1955, far outstripping inflation during the period (Chapter 2). Yet during this same period, passenger fatality rates from train accidents declined precipitously from the low levels that had obtained during the 1930s.

However, as had occurred during World War I, spectacular collisions such as the Annandale wreck woke up the ICC, and in 1947 it intervened with a major regulation intended to improve signaling and safety. On June 17, 1947, it ordered use of block signals on all tracks where freights traveled over 50 mph and passenger trains over 60 mph. It also mandated automatic train control or use of cab signals anywhere passenger train speeds exceeded 79 mph. By 1953, when the order was fully implemented, block signaling had hardly increased at all while road governed by train control had grown by about 2,400 miles. The main effect of the order seems to have been to slow the trains, thereby pushing traffic to the (much more dangerous) trucks and automobiles.[36]

Far more important were a host of changes to railroad equipment and operating practices, many of which had prewar origins. The ICC finally mandated the AB freight brake in 1945. The brake had originated in the 1930s, but the carriers had been slow to adopt it due to its expense and the economics of freight car interchange. Centralized traffic control spread to about 10 percent of all road in 1955. Companies typically introduced it on single-track, high-density lines to speed traffic and avoid the need for double tracking. Where it replaced the old Timetable and Train Order system, it sharply reduced collision risks. Companies also installed electro-pneumatic disc brakes on passenger cars, as well as tight-lock couplers that would help prevent buckling in the event of a derailment. The carriers finally embraced the diesel during these years as well. It would not blow up, which had once been a bad habit of steam locomotives. Still, if its broader impact on railroading was revolutionary, its safety effects were modest because by 1945 incremental changes had greatly reduced the dangers of the iron horse.[37]

The new electronic sensing and nondestructive testing devices developed in the prewar years continued to spread, providing low-cost information and collectively reducing risks from derailments. In the 1950s, the Sperry Corporation developed an ultrasonic version of its rail flaw detector that would catch otherwise undetectable dangers. Portable, hand-held train radio use spread rapidly, improving safety in a host of ways. Magnaflux and similar forms of tests to find flaws in axles, wheels, and locomotive parts continued to spread, and companies gradually installed hotbox and dragging equipment detectors.

Impressive as these improvements were, there remained important gaps. The carriers' eroding profitability diminished resources and slowed the diffusion of a host of improvements. A Federal Communications Commission (FCC) ruling that refused to free up radio spectrum for railroad use impeded the spread of train radio and encouraged the PRR to invest in an outdated induction-based system. There were also flaws in the approach to collision prevention, for the ICC simply mandated train control for high-speed passenger lines while the carriers focused on centralized traffic control for high-density freight lines. Thus, low-speed, high-density commuter lines fell between the stools, and several wrecks on the Long Island demonstrated both the regulatory lapses and the problems resulting from eroding finances. On November 22, 1950, an engine driver ran a signal and collided with the rear of another train near Richmond Hill, New York (Table A1.1). The speed was only 30 mph but the aging cars telescoped and the wreck killed 79 people. Hearings revealed that the lead car in the following train was 40 years old and 25 days overdue for inspection. The Long Island was also bankrupt, having been denied a commuter fare increase since 1918. With the horse out of the barn, the ICC recommended installation of automatic train control, which the Long Island's owner, the PRR, duly implemented. There were other lapses too. Few companies employed speed recorders and the ICC ignored the issue; while the carriers had pioneered the use of energy-absorbing vestibules for passenger cars in 1913, they paid little attention to the safety of interior car design or of safe egress after a wreck.[38]

Little Accidents

Passenger fatality rates fell in half between 1945 and 1955 (Figure 1.2) and, as noted, much of the gain resulted from declines in little accidents at stations and elsewhere. This reflected the continuing long-term shift of traffic to safer urban stations combined with lengthening journeys that reduced station exposure relative to passenger miles. Workers too saw sharp improvements in safety as fatality rates fell about 58 percent in the first postwar decade (Figure 1.2). As during the interwar years, all major groups of employees participated in these gains. The process was no doubt encouraged by personal injury costs, which continued to rise.

Grade crossing accidents and fatalities also declined (Figure 1.3). Some of the reduction reflected fewer train miles; yet motor vehicle travel increased sharply from 1945 to 1955, while the number of crossings declined hardly at all. The gains reflected a gradual increase in guarded crossings, better methods of guarding, and economic models to determine which guards belonged at which crossings, as well as railroad safety campaigns similar to those of the interwar years. Some companies also experimented with novel safety procedures. The Great Northern, for example, applied Scotchlite lettering to its boxcars to make them more visible at night. The technology proved unreliable, however, and not until the 1990s would it become workable (Chapter 6). In addition, trespassing fatalities continued their long-term decline as a result of postwar prosperity and the decline in train miles.[39]

By 1955, the dangers of railroading as they affected passengers and workers had been declining for decades. These gains were almost entirely the result of private enterprise operating in an environment that made most accidents expensive. Thus, richer usually is safer. In response to market forces and some public prodding, the carriers committed the resources and their technological community developed the information that underlay steady gains in both efficiency and safety. Organizational innovations such as Safety First began to develop a nascent safety culture, while accidents to trespassers and at crossings began to fall between about World War I and 1930. While these happy outcomes were largely a result of market forces, there was always a regulatory "gun behind the door." But the gun was not used often and then not always well. More important were the ICC's threats that resulted in both better freight brakes in the interwar years and the discovery of transverse fissures as a cause of rail failures. As all forms of accidents fell, the social costs of railroad transportation—fatalities per unit of transportation output—declined as well, falling about 85 percent between 1900 and 1965. By the 1950s, except for an occasional, spectacular wreck, railroad safety had disappeared as a public concern.

Yet beneath the surface all was not well. The double bargain where the railroads would be allowed to determine their own safety priorities while the ICC allowed them to earn adequate revenues was beginning to crumble. Schumpeterian innovations had not stopped with the railroads, and the new competition from trucks, automobiles, and airplanes was eroding the railroads' markets, while economic regulation impeded an effective response. The carriers' financial position had been deteriorating since the 1930s, and in the 1950s the lack of resources began to erode safety. So did another uncoordinated stampede to newer, heavier equipment that generated safety problems similar to those that had arisen with the transition to heavier equipment a half century earlier. The result, after 1955, was an explosion of train accidents. Moreover, these disasters included for the first time numbers of frightening hazmat accidents, and they occurred in a period of heightened public worries over environmental and other risks. Within a quarter century, public outcry over these matters would almost completely reverse the old formula of strict economic and loose safety regulation. In 1966, Congress removed safety oversight from the ICC to the newly created Federal Railroad Administration (FRA), which it soon blessed with sharply expanded powers, and in 1976–1980 it largely substituted the market for the ICC's economic regulations. The impact of these and other changes in railroading on the carriers' safety in modern times is the subject of the remainder of this book.

2

Off the Tracks

The Rise in Train Accidents, 1955–1978

Death may be riding our railways daily. Tragedy could be lurking on every train.

James O. Eastland, Senator from Mississippi, The New York Times, *1969*

For want of a bolt, the joint was lost,
for want of a joint, the track was lost;
for want of a track, the train was lost.

William H. Moore, Vice President, Southern Railway, Railway Track & Structures, *1970*

Train wrecks have always been newsworthy, but in 1955 railroad safety had not been much in the news for a half century. During the following 25 years, however, train accidents again became an important public issue. A spectacular derailment on the Pennsylvania Railroad in 1968 helps explain why. If the Pennsylvania had once called itself the "Standard Railroad of the World," by the 1960s that was only a memory. On January 1, 1968, an empty tank car on westbound Pennsylvania Railroad Train PR11-A, an 88-car freight, struck a broken rail and derailed near Dunreith, Indiana (Table A3.1). It was one of 5,487 derailments reported that year, and one of the first investigated by the newly created National Transportation Safety Board (NTSB).

The body of the derailed tanker lifted off its truck and swung around, where it collided with eastbound Train SW-6, a 106-car freight, traveling about 32 mph, causing a general derailment. Train SW-6 contained five tank cars of hazardous chemicals, which were punctured in the derailment and caught fire. Joe Pitcher, a police officer from nearby Knightstown, was first on the scene. John C. Linegar and Mike Smith of the Indiana State Police soon joined him. They immediately set up a security perimeter and began to evacuate townspeople. Fire trucks soon arrived, parking close to the burning tanker, and Pitcher told them to move back. As the fire truck began to move, Pitcher recalled, "I heard a big hiss and I knew that it was trouble." At that point the burning tanker containing 20,500 gallons of ethylene oxide blew up in a fireball, tossing Pitcher in the air. "The only thing I remember," Smith recalled, "was Joe started running . . . and some of the liquid from the explosion landed on . . . [him]," setting his clothes on fire and sending him to the hospital. According to the NTSB, "The dome of the exploded tank car, weighing about 1,600 pounds, was

Dunreith, Indiana, the following day. The burning tank cars can be seen in the center, and at the bottom is a trailer full of refrigerators that did not make it. The top of the image contains the remains of the tomato cannery that resulted in exploding tomato cans as it burned. (Courtesy Indiana State Police Museum)

propelled a distance of about 720 feet into the residential area." After the explosion, Linegar recalled, "we did not need to ask the media to move back."

The explosion and fire destroyed a number of businesses, including a tomato cannery. As flames engulfed the cannery "those cans of tomato juice started exploding," Linegar remembered. "It sounded like gunshots." Another tanker of highly poisonous acetone cyanohydrin was punctured and spilled into a brook, killing several animals that drank from it and polluting municipal water supplies miles downstream.

Dunreith resulted from bad track—a rail that broke in a poorly maintained, unsupported joint—while the design of tank cars allowed their truck and body to separate, also contributing to the wreck. The NTSB's report was thorough and its conclusions unsettling. Among its 10 recommendations the board urged improved track inspection and better car design. It also went on to condemn the informal arrangements whereby the Interstate Commerce Commission (ICC) had delegated responsibility for dealing with such accidents to the Bureau of Explosives, pointing out that no one from the bureau had arrived on the scene until 10 hours after the wreck.[1]

Dunreith also made national headlines. Indeed, if any single train accident doomed the old system of safety regulation I have termed "voluntarism," it was Dunreith. It inaugurated a series of spectacular hazmat accidents in 1968–1969 that propelled freight train safety into the public arena, dominating congressional hearings (Chapter 3) and resulting in the modern regime of safety regulation. Dunreith also typified much that was going wrong with railroad safety at the time, and why it was again becoming a public issue. The railroads were reporting an upsurge in train

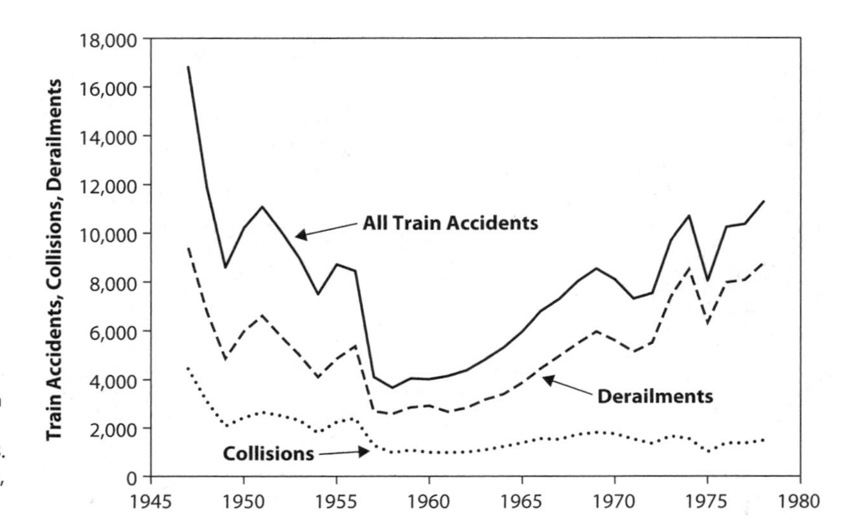

Figure 2.1. Reported Train Accidents, Derailments, and Collisions, 1947–1978. (ICC/FRA, *Accident Bulletin*, and author's calculations)

accidents—especially derailments—that began in the late 1950s (Figure 2.1). The poorly maintained track that led to Dunreith was replicated on countless miles of other carriers. In addition, Dunreith was a "system" accident; that is, it illustrated one of the many ways that equipment and track could interact to cause accidents.[2]

This chapter begins with an analysis of the statistics on the rise in train accidents—especially derailments. The following section delves into the major causes that lay behind this increase in wrecks, including the collapse in the carriers' finances and the increasing problem of track-train dynamics. The third section looks in more detail at the causes of some of these accidents and explains why the increase in train wrecks saw no corresponding rise in casualties. The following two sections investigate the increase in accidents involving hazardous materials, and the inadequacy of the carriers' response, for these, more than simply the rise in train wrecks, were what led to the revolutionary changes in safety regulation discussed in Chapter 3.

The Rise of Train Accidents

Figure 2.1 depicts train accidents as reported to the ICC and later the Federal Railroad Administration (FRA). By definition, train accidents involve the movement of on-track equipment (trains, of course, but also, for example, work equipment) and are defined by a monetary threshold. For perspective, the data begin in 1947. Only collisions and derailments are detailed; railroads also reported a stew of "other" train accidents. These included not only boiler explosions in the days of steam and crank case explosions as diesels took over, but also train-car collisions at grade crossings and—in one case—a collision with an apartment building that was being moved. Crossing accidents are discussed in detail in Chapter 6;

here the focus is on all forms of collisions and derailments. As can be seen, derailments have always outnumbered collisions and both forms of accident declined sharply down to about 1958. Thereafter, however, while collisions increased only modestly, derailments ballooned, peaking in 1978.

Counting Train Accidents

One should not take these reported data at face value. As Chapter 1 noted, both inflation and the reporting threshold influenced the number of train accidents. For example, in 1947, the threshold was $150 and the ICC treated wrecks that cost (say) $140 not as train accidents but instead placed them in the much less politically sensitive category of train service accidents (later train incidents). However, if inflation were to blow up the cost of such a small wreck to (say) $160, it would become a reportable train accident unless the commission also raised the reporting threshold. Similarly, a sharp increase in the reporting threshold without any corresponding increase in accident costs might make a number of accidents simply disappear into the train service accident rabbit hole.[3]

How important these influences were depended on both the amount of inflation and the speed with which the regulators responded to it. As a measure of inflation I use an Association of American Railroads (AAR) price index that captures the costs of materials and wages and benefits. Figure 2.2 presents index numbers of railroad prices and reporting requirements for 1947–1978. As can be seen, the reporting threshold rose more than prices from 1947 to 1957. The result was to underreport accidents during these years. From 1957 to 1974, however, rising prices along with stable reporting requirements overstated the rise in accidents. The large increase in the reporting threshold in 1975 was apparently intended to make up for inflation in the previous years, which—as can be seen—it overdid. Conveniently, this caused several thousand derailments to "disappear" from 1976 to 1978 at a time when rising accidents had put the FRA under congressional scrutiny.[4] Employing regression analysis, I estimate the impact of changes in prices relative to reporting requirements on measured derailments and collisions from 1947 to 1978, and subtract the estimated impact to obtain corrected measures of each type of accident. The results for derailments are contained in Appendix 2.[5]

For most of the years from 1947 on, underreporting was substantial. Indeed, in the early years the ICC was quite candid about this, observing in 1948 when it raised the threshold from $150 to $250 that only about 3 points of the reported 27 percent decline in train accidents that year were real. The following year after another increase it again noted that "part of the improvement is factitious."[6] A study by the Railroad Retirement Board noted that the doubling of the accident threshold in 1957 (from $375 to $750) caused a 56 percent drop in yard accidents! My calculations suggest that total reported derailments fell 50 percent while their corrected values

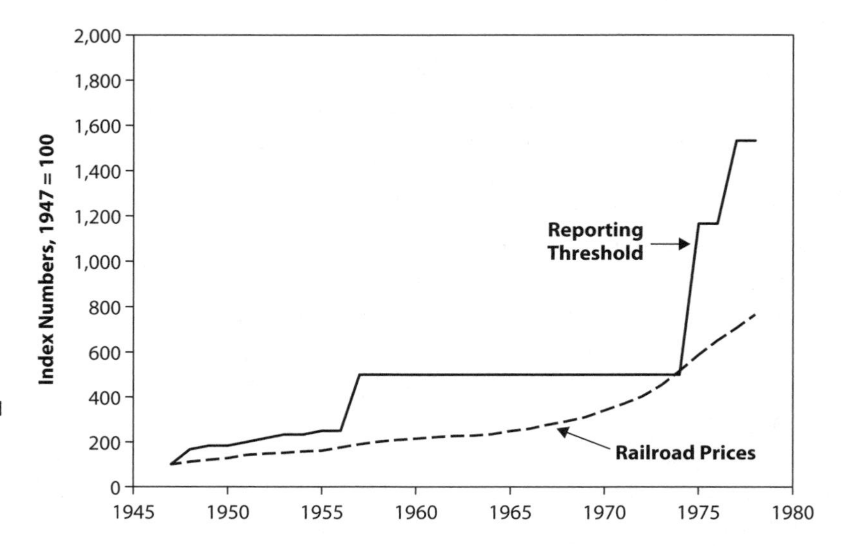

Figure 2.2. Reporting Requirements and Railroad Prices, 1947–1978. (ICC, *Accident Bulletin*, and AAR, *Railroad Facts*, various years)

declined only about 32 percent (Appendix 2). There were no changes in reporting requirements from 1957 to 1974, a time of considerable price increases, and the result was that reported derailments rose by 117 percent compared to 90 percent for the corrected figures. However, the near quadrupling of reporting threshold between 1974 and 1978 again sharply suppressed reported accidents.[7]

There are other difficulties with the train accident figures as well. While it makes sense to present such data relative to exposure, both the ICC and the FRA traditionally published collisions and derailments per train or locomotive mile. Such an adjustment is plausible for collisions, which do indeed depend on train miles, but for derailments car miles are a more appropriate measure, as a 1962 report by the Railroad Retirement Board and later complaints by *Modern Railroads* noted. Consider: if derailments were unchanged while the carriers shifted to fewer, longer trains with the same number of cars, the accident rate per train mile would rise, while the car-mile rate would be unaffected. Because trains were in fact growing longer, reporting derailments per train mile imparted an upward bias to accident rates. From a trough in 1957 to the peak in 1978, reported derailments per train mile rose by a factor of 4, but only by 3.5 when adjusted for car miles. However, for this period freight car miles showed little change; hence, while both reported and adjusted derailments relative to freight car miles are contained in Appendix 2, they are not presented here.[8]

Finances, Technology, and Derailments

In the years after World War II, the railroads' worsening financial problems increasingly undermined safe railroading. While one can make a case that these problems date from the Eastern Rate Case of 1910, the difficulties worsened in the 1950s. The Penn Central bankruptcy brought matters to a head. The Penn Central was the outcome of a 1968 merger between the Pennsylvania Railroad and the New York Central, along with the ICC-mandated inclusion of the financially troubled New York, New Haven & Hartford. It became the largest railroad in the country. Two years later, on June 21, 1970, it declared bankruptcy. While some of its problems were unique, its collapse reflected and symbolized the deeper forces that were threatening nearly all American railroads.[9]

The problems had been a long time coming. In 1947, a *Railway Age* editorial described the constellation of forces shaping railroad earnings as "The 'System' That Always Fails" and went on to claim that profits were half of what they should have been.[10] This "system," as the *Age* and others saw it, had three components: (1) the rise of intermodal competition that began about World War I, and which the carriers always claimed was heavily subsidized; (2) the railroads' powerful labor unions, which inflated costs and hindered productivity growth; and (3) ICC regulations and other public policies that raised costs, reduced profitability, and hindered adaptation to changing market conditions. There were also changes in economic activity with important consequences for the railroads that the *Age* ignored.[11]

Capitalism, as Joseph Schumpeter saw, is inherently restless, creative, and destructive, all at once. And if the automobile, the truck, and the airplane were wonders of the twentieth century, they were disastrous to the carriers' way of doing business. As early as 1912, the *Railway Age* noted that the automobile was eroding traffic on branch lines, while trucks were stealing short-haul freight traffic. By the mid-1930s, as roads and vehicles improved, there were some 220 billion vehicle miles traveled and, no doubt, more passenger miles. Some of this increase came at the expense of the railroads. In the mid-1930s, the airlines introduced the DC-3, which with a cruising speed of about 200 mph, a range of more than 1,500 miles, and a capacity of about 25 passengers was superior to train travel. By the late 1940s, aircraft cruising speeds reached 340 mph, and planes were larger with an intercontinental range. From 1930 to 1960, railroad passenger miles declined from 26 billion to 21 billion while airline passenger miles over that period rose from 73 million to 35 billion. In 1958, Patrick McGinnis of the Boston & Maine explained the source of railroads' problems to Congress. "The culprit is that wonderful thing called progress," he said. McGinnis was at least partly right. Two years later, the airlines introduced jets.[12]

Unlike the carriers' passenger service, railroad freight traffic continued to grow during these years, but at a much-diminished rate. Between 1930 and 1950, ton-miles rose from 390 billion to 597 billion, but they hardly increased at all in the following decade. Better trucks and roads, especially the expansion of the interstate system after 1956, resulted in an increase in truck traffic from 20 billion ton-miles in 1930 to 298 billion in 1960, much of which came at the expense of the carriers' high-margin, less-than-carload shipments. Eastern and Midwestern carriers, which had thrived on short-haul traffic, found truck competition especially severe. Thus, the carriers in the worst shape included not only the Pennsylvania and the New York Central, but also the Central of New Jersey, the Erie, the Lackawanna, the Lehigh Valley, the New Haven, and the Rock Island. On the New Haven, automobiles and trucks had stripped so much traffic from branch lines that by 1960 80 percent of its freight revenue came from 20 percent of its route miles. Pipeline and water traffic also expanded, stealing bulk shipments of crude oil as well as grain and other commodities.[13]

The rise of this new competition required major changes in railroading. The carriers needed to improve service with faster, on-time deliveries. By the late 1960s, in editorials, articles, and surveys, *Modern Railroads* was excoriating them for slow and unreliable service that ceded markets to trucks. They also needed to raise productivity, and their unions often impeded this process. Diesels had almost entirely replaced steam locomotives by 1955, yet railroad unions were still fighting to keep the remaining 5,000 firemen as late as 1985. In the 1890s, companies had agreed to a union demand that 100 miles of travel constituted a day's work for freight trainmen (and 150 for passenger work). And so it remained, well into the 1980s. Yard and road train workers could not do each other's jobs. The effect of such rules on the Long Island Rail Road was that engine drivers who worked only 29 hours a week were paid for 72 hours.[14]

Yet if intermodal competition and balky unions beset the carriers, the core of their problem was government. The proof lies in the years after 1980; with economic deregulation the carriers have prospered—yet they are still unionized and intermodal competition has increased. Federal and state policies contributed to the carriers' postwar unprofitability in a host of ways, only a few of which can be noted here. State full crew laws (Chapter 1) usually required 6-person freights and as late as 1965 at least 10 states had such statutes on the books. Yet by then the carriers' preferred crew was three or fewer, and with its operating unions on strike during the mid-1960s, the Florida East Coast was running trains with a crew of two. Such rules cost money: $500 million a year in 1959, the AAR claimed, and they made more frequent, shorter trains uneconomical, impeding the carriers' ability to compete with trucks.[15]

With most of their assets immobile, railroads were also ideal targets for state taxes. A study of 31 states done by the AAR in 1957 compared tax rates for railroads with that of other businesses and found that the carriers' higher rates raised payments by $141 million a year. New Jersey, for example, taxed railroad property at twice the rate of other businesses. A few states even taxed the railroads to support airport construction. Twenty years later, another study found similar although less extreme differences.

States also usually opposed abandonment of branch track and service. Typically, opposition was greatest on lines with substantial traffic, which were also those that lost the most money, and while states ultimately approved about 87 percent of passenger train discontinuances requested between 1951 and 1956, they often took a long time to do so. Alfred Perlman of the New York Central told Congress that abandoning one commuter line took three years even though it was losing $3 million a year. It would have been cheaper, he thought, to have bought each of the commuters a new Chevrolet.[16]

More than any other government entity, the ICC shaped the railroads' fate. While it is easy to criticize the commission, at least in broad terms, it was doing what Congress wished. Railroad economic regulation that began in 1887 was a response to the carriers' near monopoly, but while this condition had disappeared by about 1930, the various transportation acts never adequately addressed these changed circumstances. Yet if Congress set the broad contours of policy, the ICC had free will. The commission had power over rates, mergers, entry, exit, and financing. Its deliberations often moved with the speed of a glacier; its findings could have erred in favor of competition, but they rarely did. Along with state foot-dragging, its hearings over proposed abandonment of service or routes inevitably slowed those processes.

Control over rates hamstrung the carriers' ability to respond quickly to truck competition, and it hindered the development of intermodal transport in a number of ways. It obstructed the development of store-door shipments and prevented the carriers from entering long-haul trucking and water transport.[17] In times of inflation, simply delaying rate increases eroded railroad profitability, and in 1958 James Symes presented a table to a Senate committee that computed the cost of delay for six rate increases between 1946 and 1957 that the Pennsylvania had requested and that were ultimately granted. The revenue lost from delay, he claimed, came to $131.9 million. A later study put the cost of such delays to the carriers as a whole at $2.2 billion from 1966 to 1975. In 1961, when the Southern Railroad invested in large Big John hopper cars to transport grain to the Gulf Coast, it cut rates from $10.50 to $3.97 per ton to attract the traffic. Barge lines protested to the ICC. It took until 1965 and required two trips to the US Supreme Court before the railroad prevailed. Similarly, when the carriers introduced unit coal trains, the ICC's rate policy threatened to make them uneconomic.[18]

Collectively these forces impeded the carriers' ability to react to the rise of intermodal completion, retarded productivity, and reduced profitability. The railroads employed too many unproductive workers and had too much unproductive plant and equipment. Freight cars averaged about 50 miles a day. A 1976 study of Midwestern carriers demonstrated that 20 percent of the track carried two-thirds of the traffic. Such low

Table 2.1. Railroad Deficit from Passenger Operation, Selected Years, 1963–1975 (millions of dollars)

Year	Avoidable	Total
1963	9	399
1965	44	421
1967	138	486
1969	226	464
1970	252	477
1973	13	187
1975	16*	189

Source: AAR, *Railroad Facts*, 1977.
*Surplus.

productivity generated low profitability. Contributing to the problem was the deficit on passenger service (Table 2.1), which in 1970 amounted to at least $252 million, equaling 52 percent of the carriers' net income that year.[19] As a group, the Class I railroads' return on assets as defined by ICC accounting rules usually ranged from 2 to 4 percent, which was far less than most other industries paid, and therefore much less than the railroads' cost of capital. As a result, except for their ability to borrow on equipment trust certificates, the carriers were shut out of the capital markets. "Nobody knows how to sell a railroad equity [stock] issue," *Modern Railroads* complained in 1975, and it wondered how the carriers were supposed to refinance bonds at 14 percent when they were only earning 4 percent.[20]

In fact, actual profitability was even less than reported, for the reported figures ignored deferred maintenance. In addition, under ICC rules, the railroads could not depreciate track and roadbed. Rather, they employed "betterment accounting," which capitalized the investment, and then wrote it off as an expense only when it was replaced. With little replacement going on after World War II, the net effect of betterment accounting was to ignore depreciation—a particularly serious omission in a capital-intensive industry—thus overstating net income.[21] Finally, neither the ICC nor the railroads fully grasped the effects of inflation on railroad profitability. For example, in 1949 *Railway Age* crowed that the carrier's investment of $1.3 billion in 1948 compared favorably with the $1.1 billion they had invested a quarter century earlier in 1923. While this amounted to a 20 percent increase, the journal failed to note that the general price level had risen 37 percent over the same years, and so the carriers were investing less than in the earlier period. Thus, profitability also needs to be adjusted for inflation. After about 1967, railroads' inflation-adjusted returns were negative.[22]

Such long-term unprofitability had many consequences. By stigmatizing the carriers as a dying industry, it made attracting top talent in engineering and management more difficult. Anecdotal evidence suggests that

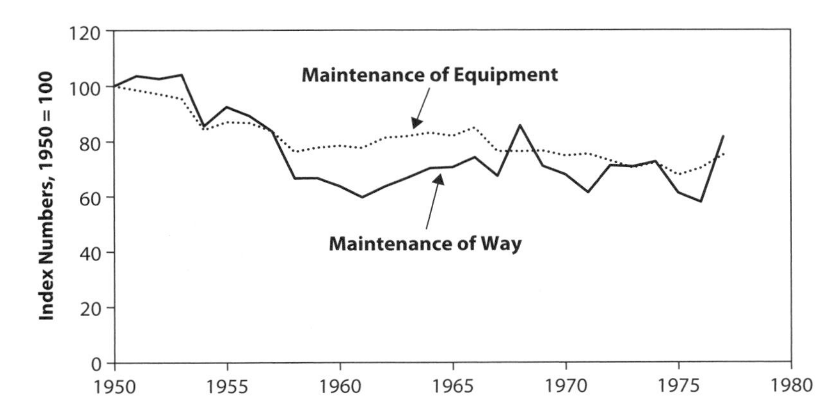

Figure 2.3. Real Maintenance of Way and Equipment Spending, Adjusted for Track, Traffic, and Productivity Growth, 1950–1977. (AAR, *Railroad Facts*, and author's calculations) Real spending on maintenance of way adjusted for track miles; real spending on equipment adjusted for freight and passenger car miles; both adjusted for productivity growth.

management of some carriers was weak. Robert Hamilton of the Southern claimed that railroads had "the worst management of any major industry." Roger Grant's research reveals that mediocrities managed the Erie on the eve of its merger with the Lackawanna in 1960. When Jervis Langdon became president of the B&O in 1961, one of his subordinates observed that "the whole B&O leadership was dead above the neck," and when new management took over the Grand trunk in 1971 they "were dismayed to find steam locomotive parts stocked . . . a decade after the demise of steam."[23] Lack of funds and unprofitability also retarded the introduction of newer, safer methods and led to deterioration of plant and equipment. The B&O began its first centralized traffic control installation only in 1958, but the project was "held up because of lack of funds." Some years later, *Modern Railroads* complained that railroad holding companies were not reinvesting revenues in the railroad: but would *you* want to invest for a minus 5 percent return? Thus, by 1950, the ICC was already estimating that the railroads had accumulated $950 million of deferred maintenance of track and equipment. Thereafter, matters got worse. As can be seen in Figure 2.3, real expenditures per mile of track and real equipment expenditures per freight car mile for Class I carriers (both adjusted for productivity gains) fell 40 percent and 30 percent, respectively, between 1950 and their low point in 1976. By 1974, an FRA study indicated under-maintenance of about $5.7 billion in rail and ties alone.[24]

In track and equipment maintenance, as in company safety work, there is a Red Queen effect: simply to maintain quality requires a high level of spending. Thus, physical measures of rail and tie renewals point to woeful under-maintenance. American Railway Engineering Association (AREA) data indicate renewals averaged about 3 percent of all ties in the early 1950s; such figures implied under-maintenance unless ties lasted 33 years (1/0.03); in fact, actual tie life was about 25 years. Throughout the 1960s, however, renewals fell to fewer than 2 percent of the total—implying

The upsurge in derailments that began in the late 1950s was due to years of under-maintenance that had resulted in massive deterioration of railroad track—as this 1967 image of the New Haven indicates. (*Trainman News*, October 23, 1967, courtesy SMART Transportation Division)

large-scale deterioration. ICC data on miles of new and secondhand rail re-laid amounted to about 6 percent of track miles in 1950, implying a rail life of nearly 16 years, but by the 1960s renewals had fallen to 2 percent of track miles or less. Unless rail life had miraculously risen to 40 years or more—which in the face of rising train weight was unlikely—the figures again suggest massive under-maintenance. The AREA also collected data on rail defects; five-year failure rates bottomed out in 1957 and rose steadily thereafter, in 1965—the last year they are available—standing at 2.4 times their 1957 level. For all rail of all ages, failure rates from all causes rose about 38 percent from 1962 to 1970—the only period for which they are available.[25]

The reader should take both the monetary and the physical data on under-maintenance with a dose of salt, for maintenance requirements also declined during these years. It made economic sense to let standards deteriorate on light-traffic branches while other lines were destined for scrap. Still, contemporaries also believed that under-maintenance was real. In 1959, Daniel Loomis, president of the AAR, admitted that there were "big backlogs of improvement projects." "Standards of maintenance . . . are not being observed," a representative of the B&O told the ICC in 1961 and went on to detail the deterioration of that company's track and roadbed. In 1964, the editor of *Railway Track & Structures* commented that the approach to maintenance seemed to be "how little can we get away with." A year later, the editors of *Modern Railroads* observed that tight budgets had led to weeds growing up in the tracks. By the 1970s, that journal was terming deferred maintenance a "crisis" and it later referred to the "cancer of maintenance deferral" that was a "time bomb in the track department." By then the bomb had been ticking for two decades. Later the journal referred to the "continuing deterioration in railroad tracks." Equipment was also crumbling: ICC data reveal a steady rise in defective locomotives and freight cars. In 1958, James Symes of the Pennsylvania spoke for most of the railroads when he told Congress "we are deteriorating badly."[26]

Track-Train Dynamics

A second cause of the rise in derailments was the wholesale introduction of much larger and heavier equipment that began in the 1950s. Moreover, the new equipment not only led to its own set of risks, it also sometimes interacted with bad track to yield wrecks that neither alone would have caused.

The carriers did not cede their markets to trucks and cars without a fight. They introduced improved service with the development of unit trains for coal and then grain and other commodities that could avoid classification yards, thereby increasing speed and equipment utilization. Piggyback (trailer on flatcar, or TOFC) and containerized (container on flatcar, or COFC) service arrived in the early 1950s. Some of these new services required new equipment. To be economic, piggyback required longer freight cars—some reached 89 feet. Other cars also grew in size: in 1950, the 44-foot, 50-ton boxcar ruled the roost and average car capacity was about 53 tons. By 1980, that average had risen about 50 percent to 79 tons and some cars were very large indeed.[27]

Among the first of these were the Big John grain cars noted above, purchased by the Southern in 1959. They carried 103 tons and weighed 22 tons for a gross weight of 125 tons. With a payload/deadweight ratio of nearly 5:1 at a time when most cars yielded about 2:1 the economics were compelling; Southern expected to earn about a 32 percent return on their extra cost. Scenting such payoffs, others quickly followed. In 1961, General American Tank Car designed a 30,000-gallon, 125-ton tank car (the "pregnant whale") to help the carriers compete with pipelines. In 1966, the Pennsylvania introduced the 38,000-gallon "rail whale." The L&N bought 100 100-ton coal cars with automated dumping capability (they would unload in 23 seconds) to be employed in a unit train. They replaced 600 conventional cars. The new cars were not only large, they were also innovative; tank cars began to be constructed without a full underframe, employing instead the tank as its own support. Soon 150-ton cars appeared and by 1965, as *Railway Age* noted, "the big cars [had] become regulars."[28]

By about 1965, they had also become big problems, for derailments had begun to rise. As noted in Chapter 1, equipment decisions were largely the domain of the Master Mechanics, which behaved like a political interest group that focused on locomotives and rolling stock. Similarly, equipment suppliers designed and tested cars and parts, but it was no one's job to test how heterogeneous trains of heavy and light, empty and loaded, cars interacted under various track conditions over long periods, and virtually no such testing was done. Yet such equipment along with track and roadbed was part of an interacting technological package—and around World War I the failure to coordinate these interactions had led to trouble as new car weight rose first to 50 tons and then 70 tons in the 1920s.

Four decades later, the problems resulting from 100-ton cars were yet another manifestation of the same old problem. Efforts to develop institutions that would integrate track and equipment design probably dated from the 1930s, when the AAR Mechanical Division and the AREA had established a Joint Committee on Relation Between Track and Equipment. Its first concern was the stresses that poorly counterbalanced steam locomotive driving wheels generated in track. In 1959, the committee recommended wheel loads of no more than about 800–830 pounds/inch of wheel diameter. Thus, a 33-inch wheel should support no more than 26,400 pounds. The 70-ton car with a gross weight of about 208,000 pounds resting on 8 33-inch wheels passed the test, as it resulted in a load of about 792 pounds/inch. Yet when the committee finally sounded the alarm on large cars, no one responded. Newer 100- to 125-ton cars resulted in a wheel load of 951 pounds/inch; even with 36-inch wheels the load was 872 pounds/inch. Some roads ran cars with more than 1,000 pounds/inch of wheel load and so the railroads were once again caught flatfooted by the epidemic of derailments that arose.[29]

John Fishwick became president of the N&W in 1971. The following year he complained to the AREA about the "lack of coordination and communication" he found at that company. "We need to look at the railroad and its work as a system . . . rather than as a series of rigidly compartmentalized pigeon holes," he argued. In 1981, John Read, general manager of the Bessemer & Lake Erie, also condemned the "provincialism" of railroad departments and he singled out the design of freight cars by mechanical departments without considering their impact on track as a good example. Yet if lack of coordination was one problem, lack of information was another. Many of the problems of the new cars took time to reveal themselves; until then skeptics could dismiss merely theoretical claims. Finally, arguments about car size were inherently economic and their critics simply did not have the cost data to back up their arguments. Finances also played a role in shaping the railroads' investment choices; the carriers could sell trust certificates for freight cars because they provided good security, whereas track and roadbed did not.[30]

The new equipment contributed to derailments in two ways, and deteriorating track and roadbed magnified both of them. The first of these reflected the effects of high wheel loads. Heavy loads, of course, required more braking power and generated more heat, which led to cracking and failure of high-carbon wheels. Heavier wheel loads might widen gauge—especially on curves of poorly spiked track. The heavier cars also crushed rail heads, and unit trains—where the same impact would occur the same way, perhaps 400 times in succession—were particularly hard on track. About one-third of main track was still composed of rail weighing less than 100 pounds/yard, and the heavier cars worsened problems of web failures at joint bars in lighter rail. Robert Sewell had been a signal maintainer on the Southern. "Every time you've got a joint you've got a point of weakness," he noted. When the wheels hit a joint they would "chew that tie to pieces [and] knock the ballast out." When the North Western merged with the Chicago Great Western in 1968 and began to run heavy trains over that line's light track, there were so many derailments the company had to label them by station and number—Cumming, Iowa, I, II, and III.[31]

Heavy loads also led to shelly rail heads that ultimately might yield dangerous detail fractures that could spread rapidly in older rail—a problem that became so important in the 1960s that the AREA rail committee began a long-running research project to find a solution. The normal friction generated by a truck with fixed axles rounding a curve resulted in rail corrugation as the wheels slipped. Corrugations reflected the high-contact stresses that resulted in surface cracks and plastic flow of metal; unchecked they could lead to rail failure. Both the Canadian National and the Canadian Pacific introduced 100-ton cars in unit trains about 1964 and promptly began to experience worsening corrugation, or "washboard rail." Almost invariably, it occurred on the inner and lower rail of a curve.[32]

The second set of problems that arose from the new cars reflected design constraints. The railroads' 56.5-inch gauge and clearances precluded widening cars; so they grew taller and longer and both generated a host of tracking problems that were magnified by the deterioration of track and roadbed. As a car rounded a curve, the ratio of lateral to vertical (L/V) forces increased. The giant hoppers and "high cube" cars had a high center of gravity, which worsened the problem. Moreover, rails were elevated on the outside of the curve to counteract centrifugal forces at high speeds. But slow-moving trains leaned into the curve. Combine this with poorly maintained track with a low spot on the inner rail and the inner wheels might climb the rail or even tip the train over. Older wheels with a sharper flange angle were less likely to climb than newer wheels. Coupling long and short cars also raised the odds of derailment on a curve, as a long car and coupler might literally pull the shorter car and coupler off the track. Newer locomotives had more powerful dynamic braking and required great skill in use; slowing the train might lead it to jackknife and derail as the slack rode in—especially when entering a turnout.[33]

The new cars also tended to rock 'n' roll. Jointed rail was prone to low spots at the joints, with staggered joints 39 feet apart. Eighty-nine-foot cars had trucks almost precisely two rail lengths apart. With such equipment or with cars having a high center of gravity, at certain speeds harmonic motion set in. Depending on the car's springing and snubbing and track conditions, wheels might lift enough to cause a derailment. The long, light flatcars were also prone to derailment from the compressive forces during dynamic braking, especially if they were located near the locomotive. At certain speeds these cars would "hunt" (weave from side to side) as well, increasing lateral forces, and the problem was worse if the gauge was slightly wide. Accidents from such conditions should not have come as a surprise. In 1964, the Mechanical Division's wheel committee reported on the decision to raise freight cars' center of gravity from 84 to 98 inches. It urged a "thorough review" of track conditions, pointing out that new cars had a tendency to rock 'n' roll and go off the rails on curves.[34]

As railroaders understood, both bad track and track-train dynamics were problems that affected the railroads as a system. Because of freight car interchange, the newer, larger cars might cause derailments on foreign as well as home lines, while bad track on any one carrier was an economic problem for all because it slowed shipments, worsened loss and damage, and generally eroded system profitability. *Modern Railroads* described the carriers with poor track as bad apples that could spoil the barrel.[35]

An Explosion of Train Accidents

The ICC and later the FRA reported train accidents under three broad cause categories: defects in track and roadbed, defects in equipment, and "employee negligence" (later given the less pejorative and more accurate name "human factors") plus "miscellaneous." Table 2.2 presents collisions and derailments from these causes for selected periods from 1947 to 1978. Because the concern here is with the relative significance of various causes of accidents, the data are not corrected for inflation or reporting changes.

Down to the mid-1960s, equipment failures were the dominant cause of derailments, typically accounting for 40 to 50 percent of the total.

Table 2.2. Reported Collisions and Derailments by Major Cause Category, 1947–1978

		Collisions (annual averages)				Derailments (annual averages)			
Period		All Causes	Human Factors	Track	Equipment	All Causes	Human Factors	Track	Equipment
1947–1950	Number	3,018	2,826	4	104	6,761	1,425	1,576	2,447
	Percent	100%	93.6%	0.1%	3.4%	100%	21.1%	23.3%	36.2%
1959–1962	Number	1,013	898	8	50	2,817	339	560	1,394
	Percent	100%	88.7%	0.8%	4.9%	100%	12.0%	19.9%	49.5%
1975–1978	Number	1,303	1,019	80	84	7,784	1,264	4,042	1,797
	Percent	100%	78.2%	6.1%	6.4%	100%	16.2%	51.9%	23.1%

Source: ICC/FRA Accident Reports, various years.

Note: Data are uncorrected for inflation or reporting changes. Miscellaneous causes omitted.

While a large number of pieces of a car or locomotive could cause an accident, the leading culprits were axles, especially journal hotboxes; bad trucks, and broken wheels. The carriers' efforts to stem hotbox accidents are detailed below. An example of what motivated this campaign occurred on April 2, 1961. A hotbox on the 39th car of Milwaukee Train 263, a 137-car freight, heading west at about 57 mph near Mauston, Wisconsin, derailed the remainder of the train, killing 3. As this wreck reveals, equipment failures often occurred at high speed because they seldom gave any advance warning, and therefore they could be especially deadly.[36]

Clearly, however, during the 1960s the mushrooming of wrecks from track and roadbed eclipsed the rise in equipment failures. In 1974, reported derailments stood at 3.3 times their 1958 level. Those from broken rails and joints were 5.5 times as high; for frogs and switches the figure was 9.5; and derailments from bad ties were an astonishing 165 times their 1958 level. Until 1975, when the FRA revised its accident codes, it is not possible to separate derailments that simply resulted from bad track from those due to track-train dynamics, for many wrecks that were placed in another category might well have involved such interactions. In 1978, however, track-train dynamics accounted for one-quarter of all derailments.

As demonstrated above, this deterioration in track and equipment and thus the surge in derailments reflected the carriers' crumbling finances. Predictably, therefore, accidents were most common on the roads in the worst financial shape. Finances mattered a lot; derailments on the weak roads skyrocketed. In 1969, the crumbling Penn Central alone accounted for 26 percent of all derailments nationwide. Statistical analysis confirms this conclusion that low profitability was a major cause of the rise in derailments.[37]

The Rise in Collisions

Like derailments, collisions dropped sharply in the early 1950s, and much of the reported decline was the result of sharp increases in the ICC reporting threshold (Appendix 2). Also, like derailments, almost all collisions involved freight trains. After 1958, reported collisions began to rise but far less than derailments. Adjusted for inflation and reporting changes, the increase from a trough in 1958 to their peak in 1969 was about 39 percent and thereafter they fluctuated until beginning to decline after 1978.

As Table 2.2 reveals, collisions largely stemmed from human factors, which accounted for more than 90 percent of the total in the 1950s. However, the railroads' deteriorating track and equipment even increased collisions, for these causes rose to account for about 12 percent of the total in the late 1970s. These stemmed, for example, from defective brakes or hand brakes, or broken couplers that might cause a train break-in-two. Defective car retarders caused only 9 accidents in 1958 but accounted for 43 wrecks in 1969. Still, human factors remained the dominant cause.

Two collisions on the Union Pacific and the Illinois Central in the late 1960s illustrate such causes.

On the Union Pacific, near Wamego, Kansas, on December 21, 1967, eastbound Passenger Train 18, consisting of 1 diesel-electric locomotive and 6 cars, took the siding to allow westbound Extra Freight 857, consisting of 3 diesel electric units, 54 cars, and a caboose to pass. The weather was clear, but it was dark; the track was straight and level and governed by block signals. For some reason the engine driver of Train 18 ran through the siding and emerged onto the main track in front of Train 857, then traveling at about 50 mph, resulting in a head-on collision that injured 79 passengers and crew. The engineman of Train 18 was 69 years old, and while suffering from diabetes, was apparently in good health and had been on duty about 3 hours. No convincing cause of the accident was ever ascertained, yet later research (see below) that investigated how fatigue might dull the reflexes of even seemingly well-rested engine drivers would begin to shine light on such apparently mysterious wrecks.[38]

A bit less than two years later, at about 4:30 a.m. on September 26, 1969, Illinois Central Extra Freight 5055 North, consisting of 3 locomotives, 128 cars, and a caboose was traveling on track governed by automatic block signals near Riverdale, Illinois. The engineman ran an approach signal that showed caution and a home sign set to stop, plowing into the rear of Extra Freight 1214 North at about 60 mph, killing three and injuring three of the train crew. The engine driver had been on duty about 3.5 hours; he had been subject to several recent disciplinary proceedings for actions that had resulted in injuries and accidents; tests revealed that he probably had a blood alcohol level of 282 milligrams per 100 milliliters when he came on duty—which is to say he was legally drunk according to Illinois law. Alcohol was a "human factor" that had plagued the railroads since their inception, but it was not until a disastrous passenger wreck in the 1980s that it would become a public policy issue (Chapter 4).[39]

Yet main-line collisions like these that resulted in deaths and many injuries were always exceptional. In 1969, a peak year for collisions, only 126—about 7 percent—of the 1,810 reported collisions resulted in any casualties. Most were yard accidents; in fact, switching accidents resulted in about 86 percent of the rise in collisions from their trough in 1958 to their 1969 peak. As noted, most such accidents reflected human fallibility: in 1969, 399 collisions reflected improper use of hand brakes while another 250 resulted from failing to have a man on a set of cars during switching.

While definitive explanations for this rise in yard collisions are impossible, some speculations are in order. Yard collisions rose at a time when the costs of loss and damage to shipments were also increasing. Much of this resulted from rough handling in yards, as workers sent cars down hump yards at too high a rate of speed or without men to guide them, which might also lead to derailments.[40] Thus, rising loss and damage, yard collisions, and the rise in derailments were all part of the same package and were symptomatic of a broader organizational breakdown. *Modern Railroads* thought that such accidents reflected deteriorating morale as employees watched their companies crumble and dangerous conditions spread (Chapter 5).[41]

An Accident Surge Without a Casualty Surge

The blossoming of collisions and derailments after the mid-1950s shows surprisingly little impact on the casualty statistics for those years. Passenger injury and fatality rates per billion passenger miles from these causes fluctuated but show no trend (Chapter 5). Similarly, while injury and fatality rates for all workers rose after 1957 (Chapter 4), train accidents were not the primary cause and train accident fatality rates for trainmen show little trend. Such data suggest that the increase in collisions and derailments was composed largely of minor "fender bender" sorts of wrecks. Several other pieces of evidence support this inference. For example, in 1958, only about 5 percent of all derailments generated casualties; by 1974, the figure had shrunk to around 2 percent. Collisions too became slightly less dangerous over this period. Similarly, an AAR study of the period 1966–1974 showed that use of a higher monetary threshold caused the increase in train wrecks during those years to disappear.[42]

The absence of any significant impact on casualties from the rise in collisions and derailments results from their causes. As noted, accidents to yard trains accounted for 86 percent of the increase in collisions between 1958 and 1969 and these were likely to be low speed. ICC data also demonstrate that yard derailments rose from about 12 percent of the total in 1958 to 30 percent in 1975 and they too were usually at low speed. Similarly, derailments that resulted from track-train dynamics were most common at low speeds. In addition, as demonstrated above, most derailments from bad track were at a location known to be dangerous because of rotted ties, light rail, or poor roadbed and so they too typically occurred at low speeds. Thus, an AAR investigation of accidents from broken rails found that most occurred at low speeds and often in yards. FRA data for 1979 reveal that of the 7,482 derailments that year, 67 percent occurred on the 2 lowest grades of track where the maximum allowable speeds were, respectively, 10 and 25 mph. The absence of casualties from the surge in train wrecks led carriers to see them as economic but not safety problems. Yet all such wrecks contributed to the totals, and probably few readers dug beyond such headlines as "Train Derailments Reported Growing." In addition, a tiny fraction of the derailments generated outsized headlines—especially when they increasingly began to involve hazardous substances.[43]

Hazardous Material Accidents

As Chapter 1 demonstrated, wrecks that involved hazardous materials (hazmat) had been a fixture of railroading since the nineteenth century and had been a concern of public policy since 1908. The railroads had shaped their own policies; the Bureau of Explosives focused on packaging of explosives and acids, and it coordinated with the American Railway Master Mechanics and tank car builders to set safety standards for such equipment. The ICC's role was largely to transform these private safety decisions into public regulations. The system seems to have worked quite well down through the 1950s: Bureau of Explosives and ICC annual reports reveal few accidents and casualties and there were few newspaper headlines featuring such accidents.[44]

During the 1960s, three developments undermined this cozy arrangement. The first was the general rise in health, safety, and environment concerns, and the increasing skepticism of business during these years. David Vogel has depicted "business on the defensive" in the years immediately after 1969 and that certainly applies to the public view of railroad safety. Representative John Culver, an Iowa Democrat, voiced such suspicions in congressional hearings on hazardous materials in 1969. "I feel that the extraordinary position that is occupied by the Bureau [of Explosives] has been unfortunate . . . Such a condition is inimitable, in my belief, to the American system and it should be corrected," he argued. This was the robber baron position stripped of its rhetoric. Second, the rise in derailments inevitably increased hazmat accidents, and as freight trains became longer and cars heavier, they contained more kinetic energy, raising the odds that hazmat cars in an accident would be punctured or ruptured. Even at slow speeds hazmat wrecks might lead to disaster, however. Recall that the wreck at Dunreith, Indiana, occurred at about 32 mph. Even if such a wreck caused no spill or casualties, its potential still might require evacuations and frighten everyone in the county. Third, after World War II, as the petrochemical industry expanded, the railroads began to carry an increasing amount of flammable, explosive, or poisonous substances.[45]

The postwar years saw long-term rapid growth in real GDP, which rose at an average rate of 3.86 percent a year between 1947 and 1970, and thus doubled roughly every 18 years. Output of many chemicals, including a whole range of petroleum and natural gas derivatives such as benzene, ethylene, and propylene, grew even faster. These were feedstocks and thus inputs into a host of final products, and all were more or less hazardous. LPG (liquefied petroleum gas) consumption rose from about 6 to 37 billion tons between 1950 and 1970. Production of some inorganics also expanded rapidly; output of ammonia increased by a factor of 10 between 1947 and 1974, while chlorine production rose 4.69 times between 1950 and 1970. While by no means all of this output moved by rail, much of it did and this implies that the carriers' shipments of hazards must have risen sharply.[46]

There are no public data on hazmat accidents or transport before 1971 and data in modern format date from 1975. However, it is possible to construct a rough index of the probability of hazmat accidents during earlier years. One measure of exposure is tank car miles. Assuming that average miles per car are the same for tank and other cars, and that all cars have the same probability of derailment, then we can construct an index of hazmat accident potential based on expected tank car derailments as follows:

$$\text{Expected Tank Car Derailments} = (\text{Derails/Car Mile}) * (\text{Tank Cars}) * (\text{Miles Per Car})$$

Of course, such a measure ignores hazardous products other than liquids and so its change is likely to be more accurate than its level. In any event, exposure (the last two terms) rose about 85 percent from 1950 to 1974, while exposure * derailments/car mile rose 165 percent. The increase from 1958 to 1974 was about 190 percent, however. Thus, the frequency of spills should have risen rapidly over this period. Moreover, if the average size of spills also rose, as it probably did because tank car capacity rose about 30 percent, this would have led to a disproportionate increase in the number of large spills. An investigation of the stories in major newspapers supports such inferences. Few spills and fewer disasters made the news during the 1950s, but in the 1960s and especially the 1970s the number grew rapidly, reflecting the growth in chemical shipments and the surge in train accidents.[47]

The causes of these major wrecks mirrored those of other train accidents during these years. The reader will recall that the culprit at Dunreith was bad track and a broken rail. A little more than a year later, on January 25, 1969, a broken wheel caused the derailment of 15 cars of LPG on a Southern Railway freight in Laurel, Mississippi (Table A3.1). In addition to the LPG, the train also carried one car each of ammonium nitrate, anhydrous ammonia, hydrocyanic acid, isopropanol, and benzol benzene. The LPG exploded, and "pieces of tank cars ranging in size from three quarters of a tank . . . were hurled up to 1,600 feet from the wreck, igniting dwellings." Like the earlier explosion at Dunreith, this was what the NTSB later called a BLEVE—a boiling liquid expanding vapor explosion—and it was much like an old-fashioned steam boiler explosion, as it was the vapor expansion, not the burning propane, which propelled the car. The blast killed 2 people, injured 39 others, destroyed 58 houses, and caused the evacuation of about 100 people. Again, like Dunreith, Laurel was of more than ordinary significance, for the NTSB investigation raised disturbing questions that undermined the carriers' claims that

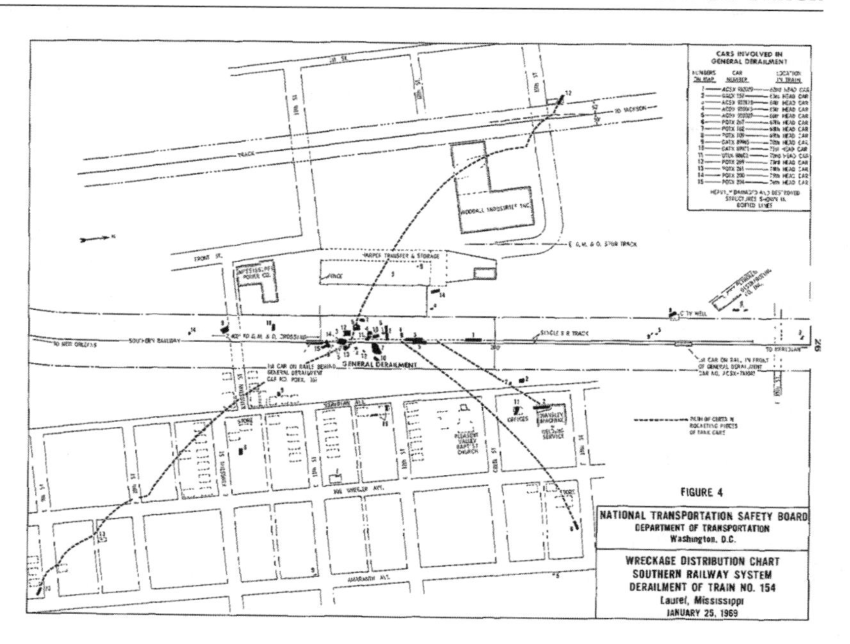

The derailment of a freight train carrying LPG at Laurel, Mississippi, on January 25, 1969, resulted in a massive fire and BLEVE that blew wreckage hundreds of feet (dotted lines), killing two and resulting in much property damage. The accident was caused by a broken wheel, but it revealed a host of problems with railroad safety procedures. (National Transportation Safety Board)

The offending wheel that caused the BLEVE at Laurel, Mississippi, on January 25, 1969, had been so poorly machined that the defects were visible to the naked eye, raising the question of whether the carriers were capable of self-regulation. (National Transportation Safety Board)

they were capable of self-regulation (see below). Both the AAR and the FRA immediately began a review of tank car safety (Chapter 3).[48]

The fire in Laurel was barely out when, on February 18, it was the turn of Crete, Nebraska (Table A3.1). A 96-car freight of the Chicago, Burlington & Quincy Railroad derailed, striking another train containing 3 cars of anhydrous ammonia, one of which fractured and exploded, releasing about 29,000 gallons of ammonia that formed a cloud of gas, killing 8, seriously injuring 28, and causing most of the town to evacuate. Such an accident in a major city might have been catastrophic. Bad track was again the culprit; the NTSB concluded that maintenance was not up to "normally acceptable" standards. Later testing by the AAR discovered that the hazmat release resulted because the tank car steel was brittle and therefore fracture-prone. There were 11,000 such cars in service, the NTSB pointed out, many of them hauling LPG. They represented, therefore, the potential for 11,000 BLEVEs such as that at Laurel.[49]

After a seven-month lull, a third disaster struck. On September 11, 1969, Illinois Central Train 2d 76, a 149-car freight traveling through Glendora, Mississippi, buckled when the engineman hit the emergency brake to avoid a pedestrian (Table A3.1). The train makeup included a number of empty cars toward the front with loaded cars, including 10 that carried vinyl chloride, in the rear. Such an assembly was almost perfectly designed to derail in an emergency, for the brakes stopped the empty cars much more rapidly than those that were loaded, resulting in enormous compressive forces on the unladened cars—a problem, the NTSB noted, that had worsened with the arrival of very large equipment. As the brakes were applied, they compressed an empty boxcar so violently that its frame buckled, throwing it and another 14 cars off the track, including 8 that contained vinyl chloride. Several cars caught fire and burned overnight, but the real disaster did not come until the following day.

> About 6:45 a.m. the following day, GATX car 85069 containing vinyl chloride exploded violently. The tank separated near one end . . . [and] a section of the tank, about 20 feet long, landed about 350 feet southeast of the site . . . A part of the opposite end, including the underframe, was propelled northwestward. This piece, after attaining the height of 50 to 60 feet, hit the ground 725 feet from the track and ricocheted over a 50-foot tree for another 100 feet . . . The other tank head landed 650 feet southeast of the track in a tenant house yard. Another section, measuring 7 by 18 feet, went 850 feet south of the point of explosion.

The wreck burned for 12 days. There were 2 injuries, but worries over phosgene gas (a result of the combustion of vinyl chloride) led to the evacuation of between 17,000 and 21,000 people, including Senator James Eastland, who was "routed from his plantation home."[50]

On June 21, 1970, in Crescent City, Illinois, a hotbox caused the derailment of Train 20, a 109-car freight of the Toledo, Peoria & Western, dumping 10 tank cars of LPG, which blew up, yielding 3 BLEVEs, injuring 66, and wrecking most of the downtown business district. On October 8 that year, near Sound View, Connecticut, a broken truck derailed a Penn Central freight that then collided with an oncoming passenger train, dumping about 1,100 gallons of LPG that caught fire, injuring 7 crew. Some spills resulted from collisions or derailments during yard switching, and these were particularly dangerous because they were often located near populated areas. One such wreck on the Norfolk & Western occurred on July 19, 1974, in Decatur, Illinois, when a switching locomotive pushed a

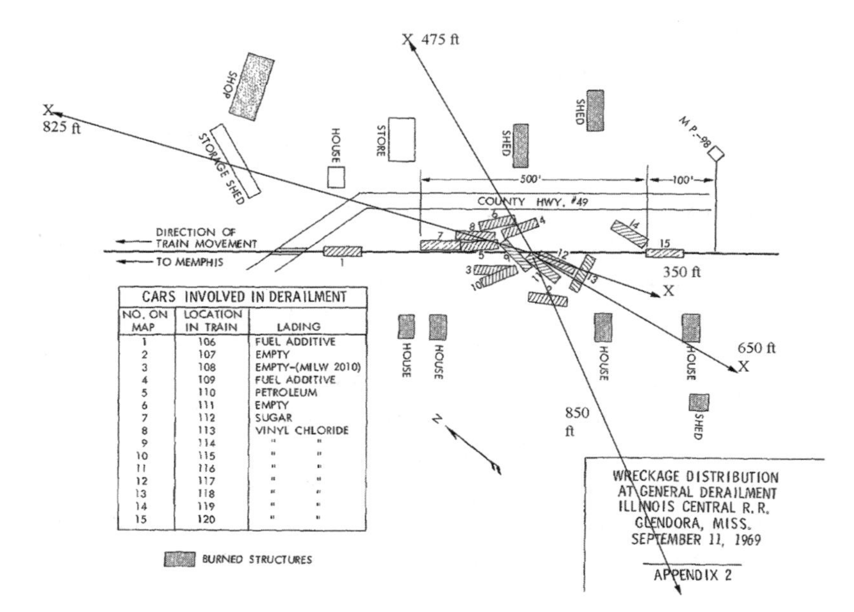

CARS INVOLVED IN DERAILMENT		
NO. ON MAP	LOCATION IN TRAIN	LADING
1	106	FUEL ADDITIVE
2	107	EMPTY
3	108	EMPTY-(MILW 2010)
4	109	FUEL ADDITIVE
5	110	PETROLEUM
6	111	EMPTY
7	112	SUGAR
8	113	VINYL CHLORIDE
9	114	" "
10	115	" "
11	116	" "
12	117	" "
13	118	" "
14	119	" "
15	120	" "

The derailment at Glendora, Mississippi, on September 11, 1969, was the result of an emergency application of the air brakes that caused the train to buckle. The wreck punctured tank cars filled with vinyl chloride. The resulting fire and BLEVE threw pieces of the tank car hundreds of feet (arrows) and resulted in the evacuation of 21,000 people. (National Transportation Safety Board)

cut of 5 jumbo (30,000-gallon) tank cars containing isobutane onto a different track without benefit of car retarders. Going too fast, they collided with another car, puncturing one of the tankers and resulting in a BLEVE that killed 7 and injured 349.

Every year brought more such spills. At their peak, the railroads generated at least a dozen accidents during 1976–1978 that led to major hazmat spills. These spectacular incidents and even more minor spills of exotic, lethal materials were significant beyond the comparatively few casualties actually generated, for as everyone understood, this was luck. The derailment of a 21-car Burlington Northern freight near Callao, Missouri, on January 8, 1971, provides a good example of these near misses. A loose wheel on the third car caused a 14-car derailment, wrecking a bridge and tumbling the train into a creek. The derailed cars included five of anhydrous ammonia and one of LPG. A small fire caused a car of ammonia to explode, sending it 600 feet, but the wreck was in a rural area. There were no casualties, but authorities evacuated 20 families.[51]

As Chapter 1 argued, train wrecks have characteristics that may make them seem especially horrifying, and hazmat accidents are in a class of their own. Not only are they uncontrollable and involuntary, they often involve new or frightening substances such as explosives, chemicals, radioactive materials, and poison gases. And as they had after 1900, media stories played on and probably amplified public worries. Thus, in the late 1960s, headlines began to appear with pictures of fireballs from spills of petroleum distillates and clouds of poison gas. The *Chicago Tribune* headlined Laurel "Train Blast Rips Part of Town," while the Crete story featured "Ammonia Gas from Leaking Tank Car Kills 8 in Nebraska," and the articles included illustrations of blasts and workers in hazmat suits. After the Crescent City wreck, that paper's front-page headline, in large type, was "Blast Sets Town Ablaze." Because these and similar stories came largely from Associated Press or Universal Press International reports, they appeared in nearly every newspaper in the country. Such disasters made the rise of train accidents seem especially scary. Together, the growing number of train accidents and hazmat spills began to focus the concerns of the railroad technological community, but their response would be too little and too late.[52]

Private Enterprise and Train Accidents

As we have seen, there had been a run-up in collisions and derailments before 1907 and again during World War I; these had generated both public outcry and a number of regulations. In the main, however, the response to earlier safety problems had been left to the carriers working through the railroad technological community. In the 1950s and 1960s, this same community gradually came to focus on the worsening problem of derailments, but they were slow to do so. With the eruption of hazmat wrecks after 1968, events overtook their efforts. A shortage of funds also slowed the diffusion of new and safer technologies.

Research, Technology, and Derailments

In 1962, the National Academy of Sciences (NAS) surveyed railroad research. They—discovered that the carriers and the AAR typically spent about $6 million (0.06 percent of revenue) yearly on research—of which only about 12 percent was on operations research and 0.5 percent was basic. By contrast, a survey of 16 major manufacturing industries at about this time found research spending constituted 1.74 percent of revenue. A *Railway Age* survey of 1966 found much more activity. Twenty-two carriers reported research budgets totaling $42 million and $45 million in 1965 and 1966, respectively. Again, most of the work was applied research and development and testing, but the survey also discovered increasing interest in systems research and marketing. The AAR itself maintained a research facility on the Illinois Institute of Technology campus, and it occasionally employed university researchers as well. Some of its projects came from committees of its Mechanical Division and the AREA. In 1969, the AAR furnished the latter group with $329,000 for such work, but its contributions had been declining in real terms for a decade. Individual carriers often cooperated in AAR projects, employing their own tracks for testing, and some maintained their own laboratories, including the Canadian National, the Southern Pacific, the Southern, the New York Central, the Denver & Rio Grande, the Seaboard, the C&O, and the Pennsylvania.

Railroad equipment suppliers did more. These included such companies as American Brake Shoe, General Electric, Griffin Wheel, Union Switch & Signal, Westinghouse Air Brake, and GM's Electro-Motive Division. They had their own trade associations such as the Railway Wheel Association, as well as an umbrella group, the Railway Supply Institute (later Railway Progress Institute, or RPI). The NAS survey of 45 percent of them reported R&D spending of about $15 million annually, or about 1.7 percent of sales. The *Railway Age* survey noted above for 1965 and 1966 revealed R&D equal to 2.73 percent and 3.1 percent of sales for 61 suppliers. Findings of all these groups appeared at meetings and in proceedings of technical societies—predominantly the AAR Mechanical Division and the AREA, both of which set standards. Finally, the railroad technical press, although shrunken in size, continued to disseminate technical and other developments and provide informed commentary on all aspects of railroading.[53]

Critics found this thin gruel. In 1967, Aaron Gellman, vice president of planning at Budd, a railroad supplier, termed the lack of large-scale cooperative research a "scandal." Yet despite the lack of resources, together these groups contributed to a steady increase in the carriers' efficiency in the 1950s and 1960s. One study found that output per unit of input rose at about 1.5 percent a year from 1951 to 1974, faster than the rest of the economy. Considering the carriers' economic problems and excess capacity, that was a remarkable achievement.[54]

As has been stressed, most technological progress improved both safety and productivity. Air brake improvements and the responses to hotbox problems during the 1950s and 1960s provide good modern examples. The longer, heavier trains that the carriers were dispatching in the 1960s required better brakes. The release on the old Westinghouse AB brake was so slow that it took as long as 80 seconds to reach the caboose of a 150-car freight. Thus, a release while the train was running might well break it in two, while an emergency application could cause buckling, as happened at Glendora. In the 1960s, Westinghouse developed an accelerated release valve (type ABD) that might cut the time to 13 seconds, thereby improving train handling and reducing stopping distance. The ABDW brake came in the in the 1970s; it yielded faster application and release compared to the ABD, as it could stop a heavy freight on a 1 percent down-grade in 22 percent less distance. It was required on all new equipment after 1977. Help had also arrived with the diesel, which brought dynamic braking; it also allowed use of a radio-controlled remote locomotive two-thirds of the way back, so the engineman in the head locomotive could operate both units and effectively control a 150-car freight like a 50-car train. Distributed power, as such operations were termed, not only shortened stopping distances, it also saved fuel, reduced wear on curves, and virtually did away with break-in-twos. In addition, somewhat earlier, Westinghouse and Johns Manville had developed composition brake shoes that had both a higher and more uniform coefficient of friction than the older cast iron shoes, which improved stopping power and reduced the likelihood of wheel sliding. Controlling long trains with this new equipment remained difficult and so companies developed engine driver training programs; in the 1970s, some of them such as the Santa Fe and Southern Pacific began to use locomotive simulators.[55]

Hotbox problems provide another example of new technologies that yielded multiple payoffs. As we have seen, hotboxes had plagued the railroads from the beginning. They disrupted schedules, hampering competition with trucks, and caused many sensational accidents when an overheated axle broke at speed. Hotboxes resulted from a lack of lubrication, a misaligned bearing, dirt, or "waste grab," where the lubricating material wrapped around the axle journal. In the 1950s, hotboxes became the leading cause of train wrecks, accounting for one-third of all equipment-related derailments in 1956. The carriers finally began to focus on the problem. In 1959, the president of the Santa Fe called them "today's biggest operating problem." The derailment of New York Central Extra Freight 1087 on August 25, near Syracuse, New York, is typical of the thousands of such accidents. A hotbox burned off the end of an axle when the train was traveling about 38 mph, causing it to derail. The cars then hit a signal tower, killing its occupant, and went on to demolish a bridge.[56]

Help, however, was already on the way. In the 1950s, "lube pads" arrived; these were simply lubricated containerized pads that did away with waste grab. Reports from the Norfolk & Western and other roads revealed that they cut hotboxes by 40 percent. Roller bearings were a better solution and Timken had developed one for freight cars in the 1930s; an improved, simpler-to-apply model came out in 1954. Not only did it reduce hot bearings, it also saved fuel and labor as well, as the company demonstrated in several widely published cost-benefit analyses. The problem, however, was that the high initial cost combined with the externalities of freight car interchange reduced individual companies' payoff to such innovations. To equip cars that would be lost in interchange with better bearings "would be entirely charity on our part" an Atlantic Coast Line representative explained, and this logic largely confined such bearings to cars that stayed on home roads. Moreover, while AAR interchange rules required roller bearings on all new cars after January 1, 1969, "complete conversion will take years," one journal noted, because the freight car stock turned over slowly. In 1957, Servo Corporation and several railroads developed and installed the first wayside hotbox detector that might catch the problem before a wreck. They were expensive and spread slowly; in 1962, the railroads had 400 of them in use—1 for every 943 miles of track—and in 1969 there were only 1,500 on Canadian and US roads combined.[57]

By the early 1960s, hotboxes had become the central focus of AAR research. The AAR tested and approved different pads and lubricating oils and banned the old waste-lubricated bearing in 1960. *Railway Age* and other trade journals published numerous articles on the hotbox problem and the AAR Mechanical Division compiled statistics on both car miles per hotbox and the spread of roller bearings. Such actions contained the problem. In 1962, one publication proclaimed, "Hot Boxes Dropping to an All Time Low." Banning old bearings in the 1960s cut hotboxes nearly in half; thereafter, they continued to decline with the gradual spread of roller bearings and better journals, oils, and maintenance procedures. Despite such successes, accidents from hotboxes still ranged around 250 a year by the late 1970s, in part because the new oils did not smoke when hot to warn of impending trouble, while longer, faster freights also diluted inspection quality. Some of these wrecks were on trains carrying hazmat substances.[58]

In the late 1950s, derailment problems shifted from hotboxes to track and roadbed and track-train dynamics, as roadbed continued to deteriorate and the new heavyweight cars came on stream. In 1961, the AAR's head of engineering research warned "What Today's Heavier Wheel Loads Are Doing to Your Rail." In response, the railroad technological community increasingly began to focus on these areas. Each year the AREA reported ongoing rail research and companies discovered that rail grinding could solve corrugation problems, while high-cant tie plates reduced shelly rail, both of which extended rail life and reduced accident risks. There were also endless experiments with concrete cross ties to improve roadbed. Flange lubricators reduced rail wear as well as the dangers from bad wheels. Companies employed hand-held ultrasonic detectors that would find flaws within rail joints and Sperry's motorized rail detectors added that feature too, while the spread of continuously welded rail (CWR) also reduced joints. Computerizing rail defect data allowed companies to prioritize problems and respond more quickly. Use of remote sensing detectors also spread, to include not only hotboxes, but also dragging equipment, broken wheels, earthquakes, bridge fires, and falling rocks. *Railway Age* claimed there were more than 8,000 detectors of all kinds in use in 1966. A host of new machines, along with products such as panelized turnouts, reduced the cost of track work, while portable radios and computerized scheduling improved efficiency. About 1966, the AAR instituted studies of track-train dynamics using a test track on the Louisville & Nashville. Its findings resulted in longer shank couplers, information on the importance of center plate lubrication, and various snubbers that would reduce rock 'n' roll.[59]

Yet despite these considerable efforts, the carriers' response to rising derailments fell short in two respects. The first involved money; maintenance of way and equipment budgets declined in real terms during the

1950s and 1960s, even accounting for rapid productivity growth. Thus, roller bearings spread at a glacial pace. Hotbox detectors could prevent many accidents and save millions, *Railway Age* claimed, "if used more widely." CWR advanced about 3,000 miles a year. And so, despite the new technologies, accidents soared.

Second, perhaps because the railroads saw train wrecks primarily in economic rather than safety terms, there was a "business as usual" quality to the carriers' actions and they failed to grasp the political ramifications of the rise in wrecks—especially those like Dunreith that involved hazardous materials. Even as late as 1977, George W. Way of the AAR Research and Test Department was proclaiming that "derailments . . . are an economic problem; not a serious threat to life and limb." The proceedings of the AREA, whose members' job it was to maintain the track and roadbed, show little concern with the cascade of derailments during the 1960s. Thus, until the NTSB rather pointedly called for better track standards in its Dunreith report, the men responsible for track and roadbed seem to have been unaware of the storm that was building. Nor was the AAR Mechanical Division any more alert to the political dangers that arose from derailments. In 1968, Rex Manion, a vice president of the more politically attuned AAR, gave a talk to the mechanical men entitled "Up to Our Neck in Bugs." It was the first time that the organization had discussed the seriousness of the derailment problem. "If we don't come through," Manion warned, "we are sure to have people outside our industry coming in to tell us how to do our jobs."[60]

By the time Manion warned the mechanical men, Dunreith had put tank car design in the news. The following year (1969) brought the disaster at Laurel, in which—the reader will recall—a derailment from a broken wheel ditched 15 cars of LPG that blew up. The NTSB report ran to nearly 75 pages and raised a host of questions. The wheel broke from stress concentrations in an area so poorly machined that the defects were visible to the naked eye. Why had inspectors not condemned it? The tanks themselves mostly failed from coupler punctures that resulted when the train came apart—yet tight-lock couplers had been available for a generation.

In this and a separate report the NTSB raised a series of questions about the design of the new, jumbo tank cars intended for shipping LPG. They had twice the capacity of older cars. They were heavier and lacked both insulation and center sills. How would they respond to the compressive forces in a derailment? The new equipment designs had been approved by the AAR tank car committee and the ICC. Yet there were no requirements to test the behavior of pressurized vessels under accident conditions; indeed, as the NTSB pointed out, "the entire process lacked any documented safety methodology . . . We cannot understand the lack of testing." Such tests, for example, might have discovered the inadequacy of safety valves. At Laurel the safety valves worked perfectly; yet the cars failed from the buildup of internal pressure, resulting in a BLEVE that had rocketed them around the area for distances up to 1,600 feet.[61]

These hazmat disasters raised the ultimate question: how well were the men of the Mechanical Division doing their jobs? As *Railway Locomotives and Cars* bluntly put it, the carriers' self-regulation had come under scrutiny—and in the eyes of critics, it had failed. The result was a burst of activity. Laurel especially seems to have electrified the AAR. Before the NTSB report became public, the Mechanical Division had proposed that after January 1, 1970, new tank cars be equipped with type-F (tight-lock, passenger train-type) couplers so they would not come apart and puncture adjacent cars, and it began metallurgical investigations of tank car steel. Within six months of the wreck, the division proposed ultrasound tests and precise machining standards for all new wheels and that tests be made on safety valves and cars without center sills. It also undertook a thorough inspection of all of the plants of its wheel makers and began to investigate ways to reduce puncture risks. The RPI also reviewed the experience of the 8,700 jumbo (33,000-gallon) non-insulated pressure cars from January 1961 on and claimed to be "far from satisfied" with what it found.[62]

Finally, in 1969, the AAR made a major commitment to research and the following year it hired William Harris, a non-railroader with a PhD in metallurgy from MIT, to be its research director. Harris immediately formed a Joint Committee on Tank Car Safety drawn from the AAR and the RPI. The AAR also announced a stepped-up budget, and strengthened and centralized its Engineering Division so that instead of having recommended practices coming from a committee of the AREA they now came directly from the AAR itself. In January 1970, it issued new recommended practices for track inspection. In September 1970, it issued its first ever recommended standards for track maintenance, the purpose apparently being to get the standard out before the passage of the Federal Railroad Safety Act of 1970, which would require any federal regulations to "consider . . . relevant standards." Yet all this was too little and too late. The rise of derailments combined with hazmat accidents had made train wrecks a popular issue at a time when the public was far less willing to rely on private enterprise than had once been the case. As Chapter 3 details, the result was an outpouring of concern and a major expansion in federal regulation, followed—somewhat belatedly—by the dawning realization that economic deregulation might also prove necessary to safety.[63]

The evolution of train accidents during these years reflected the interplay of incentives, organizations, technology, resources, and regulation. Beginning in the mid-1950s, train wrecks—especially derailments—began a long upward climb. This was the outcome of the carriers' deteriorating economic conditions and their wholesale, uncoordinated jailbreak into jumbo equipment—a combined result of regulatory and organizational failures. The railroads were victims of the normal workings of competition—creative destruction—and inevitably their adaptation led labor to dig in its heels. Yet their inability to adapt to the new competition from cars, trucks, and airplanes was largely a result of regulation by the ICC and others. It is impossible to know—and equally impossible not to wonder—how many thousands of railroad accidents resulted from the perpetuation of these unwise public policies.

Even as their track and roadbed deteriorated, the railroads stampeded into very large freight equipment in a desperate effort to cut costs and win back markets. Yet their technical societies remained ill equipped to cope with track-train interactions and as had happened in the past, much larger equipment generated unanticipated problems. These interacted with deteriorating track and roadbed to yield an upwelling of derailments after 1956. The rise in derailments coincided with an expansion of traffic gleaned from the rapidly growing chemical industry, and resulted in large numbers of hazmat spills beginning in the late 1960s. The carriers responded to this surge in accidents as they had in the past with a host of important innovations that contained hotboxes and began to focus on the problems of heavy cars and track-train dynamics. But the days when they might have been allowed to solve their own problems with only prodding from the authorities were as long gone as high-button shoes. Railroad safety was transitioning from a largely private matter to a public responsibility.

3

On the Right Track

The Long Campaign Against Freight Train Accidents, 1965–2015

It seems these days that there is a great suspicion of an industry that regulates itself.

Thomas Goodfellow, President, Association of American Railroads, The New York Times, *1969*

Science is pure. Engineering is contaminated by economics, or worse yet, commercialism.

George H. Way, Association of American Railroads, AREA Proceedings, *1980*

The rise in train accidents after the mid-1950s did not immediately generate a public response. Rather, the origins of concern lay with the Johnson administration and Congress, which in 1966 created the Department of Transportation (DOT). In the process, they transferred railroad safety regulation from the Interstate Commerce Commission (ICC) to the newly created Federal Railroad Administration (FRA) and established the National Transportation Safety Board (NTSB) as well. In the late 1960s, the new agencies helped sound the alarm, both responding to increasing popular concern with train derailments and hazmat spills and stoking these same fires. Congress also became much more involved in railroad safety after 1965. Previously, legislative interest had focused largely on worker and passenger safety, but the rise in hazmat wrecks and grade crossing accidents extended congressional concerns to freight train accidents. The result was the Federal Railroad Safety Act of 1970, which ended the old regulatory regime of voluntarism, giving federal regulators sweeping powers over railroad safety. Yet for a decade, the plague of train accidents continued unabated. Finally, about 1978, the carriers turned the corner and train wrecks began a long retreat, almost—but not quite—disappearing from public view.

This chapter begins with the change in safety regimes that culminated in the Federal Railroad Safety Act of 1970. The following section chronicles the FRA's early response to rising accident rates, which took the form of track and equipment regulations. This section also charts how the failure of train accidents to recede, along with highly publicized wrecks, contributed to the economic deregulation of the carriers between 1976 and 1980. The FRA's major regulatory achievement during these years was to mandate important tank car modifications that began to reduce hazmat spills after about 1980. The third section reviews the decline in train accidents after 1978. Its origins lay in the attempts of an increasing number of carriers to improve service, which dated from the 1960s. Economic deregulation supported and reinforced these efforts because it sharpened safety incentives and helped supply funds for needed improvements. Safety also benefited from a research renaissance that was partly federally funded and that contributed to an immense wave of technological change. The final section reviews the sharp refocus of regulation that began in the 1990s. In sum, this portrait suggests that while federal research and market-augmenting activities and to a lesser extent regulation have been important, private-sector capitalism was the dominant source promoting the decline in train accidents after 1978.

The New Regime in Railroad Safety

By 1965, it was clear to nearly any observer that the ICC was no longer an aggressive force for railroad safety. In its youth the commission had publicized its accident investigations and sometimes used its bully pulpit to push the carriers to improve worker and passenger safety, but senility had long since set in. A 1962 report by the Railroad Retirement Board noted that the commission's accident reports had become perfunctory and failed to deal with newer risks. The laws under which the commission operated had become obsolete, and it petitioned Congress from time to time to extend their coverage modestly. Yet it never sought wider powers over other aspects of safety. A congressional review in 1965 was scathing. It noted that there had been no significant decline in train accidents in a decade; that many such wrecks reflected employee negligence; and that the commission knew nothing of the company safety programs that might help avoid such negligence. Similarly, while faulty equipment was a major cause of wrecks, ICC inspections had not increased over a decade even as the inspection force—especially the number of supervisors—had grown. The commission, it seemed, was content to preside over a narrow set of safety laws but showed little interest in overall railroad safety.[1]

Yet when Congress and the Johnson administration moved railroad safety enforcement from the ICC to the FRA in 1966, the motives apparently had more to do with administrative efficiency than with the commission's failings. The Civil Aeronautics Board (CAB)—an independent agency like the ICC—also lost its jurisdiction over safety, which was moved to the newly created Federal Aviation Administration (FAA). In addition, Congress also created the NTSB in 1966—an agency that while (temporarily) within the DOT, was to be independent of it—and thus not subject to agency pressures. These changes also reflected increasing congressional interest in the details of railroad safety that was part of the broader wave of new safety and environmental regulation during the 1960s and 1970s.[2]

Together, the transfer of railroad safety regulation to a separate agency and the creation of the NTSB would have a powerful influence on railroad safety. The NTSB, which was initially staffed with ex-CAB investigators, came to railroad safety with fresh eyes—and its investigators were often horrified by what they saw. David Morgan, editor of *Trains*, compared the new agency with Nader's Raiders: "one reading of an NTSB report told us that a new cop was on the beat," he said. As Morgan suggested, its accident investigations probed beyond the initial causes of the accident; they were far more searching than the ICC reports had been. Thus, the NTSB was the intellectual heir of Charles Francis Adams and sunshine regulation. This was a market-augmenting activity; the reports disseminated accident information and made recommendations to the railroads and the FRA, pressing them to step their games up. Nor was the NTSB shy about making its views public. In speeches and open letters reported in *The New York Times* and other major newspapers, it warned of the rise in accidents and urged new legislation.[3]

The FRA, on the other hand, provided the "gun behind the door." Without the many other duties that consumed the energies of the ICC, the focus of the FRA would be largely on safety, and the focus of its critics would be on how well it was doing, which must have had a salubrious effect on management. By 1968, accident reports from the FRA were pointedly noting that its gun needed more ammunition; there were no government regulations covering the majority of railroad accidents—an insight that had somehow escaped the ICC. Together these new institutions generated a powerful impetus for expanded regulation and better safety. And they did so at a time when older ideas of federalism and limited government were under assault and public suspicions of business were at a level not seen since the days of the robber barons.[4]

Congress Makes a Law: The Federal Railroad Safety Act of 1970

Congress took up railroad safety in 1968 with Harley Staggers's (D West Virginia) Committee on Interstate and Foreign Commerce holding hearings on H.R. 16980, a bill drafted by the DOT that would have greatly broadened the FRA's jurisdiction. It was a central moment in railroad safety. Voluntarism, which had resulted in a hodgepodge of narrow rules but left the broad contours of railroad safety up to the carriers, was under assault.

After the outburst of public concern with passenger safety during the Progressive era (Chapter 1), railroad safety had largely been an inside-the-beltway interest. The list of participants in the 1968 hearings reflected these politics but with hints that the matters were becoming of broader interest as well. Staggers introduced the hearings, emphasizing the rise in train accidents; next to testify was A. Scheffer Lang, head of the FRA and a recent alumnus of the New York Central, who testified for the administration. Lang noted that even after adjusting for inflation, train accidents had risen 66 percent in the past five years and claimed that 95 percent of them were due to conditions over which the FRA had no jurisdiction. Alan Boyd, secretary of the DOT, also lent his weight to the administration bill. Al Chesser spoke for his own union (the Trainmen) and the Railway Labor Executives, while Harold Crotty represented the maintenance of way workers. Both groups enthusiastically supported the act, and urged a long list of rules, many of which, not coincidentally, would have mandated increased employment. Representatives of state railway commissions appeared to ensure that federal preemption did not go too far and to see as well if there might be any federal money for their work.

The railroads opposed the bill, supported by a small group of shippers that worried about its impact on transport costs. Louis Menk, president of the Northern Pacific, and Thomas Goodfellow, head of the Association of American Railroads (AAR), spoke for the carriers. They blamed much of the rise in accidents on inflation and claimed that passenger and worker safety was improving. Here again, the carriers depicted rising collisions and derailments as primarily an economic—and therefore a private—matter, rather than one of safety and a concern of public policy. The hearings spent little time delving into the causes of rising accidents—although in a phrase that revealed the sway old ideas still held over the imagination of some—the unions blamed the problem on "financial officials." That the carriers were going broke apparently did not seem relevant to anyone.[5]

For Lang and Staggers, and most of those who appeared before the committee, the emphasis was on train accidents and worker casualties—a traditional focus of regulation. But newer worries were intruding. For the unions especially, accidents at grade crossings were of peculiar concern because they sometimes involved collisions with fuel trucks, immolating all involved. For years, the *Locomotive Engineer* had been full of stories such as "Fiery Crash Kills Fourteen" and "Photos Cry Out for Action to Stop Crossing Crashes." Hazmat concerns arose in other ways too. Congressman Richard McCarthy (D New York) testified to two recent wrecks in his district near Buffalo, one of which involved methyl acetone, while another dumped 10,000 gallons of styrene that polluted the water supply and caused evacuations. No one was hurt in either accident, but McCarthy was worried. We need to "protect . . . the general public," he informed the committee. Joseph O'Donnell, chief of the NTSB, also testified for the bill. He too noted the rise in accidents and he warned specifically of the increase in wrecks involving hazardous materials, which he thought "raised the probability of catastrophic occurrences." And while the carriers stressed the statistics of passenger and workers safety, O'Donnell, like McCarthy, emphasized that "bystanders" were now at risk. In short, hazmat disasters had transformed railroad safety from a concern of railroads, unions, and passengers into a broader public issue.[6]

H.R. 16980 died in committee but railroad wrecks such as the earlier disaster at Dunreith and the derailment and LPG explosion at Laurel, Mississippi, on January 25, 1969, that killed two kept the issue in the news. So did the wreck in Crete, Nebraska, which *The Boston Globe* reported as "RR Wreck, Gas Cloud Kill Eight." Nearly every major newspaper featured similar stories. Such articles, complete with lurid headlines, pictures of fireballs, names of scary, unpronounceable chemicals, and casualty lists amplified public worries. There had been an election in 1968 and in February 1969 the new secretary of transportation, John Volpe, began to assemble a small task force consisting of Reginald Whitman, administrator of the FRA and late of the Great Northern, Thomas Goodfellow and other representatives of the railroads, Al Chesser and other labor representatives, and state railway and utility regulators. Meeting from April through June, they were able to hash out a general report that concluded, among other things, that the FRA should have authority to issue regulations governing "all areas of railroad safety."[7]

A month earlier, in May, with hazmat stories still making the headlines, Senator Vance Hartke (D Indiana) introduced his own bill that also conferred broad powers on the FRA. Hartke was chair of the transportation subcommittee of the Senate Commerce Committee; his constituents included those surviving the near miss at Dunreith. While the hearings would include the same interest groups, largely repeating the same testimony, the administration, not yet ready with its own bill, remained noncommittal. Hartke revealed that he understood the new public concerns: "With increasing frequency train wrecks threaten whole communities with flames, explosions and contamination by poisonous chemicals," he warned.[8]

The army also gave the cause of regulation a boost, making headlines just before the hearings by proposing to ship trainloads of phosgene gas across the country. Representatives of the army then appeared before the committee, testifying on the extent to which the military routinely shipped other equally frightening substances. Both senators from New Jersey testified in opposition to the army plan. John Reed, the new head of the NTSB, reinforced these worries. He reviewed the spate of recent hazmat spills, complete with graphic slides of fireballs and wreckage, and rehearsed the board's conclusion that new legislation was needed to control such dangers. Further testimony revealed that the FRA had essentially no expertise or authority in this area. The railroads, again represented by Louis Menk, initially opposed broad safety legislation, making the same claim that the carriers were competent to regulate their own safety. In an exchange that caught the difference between the old, voluntary approach and the newer desire for federal control, Hartke asked him to account for the Bureau of Explosives' lapses in the Dunreith accident. "You contend you are capable of self-regulation and the . . . [NTSB] says otherwise,"

Hartke asserted. Menk gave the carriers' response: "I am not denying that we have accidents, but this [problem] isn't anything that is going to be corrected by legislation."[9]

In July, Hartke moved the hearings to his home state of Indiana. Labor again supported the legislation but with an even larger contingent that included the rank and file as well as leaders. The *Locomotive Engineer* followed the developments closely, quoting Grand Chief Engineer C. J. Coughlin that "this Organization is 100 percent behind this vital measure." Hartke had also reinforced his position by inviting a number of citizens who had experienced the hazmat wrecks at Dunreith and Laurel. A representative of the Indiana State Police testified, as did the mayors of Greenfield and Richmond, Indiana, and the fire chiefs of Dunreith, Middletown, and Richmond. The railroads could not have missed the import of this broader focus: train accidents that involved hazardous substances were no longer simply a private matter.[10]

The May and July 1969 hearings adjourned with little accomplished. Yet by now the carriers had realized that opposition was a lost cause. September 1969 brought yet another hazmat disaster—this one at Glendora, Mississippi, in which a vinyl chloride spill caused a large-scale evacuation that—as noted above—included Senator James O. Eastland. "30,000 Escape Menace of Gas after Derailment in Mississippi" was the not very reassuring headline in *The New York Times*. By October, when the hearings resumed, the administration finally had a bill of its own, and Volpe himself represented the administration. Still, the participants remained divided, with the railroads favoring federal preemption that the state commissioners opposed, with labor wishing to retain the older safety regulations, and with the carriers hoping to wipe the slate clean. Despite reservations, Thomas Goodfellow of the AAR testified for the bill as it promised to subsidize railroad research and do away with certain provisions in earlier bills that the carriers disliked. Labor split, however, with the Locomotive Engineers favoring Hartke's bill. And when Hartke tried to blend the two bills he excluded federal preemption, which led to an explosion by the railroads with Benjamin Biaggini, president of the Southern Pacific, terming it "the worst possible type of safety legislation."[11]

The year 1970 brought more headlines and more hearings on yet another proposed railroad safety bill during which largely the same cast of characters delivered the same lines. It also finally brought a compromise, and the law that passed on October 16, 1970, contained something for everyone. Its purpose was to "promote safety in all areas of railroad operations and to reduce . . . accidents involving . . . hazardous substances." Accordingly, it authorized the establishment of an office with authority over shipment of hazardous materials. The law maintained older safety legislation, and it required the FRA to conduct a "comprehensive analysis of grade crossings" within one year, both of which pleased labor. It reaffirmed the role of states in railroad safety—along with a commitment of federal financial support—but made federal preemption clear. The railroads achieved federal money for research, for the law instructed the DOT to "conduct . . . research, development, testing, evaluation and training for all areas of railroad safety." As seen in Chapter 2, the law also included language instructing the DOT "to consider relevant existing safety data and standards," implying that any regulations would at least begin with AAR rules. This was part of a long tradition: the famous Safety Appliance Act of 1893 had simply mandated that the carriers adhere to their own industry-wide standards for air brakes and couplers. Finally, the heart of the bill empowered the DOT to "prescribe . . . rules, regulations, orders, and standards for all areas of railroad safety."[12]

For more than a century, federal and state railroad safety rules had been narrowly focused. They had mandated safety appliances on freight cars; required inspection of locomotives, signals, and brakes; flirted with train control; and regulated some other matters too. Yet the railroads still controlled the main contours of railroad safety. By 1970, however, the idea that private corporations would make responsible safety decisions without public supervision was as old-fashioned as a horse and buggy; all aspects of rail safety were now the concern of public policy. Still, in light of the carriers' well-publicized financial woes, it is striking that virtually no one who testified seemed able to relate finances to the accidents that so concerned them. Outsiders, however, had begun to connect the dots. In an editorial, the *Chicago Tribune* thought it deeply appropriate that the Penn Central bankruptcy and the hazmat disaster at Crescent City, Illinois, occurred on the same day—June 21, 1970. Yet there is an old adage: if the only tool you have is a hammer, every problem looks like a nail. Congress seemed to think that the cure for the carriers' safety ills was a larger dose of regulation. But the problem would prove resistant to that cure.[13]

Regulation and Train Accidents

The public and congressional outcry over derailments ensured that track standards would be the FRA's first exercise of its new powers. The new law required it to propose "initial . . . rules" within one year of the enactment of the law, making the deadline October 16, 1971. To provide what it hoped would be guidance, the AAR issued its first ever recommended standards for track maintenance in June. When Mac Rogers, head of the FRA's Bureau of Safety, told the American Railway Engineering Association (AREA) that track would be its first target, he promised to follow the format laid out by the AAR's standard but with "more rigid" numerical values. The final standards were adopted on October 20, 1971; they were to be phased in and fully implemented by October 16, 1973. An AAR cost-benefit analysis claimed that, as originally proposed, the rules would have had a capital cost of $2.2 billion and annual costs of $141 million. Modifications obtained by the carriers brought the costs of the final regulation down to $1.6 billion and $116 million, respectively, the carriers claimed.[14]

Federal Standards for Track and Equipment

Table 3.1 contains two of the many requirements from the AAR rules and the FRA final regulation. Consider those governing spiking of ties. The AAR proposal for the lowest class of track allowed 6 out of every 10 ties to have no spikes while the FRA regulations required a minimum of 2 per tie. For ties themselves, the FRA required at least 5 "non-defective" ties per (39-foot) rail, and they could be no more than 100 inches apart. Since the carriers tended to space ties 23 to 29 inches on center, this implied no more than 3 to 4 defective ties in a row while the AAR allowed 5 consecutive ties to be "broken through." The FRA also had a less restrictive definition of what constituted a defective tie. The most expensive section of the regulations, according to the AAR, dealt with rail defects. Essentially, the FRA required speed limitations of 10 to 20 mph over any rail discovered to contain a listed defect, no matter how tiny. Since the carriers often simply ignored small defects, the result would be either a lot more rail replacement or a lot more—expensive—slowdowns.

In fact, the new government track standards seem to have occasioned comparatively little protest from the railroads. In part this was because

Table 3.1. AAR Rules and FRA Regulations for Spikes and Ties

AAR			FRA		
	Track Spikes	Ties		Track Spikes	Ties
Class of Track	No. of Ties Allowed with None[a]	Max. No. Consecutive Defective[b]	Class of Track	Min. No. per Tie[c]	Min. No. Non-defective Per Rail
			6 (110 mph)	3	14 (45-inch spacing)
A (61–80 mph)	3	3	5 (80 mph)	2	12 (45-inch spacing)
B (46–60 mph)	3	3	4 (60 mph)	2	12 (56-inch spacing)
C (31–45 mph)	4	4	3 (40 mph)	2	8 (70-inch spacing)
D (16–30 mph	5	4	2 (25 mph)	2	8 (70-inch spacing)
E (0–15 mph)	6	5	1 (10 mph)	2	5 (100-inch spacing)

Source: "AAR Adopts Recommended Minimum Track Inspection Standards," RLC 144 (January 1970): 26, and "FRA Track Standards," RTS 67 (November 1971): 30–35.
[a]Out of 10 consecutive ties; tangent track through 2 degree curve.
[b]For tangent (straight) or 3 degree curve track.
[c]Tangent track through 2 degree curve.

the carriers had been involved in their creation and because the FRA proved willing to make minor modifications in the regulations when problems arose. For example, since many carriers spaced ties 23 inches on center, a spacing requirement of 45 inches essentially allowed no defective ties for Class 5 and 6 tracks (Table 3.1). A later modification changing the requirement to 46 inches solved the problem. The AREA 1972 convention revealed some support for the new regulations, as they would provide maintenance engineers both leverage and respect. But there was little protest. Some thought the result would be to reduce train speed and take some marginal track out of service and some of the provisions would require a bit of upgrading. Still J. A. Barnes of the Chicago & North Western probably spoke for the majority when he concluded that while the company would have some problems with yards and branches, "I really can't say that the FRA standards are going to hurt us."[15]

Even as it was revising the new track standards, on September 27, 1972, the FRA proposed new safety standards for freight cars that were finally issued on November 12, 1973, and went into effect January 1, 1974. While the agency had inherited from the ICC safety appliance standards that had been in force for many years, most of the freight car—its wheels, axles, trucks, and car body—had been unregulated by the government. However, over the years the AAR Mechanical Division had built up a body of rules governing virtually all aspects of freight cars and, unlike the AAR-AREA recommended practices that governed rail and track, these rules had teeth, for cars that failed to measure up would be banned in interchange. Thus, they were required practice, not simply "good ideas." Again, the FRA regulations followed AAR practice closely and the main complaints seem to have involved the complexities of getting cars inspected and the length of time needed to comply. *Railway Locomotives and Cars* concluded, "in general, the new safety standards aren't all that tough to live with."[16]

Nor, it seems, did either set of regulations have much impact. Both regulations became fully effective in 1974 and the FRA has modified them in numerous ways since. As can be seen (recall Figure 2.1 and Table 2.1), neither reported nor adjusted derailments show any discernable effect. Nor is there any imprint on the relevant physical or monetary data. New ties per track mile had been rising gradually since 1961 with no significant jump about this period. The NTSB reviewed the track standards in 1979. "There is currently no indication that the FRA track safety program is reducing the number of track-caused derailments," it concluded.[17]

This failure of the new regulations probably reflected a combination of causes. As noted above, they made only modest changes in the status quo. Testifying before Congress, John Ingram, head of the FRA, also argued that there would be a lag before the standards would show any impact. Yet while we might expect that the rules' safety effects to have lagged, their failure to show up in the monetary and physical measures is harder to explain. Labor complained that inspections were too few and poorly targeted, while fines were too small. Indeed, the General Accountability Office (GAO) later issued several scathing assessments of FRA enforcement, bluntly concluding in 1990 that the "FRA's safety inspection program does not provide assurance that the nation's railroads are operating safely." The FRA acknowledged widespread noncompliance, admitting that the cost of fines was often below the cost of upgrades. Not until the late 1990s did the agency begin to implement an inspection program that focused on risks rather than defects. An FRA official speaking to the AREA in 1974 warned his audience that "our inspectors can't fix the track for you but they can close it down." Sometimes they did, but this was not as easy as he made it sound; closing down even a branch line might yield howls of protest from shippers. When the Penn Central threatened to shut down because it had tracks even poorer than the lowest class, the FRA promptly granted the company a waiver to keep on operating.[18]

Some companies complied by reducing train speeds which, given the complexities of track-train dynamics might have had little impact, and in at least one case the regulations probably increased risks. As noted, the rules mandated speed restrictions when rail defects were found. The Sperry Corporation discovered that railroads were limiting use of its flaw detection cars to avoid finding more defects than could be fixed quickly, which would then trigger a speed limitation. However, the difficulties ran deeper still. Train accidents were not entirely the result of companies' indifference to safety that regulations might cure, as members of Congress seemed to imagine (see below). Absent the necessary cash for improvements, new rules could not work miracles. Moreover, these were specification rather than performance standards—that is, they focused on means (ties per rail) rather than ends (accidents). Yet track and equipment constituted parts of a technological system and, as the NTSB concluded in a review of the proposed standards, this required a "systems approach." Thus, the adequacy of a given number of ties per rail or spikes per tie depended not only on train speed, but also on train weight, rail weight, ballast, and much else. In short, the standards focused on fixing defects rather than preventing accidents.[19]

In response to these problems and to congressional grumbling (see below), in 1974 the FRA instituted special inspections of the 10 railroads with the highest accident rates relative to train miles. The worst improved relative to the average. Perhaps the improvement resulted because the FRA took them to the woodshed, but that seems unlikely, for virtually all the gain occurred between 1974 and 1975, when the FRA raised the accident-reporting threshold by 133 percent. Of the 10, at least 4 were primarily switching roads—the P&LE, the Elgin, Joliet & Eastern, the Indiana Harbor Belt, and the Houston Belt—and they had many low-cost, minor yard accidents of the sort that would be likely to fall below the new threshold. For example, the cost of derailments on the P&LE was about 40 percent of the national average and so it is not surprising that its reported accident rate fell from 52 to 25 between 1974 and 1975 when the reporting threshold rose. Thus, most of the "improvement" of the 10 worst was likely a statistical illusion.[20]

A second round of intense inspections in 1978 that was motivated in part by the L&N hazmat spill at Waverly, Tennessee (Table A3.1), focused on four particularly bad actors: the L&N, the D&H, the Rock Island, and the ICG). In 1982, in testimony before Congress the FRA claimed success for this venture, but the program had no discernable impact on the D&H or the Rock Island, which was gone by 1979. Rates on the ICG and the L&N did fall much more than the national average between 1980 and 1981, but the latter company at least claimed that its bad accident rate had been an aberration reflecting rapid changes in traffic patterns.[21]

The third important topic of FRA regulations in the 1970s was tank car safety. This had always been the province of the Tank Car Committee of the AAR's Mechanical Division—composed of railroads, shippers, and car manufacturers. Its first standards dated from 1903; after 1927, the ICC and later the FRA were required to approve their specifications. By 1970, there were about 180,000 tank cars on the road (roughly 10 percent of all cars), nearly all of them owned by shippers. Later research suggests that perhaps 5 percent carried hazardous materials. There were a number of different classes of car, depending on their use; the most important were the DOT-105 series, which were insulated pressure cars; the DOT-111 series, which were non-pressure cars; and the DOT-112-114 series, which were also pressure cars but without insulation. The 105 cars were often older; they were also usually smaller—less than 20,000 gallons—and they routinely carried chlorine. The 111 cars might carry sulfuric acid or alcohol. The 112-114 class of cars was a product of the giant-ism that struck the railroads in the 1960s—some carried as much as 40,000 gallons. They hauled low-density cargo such as LPG.[22]

Chapter 2 demonstrated that hazmat accidents almost surely rose sharply after the mid-1950s, but there seems to have been little awareness of this until the disasters at Dunreith, Indiana, and Laurel, Mississippi, in 1968 and 1969, which thoroughly alarmed the public. The NTSB investigations of these wrecks were searching; along with the public outcry they woke up a seemingly sleepy Tank Car Committee and the FRA as well. The NTSB recommendations resulting from these accidents also provided a research agenda to deal with hazmat risks. In addition to the specific causes of each accident—a broken rail at Dunreith and a shattered wheel at Laurel—the NTSB focused on tank car design, including hull penetration and the ways to prevent it, the apparent need for insulation, and the adequacy of safety valves.

The Laurel disaster aroused the AAR Mechanical Division, and about this time the FRA became involved. It urged use of type-F (passenger car, tight-lock) couplers on all tank cars (not just new ones as the AAR had proposed) and urged study of insulation and puncture protection. In early 1970, the FRA also began research on tank cars, employing outside contractors and committing about $900,000 a year to the subject. By September 1970, it had contracted with the AAR to study the application of head shields (metal plates welded to car ends) to prevent puncture of tank cars during derailment. William Harris, the reader will recall, had been hired about 1970 to head up the AAR's newly expanded research program. Under his urging, the AAR, along with the Railway Progress Institute (RPI), chipped in about $1 million to inaugurate what would prove to be a long-running Tank Car Safety Project, and began to study thermal protection and relief valves.[23]

These studies yielded quick and conclusive results. Comparison of the DOT-105 insulated cars with the non-insulated DOT-112-114 series demonstrated that the former were less likely to explode during fires. Research also indicated that safety valves were too small; when an uninsulated car of LPG was subject to heating, even if the valve worked perfectly pressure would build until the car ruptured, yielding a BLEVE. Experiments suggested that head shields would also be cost-effective. Similarly, while the standard type-E freight couplers would separate vertically in a derailment, with each becoming a potential puncture threat, welding a top and bottom shelf to them could prevent separation and would be cheaper than installing a type-F coupler. These resulted in a regulation, issued September 15, 1977. All tank cars were required to be equipped with shelf couplers by December 31, 1981; DOT-112-114 cars carrying ammonia needed head shields while those carrying flammable gases required both shields and shelf couplers, again by the end of 1981. An oversight, long lamented by the NTSB and not corrected until very recently, was that DOT-111 cars, used to transport alcohol and crude oil, were not required to have head shields or thermal protection.[24] The disasters continued to generate bad publicity throughout the 1970s. After the Crescent City wreck that derailed a group of 10 LPG cars, 3 of which blew up, the FRA considered limiting tank car size while the NTSB wrote to railroad executives suggesting that they consider reduced speeds and separation of hazmat cars. It also hinted at the possibility of banning the jumbo tankers. In 1974, the reader may recall, a collision in the Norfolk & Western yard near Decatur, Illinois, ruptured a 30,000-gallon tank car of butane, resulting in a fire and explosion that killed 7 and injured 349. "133 Hurt in Decatur Tank Car Blast," was the resulting headline in the *Chicago Tribune*. The FRA promptly issued an emergency order prohibiting free switching (cars unattached to a locomotive) of tank cars containing compressed gases. Two

The Waverly, Tennessee, hazmat disaster of February 22, 1978, was caused by an incorrectly installed brake that led to overheating and cracking of a high-carbon wheel. The train dumped a load of LPG that killed at least 15 people and led to an FRA emergency order banning such wheels. (http://web.archive.org/web/20060521030601/http://www.tnema.org/Archives/Waverly/Waverly7.htm)

disasters in 1978 also made headlines and speeded up the regulatory timetable. An improperly replaced brake caused a defective high-carbon wheel to break, derailing an L&N freight near Waverly, Tennessee, on February 22, 1978, and causing a propane BLEVE that killed 15. Four days later, near Youngstown, Florida, a derailment punctured a DOT-105 car of chlorine and the resulting cloud of gas killed eight (Table A3.1). "Motorists Die in Chlorine Cloud" was a *Washington Post* headline. *The New York Times* lead was "Gas in Derailed Car Kills 8,"which accompanied a picture of the wreck enveloped in clouds of chlorine gas. Two months later, the *Post* editorialized on "Tank Car Safety" while a *Los Angeles Times* editorial lambasted "the failure of U.S. regulatory authorities to do their job." The FRA issued another emergency order that year, banning high-carbon freight car wheels that were inclined to fracture when overheated, and it accelerated the timetable to equip cars with head shields and shelf couplers to the end of 1980.[25]

In 1980, after the worst had passed, the *Washington Star* ran a three-part feature on the dangers of hazmat accidents in yards with a map showing evacuation areas for a potential spill in the Potomac (Virginia) yards, which were just south of Washington, DC, and close by the Pentagon and National Airport. That same year, ABC News's *20/20* aired "An Accident Waiting to Happen," a documentary depicting track and tank car conditions that were no longer current. "ABC News bears the same relation to journalism that Charlie's Angels bears to police work," an AAR executive grumbled. Accurate or not, these stories kept the pressure on. In 1984, additional regulations to become effective in 1986 extended requirements for better insulation, head shields, and better safety valves in DOT-105 cars.[26]

Dunreith had revealed major lapses in the carriers' response to hazmat accidents as well. Officers Joe Pitcher and John Linegar recalled that when they arrived on the scene, no one knew what was in the tank cars. Officer Mike Smith thought that had the police known the risks, they would have moved people farther back. In response to Dunreith and the other disasters, the carriers, the shippers, and the FRA worked to improve the information that would be available in the case of a hazmat accident. The Federal Railroad Safety Act of 1970 had given the federal government broad powers to collect information and evaluate hazards. In 1974, the Hazardous Materials Transportation Act centralized these efforts in a Materials Transportation Bureau (MTB). The bureau adopted a UN hazardous material coding system that attached a specific number to each hazard. In an accident, first responders could use the number to obtain information on the chemicals involved. About this time the AAR developed its own coding system based on the Standard Transportation Commodity Codes but the MTB rejected it. The Manufacturing Chemists Association established CHEMTREC in 1971; it was a centralized information center, the telephone number of which would be on the paperwork attached to the cargo and which could provide technical information on chemicals in a spill. In the 1990s, with FRA assistance the carriers developed Operation Respond, which, as it has evolved, will employ smart phones to provide real-time information on railcar contents and location, car diagrams, and emergency contact information. In 1994, Conrail began a "safety train," now taken over by CSX, which travels the system providing instruction on hazmat emergencies; other companies provide similar services.[27]

These reforms had a dramatic, if delayed, impact as revealed in aggregate statistics and individual studies. Figure 3.1 shows the total number of "consists" (trains) carrying hazardous materials in all forms of train accidents, per 100 million train miles, and the proportion of all hazmat cars in an accident that leaked. As can be seen, both declined sharply from the late 1970s on to about 1987. Thereafter, hazmat accidents rose relative to train miles. Probably some of the increase reflected the expansion in petrochemical shipments; it also reflected traffic shifts as relatively low-risk yard traffic fell relative to main-line shipments. The percent of hazmat cars in accidents that leaked declined throughout the entire period.[28]

In 1978, the NTSB examined a major hazmat wreck that occurred on the Southern Pacific in Paxton, Texas, and involved tank cars equipped with the new couplers and other devices along with cars not so equipped. It concluded that the new equipment worked. A later study by the Transportation Research Board reported that head punctures had declined by 91 percent from 1965 to the early 1980s. Perhaps equally important, the combination of continuing NTSB investigations and institutionalized research by the FRA and the AAR-RPI resulted in ongoing assessments of

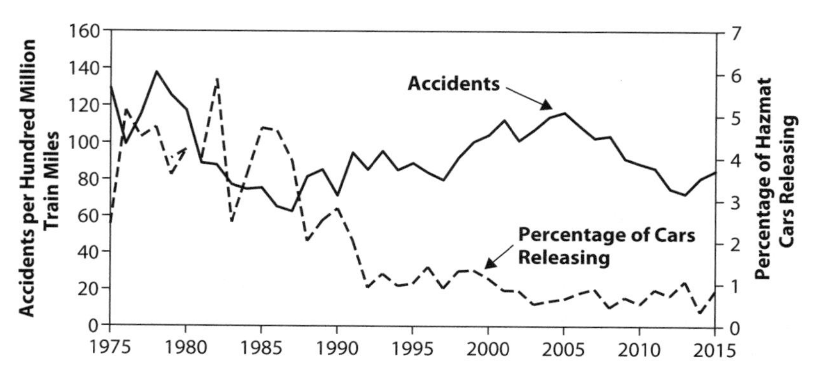

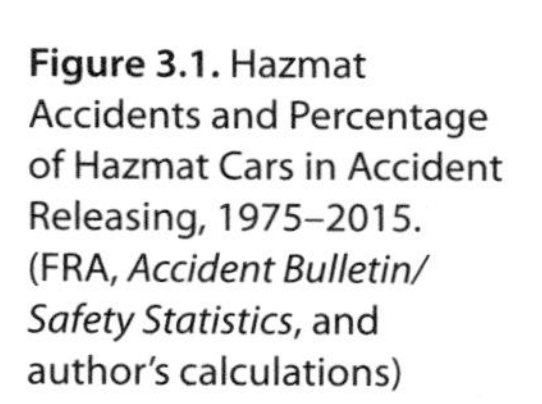
Figure 3.1. Hazmat Accidents and Percentage of Hazmat Cars in Accident Releasing, 1975–2015. (FRA, *Accident Bulletin/ Safety Statistics*, and author's calculations)

hazmat problems that involved complex risk assessments, metallurgical studies, and full-scale crash testing. In 1983, the tank car committee began to investigate the problem of brittle steel fracture in pressurized tank cars (e.g., those that might contain chlorine or anhydrous ammonia). After much study and debate, the committee ordered that all cars built after 1988 employ normalized (heat-treated) steel, which improved low-temperature resistance to brittle fracture. Research has also demonstrated the importance of car placement, train speed and length, and accident type for spill probabilities, as well as the safety gains from rerouting shipments to lower-risk journeys. These efforts have caused the hazmat problem to recede from public consciousness until the twenty-first century. Overall, it is clear that the FRA and the NTSB played constructive roles in these developments, for while improved tank cars would probably have evolved without public prodding, the involvement of these agencies surely speeded up the process.[29]

The Limits of Safety Regulation

These were the major, although by no means the only, federal regulatory efforts intended to deal with train accidents during the 1970s. In addition, the FRA issued regulations requiring companies to file operating rules (1974) and operating practices (1979), but it never attempted to set federal standards, probably because the problem was dauntingly complex while the real difficulties were with enforcement rather than ambiguous or dangerous rules. In 1976, the FRA responded to a congressional mandate (P.L. 94-348) requiring a conspicuity device on the rear of all freights intended to make them more visible to following trains. It is unlikely that this had any significant effect on rear collisions, for it was impossible to stop a long freight in time for such devices to provide adequate warning. As the decade lengthened, the collective failure of these regulations to have any demonstrable impact on the collision and derailment totals resulted in an increasing volume of criticism of the FRA, as Congress held hearings on some aspect of railroad safety nearly every year of the decade.[30]

The safety hearings also occurred against the backdrop of worsening financial conditions for all the railroads, but especially those in the Northeast and middle West. As Chapter 2 noted, it is convenient to date the beginnings of these financial train wrecks with the Penn Central's bankruptcy filing on June 21, 1970—some four months before passage of the Federal Railroad Safety Act of that year. Thereafter, railroad financial problems were almost continuously in the news and before Congress for much of the decade. The bankruptcy of the Central Railroad of New Jersey had preceded that of the Penn Central, which was soon followed by the Lehigh Valley, the Erie-Lackawanna, the Reading, and others. In 1970, Congress responded, passing the Rail Passenger Service Act setting up Amtrak to take over passenger service that had been a financial drain. Robert Gallamore and John Meyer have proposed that one of Amtrak's important benefits was to free the freight railroads of the passenger traffic albatross. In a backhanded way, therefore, because it freed up their resources, it also contributed to their safety. The Regional Railroad Revitalization Act ("Three-R Act") of 1973 established the US Railway Association and authorized substantial funding to plan to create a new line—Conrail—from the wreckage of the older carriers. In 1976, the Railroad Revitalization and Regulatory Reform Act ("Four-R Act") set up Conrail and partially deregulated the railroads as well.[31]

These tumultuous events helped reshape congressional ideas on safety. Hearings over the course of the decade reveal a deepening understanding that like so much else that was wrong with the railroads, the worsening epidemic of train accidents stemmed in large measure from inadequate finances that only economic deregulation could remedy. Four years after the "Four-R Act," with railroad financial problems still in the news, the Carter administration signed the Staggers Act into law, largely ending economic regulation of the carriers.

The Federal Railroad Safety Act of 1970 had reflected an abiding distrust of private risk management and faith in the ability of regulation to solve safety problems. Some congressional representatives erupted in outrage when reality failed to confirm their views. Reported train accidents had risen about 20 percent per train mile between 1970 and 1973 and derailments even more. In 1974, Joe Skubitz (R Kansas) of the House Commerce Committee grilled John Ingram, head of the FRA and a recent employee of the Illinois Central, complaining of inadequate inspections and fines, especially of the floundering Penn Central. Skubitz was fresh from continuing hearings on the northeastern rail financial problems, but he apparently failed see to any connection with safety. Labor representatives too blamed problems on inadequate enforcement. Representative John Dingell (D Michigan) seemed to believe that, like the robber barons of yore, railroad managers were indifferent to safety. He urged tougher enforcement, fuming, "we have more inspectors out chasing violators of the Migratory Bird Act," to which Ingram responded that with the government as creditor, fining Penn Central was like fining oneself. Dingell then suggested that "when all else fails maybe you should try the jail house," only to be informed that the likely result would be an exodus of management from any such company. Earlier Ingram had rather delicately raised the problem of finance, asserting that the government "must take the appropriate steps to insure [*sic*] this nation of a safe and *economically sound* rail transportation system" (emphasis added). Unsurprisingly given Ingram's background, this was the position of the railroads. "The real problem with accidents and track conditions," Stephen Ailes, head of the AAR, told the committee, "is the problem of money."[32]

The persistence of spectacular hazmat accidents provided a continuing spur to congressional concerns. By 1975, safety problems were also beginning to merge with the general financial difficulties surrounding the northeastern railroads. In April that year, Fred Rooney (D Pennsylvania), chairman of the transportation subcommittee of the House Committee on Commerce, revealed that he saw the connections. He introduced the hearing, observing that "what concerns me is that the railroads, many of whom are experiencing financial difficulties, may defer track maintenance at the expense of safety." Yet his thinking remained confused; later he worried that the "rail industry itself must be convinced that railroad safety is cost-effective," as if the problem stemmed from management ignorance. Asaph Hall, head of the FRA, emphasized finance. Although he had no railroad background, Hall was an engineer; he estimated deferred maintenance and capital expenditures at about $7 billion, which might have been low given that there were proposals to spend about $4.6 billion on the Penn Central (soon to become Conrail) alone. Speaking for the FRA, and indeed the Ford administration, Hall took the position that economic deregulation was the key to better railroad safety.

> Increased regulation and inspection efforts . . . will not reach the root causes of the rail safety problem . . . The industry does not have enough money for the current maintenance of its equipment and roadbed . . . Unless these improvements are made, we do not see meaningful improvement in the immediate future. The administration will very shortly submit to the Congress legislative proposals which will provide regulatory reform and financial assistance necessary to enable the railroads to . . . maintain their properties at a safe and efficient level.

Testimony by representatives of Amtrak reinforced these concerns, pointing out the impact that deferred maintenance, derailments, and the increasing number of slow orders—especially on the crumbling eastern and Midwestern lines—had on that railroad's costs and service.[33]

In 1978, Rooney was still lamenting the rise in train accidents. His concerns no doubt reflected the headlines. On February 22, 1978, just before the hearings, the wreck on the L&N near Waverly, Tennessee, had generated a bevy of alarming stories in major newspapers as did the Youngstown, Florida, derailment four days later. After Waverly and two other wrecks on the L&N in the same week, Selma, Alabama, tried to ban the railroad from town. "Death Rides the Rails," reported *The Boston Globe* in an article that questioned tank car safety.[34] Rooney again lambasted the FRA for insufficient inspections but—perhaps recalling testimony supporting the "Four-R Act"—he also acknowledged the role that inadequate funding and deferred maintenance played in the continuing rise in accidents. Harley Staggers, chair of the Commerce Committee, agreed. He told of walking the tracks in his district and seeing loose rails and spikes sticking up. He worried that if conditions didn't improve, nationalization would be necessary.[35]

In 1976, Congress had commissioned the Office of Technology Assessment (OTA) to evaluate railroad safety, and its director, Russell Peterson, reported its findings in 1978. "The accident rate does not appear to have been affected by the increased Federal inspection activity," he observed. In addition, when Rooney asked him what to do, he sang from the AAR's hymnbook, claiming, "the primary thing to be done is to provide the funds by some means to maintain the tracks."[36]

Thus, between 1970 and 1980, a rough consensus emerged among important groups on how to combat train accidents. The railroads had always claimed that the problem was primarily a matter of finances, a point they were able to document convincingly, and at every opportunity they linked safety to their financial woes. These claims received powerful support as Congress also grappled with the need for industry restructuring to cope with the carriers' financial problems. The AAR also gained important allies at the FRA, among labor groups, at the OTA, and among important members of the House Commerce Committee including its chair, Harley Staggers. When the subcommittee on transportation met again in March 1980 its new chair was James Florio, who would play a major role in the passage of the Staggers Act later that year and floor-managed the bill in the House. Florio's opening remarks on safety revealed how much had changed in a decade. "One of the best ways to ensure safety is to give carriers the tools to better their financial performance," he said, "perhaps through elimination of much of the economic regulations that have stifled the industry." Seven months later, Congress passed and President Carter signed the Staggers Act, ending the era of detailed economic regulation—although there have been recurring efforts by groups of shippers to return to the good old days. While Staggers was symptomatic of a larger swing of the political pendulum toward business interests during these years, for safety advocates it was also simply recognition of reality.[37]

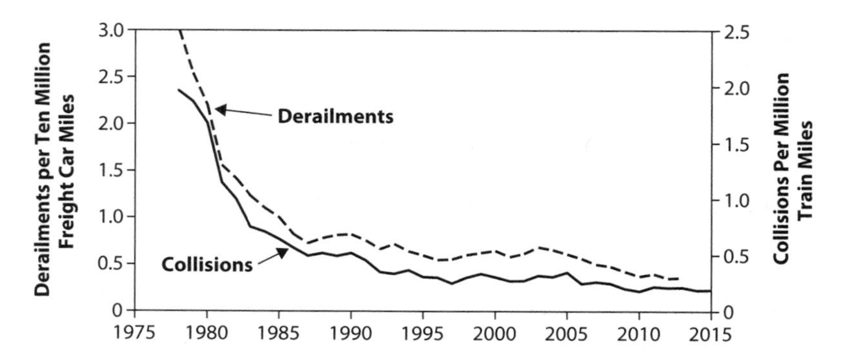

Figure 3.2. The Decline in Train Accidents: Collision and Derailment Rates, 1978–2015. (FRA, *Accident Bulletin/Safety Statistics*, and AAR, *Railroad Facts*)

The Decline in Train Accidents

In 1978, when Congressman Rooney opened that year's hearings on railroad safety, he could not have known that the worst was already over. Despite the rise in the reporting threshold, which increased at more than twice the inflation rate between 1974 and 1978 (Figure 2.1), reported derailments still rose slightly, peaking in 1978. As Figure 3.2 reveals, thereafter they receded sharply to about 1990 and then much more gradually. Overall, from 1981 to 2015 derailments per freight car mile fell at an average rate of 4.1 percent a year while collisions per train mile declined 4.6 percent a year over the same period.[38]

As discussed above, FRA regulations probably made only a modest contribution to these gains. The driving forces were three broad, interrelated changes in railroading that began in the 1960s and were reinforced by the partial economic deregulation from 1976 on. These included a growing awareness on the part of railroad managers of the need for better marketing that could not be achieved without reducing train accidents; a sharp increase in maintenance and capital expenditures on track; and rapid technological improvements in virtually every aspect of railroading. As we will see, the deregulation of entry, exit, and rates that began in 1976 deeply influenced these events. The initial precipitous decline in train accidents largely resulted because the carriers made up deferred maintenance and invested in safer technologies that had long been available. Once they achieved these gains, further progress was more modest because it required additional innovation and a shift in focus away from accidents that resulted from mechanical problems to human failures.[39]

The Imperatives of Marketing

Beginning in the 1960s, some far-seeing railroaders realized that the carriers needed to undertake dramatic changes in order to compete in the new intermodal world. In particular, they needed not only to reduce costs, but also to make quantum improvements in service. Customers complained bitterly of goods lost or damaged in shipment and of service that was slow, unreliable, and inflexible. No one was more convinced of the need for better marketing than the editors of *Modern Railroads*. Beginning in the late 1960s, that journal hammered on the need for what it called "high capacity railroading." Each year it offered the "golden freight car award" for innovative marketing. "Speed and consistency" were the keys, it urged, and it featured surveys and articles critiquing company efforts. Haltingly and gradually, the carriers responded. In modern business terminology, they began to develop a culture of service. Loss and damage claims had been rising steadily throughout the 1960s, reaching 2 percent of revenue in 1970. This was not only an expense, it also drove customers away, and the AAR began a major assault on the problem. The causes were many, but a 1968 survey revealed that the second leading cause after improper handling was train accidents. The costs of accidents, because they eroded service and drove away customers, went far beyond their immediate monetary value.[40]

Concern with loss and damage led some companies to redesign organizational structures. Having freight claims and train operations in separate departments made coordination difficult and impeded a broader accounting of accident costs. In the language of economics, these were agency problems that resulted in built-in inefficiencies. To overcome such difficulties, in 1968 the Chicago & North Western created an Accident and Loss Prevention Department that grouped together freight claims and all forms of accidents. A few years later, the C&O instituted a similar reorganization. The effects of loss and damage on costs and service thus focused management on accidents, while such a system-wide approach ensured that these costs would be conceived broadly. The Union Pacific responded to the new pressures by flattening its management structure to make it more responsive to customer needs. Speed and reliability were also essential to marketing and here the Denver & Rio Grande was one of the innovators as it shifted to shorter trains, run more often, at higher speeds. In 1967, the Rio Grande set up a Transportation Management Center to improve service and marketing. The goal was to coordinate departments so as to combine the proper package of equipment, routing, priorities, and communications to enhance service. "Management has dedicated the whole railroad operation to the marketing concept," *Modern Railroads* enthused. Several years later, *Railway Age* noted that "catering to the customer is good business" and termed the operation "Blending Service and Efficiency for Profit." CEO Gus Aydelott explained how such a focus had led the company's track and equipment to exceed federal standards: "we're trying to run a railroad at a profit . . . and you can't do that if . . . your equipment . . . [isn't] in good, safe operating condition," he explained.[41]

Other responses to the demands for better service involved innovations in traffic, one of which was the rapid growth of containerization

(TOFC, or trailer on flatcar, and COFC, or container or flatcar). Intermodal loadings rose by a factor of 5 in the decade after 1957 even as revenue ton-miles grew only about 10 percent over the same period. By 1989, Conrail was the largest intermodal carrier, and 98 percent of its boxes were in dedicated trains, which avoided much yard switching. In 1976, the Chicago & Eastern Illinois also began point-to-point piggyback service in conjunction with the Southern to speed up delivery times by avoiding yards. In addition, in the 1970s, as computerization made car scheduling easier, the carriers increasingly grouped cars in blocks to minimize switching and speed up schedules. The St. Louis, San Francisco, was among the most active of lines in blocking and arranging run-through agreements that speeded traffic by avoiding yards, demonstrating what *Railway Age* described as "an almost fanatical dedication to service." By 1970, the Southern Pacific was employing a computerized TOPS (Total Operations Processing System) to develop complex blocking and run-through arrangements. The ultimate in blocking was, of course, unit trains; in the late 1970s, their use spread from coal to grain and a variety of other bulk products, and they too typically ran point to point. In 1978, when the Missouri Pacific instituted blocking and run-through agreements with a number of other carriers, it knocked two full days off some schedules, and by avoiding yards it reduced accident risks.[42]

After 1976, economic deregulation contributed to these organizational changes in several ways. By increasing competition and allowing rate flexibility, it raised the potential return to improved marketing. Yet if the railroads were becoming more market driven they also began to develop long-term relationships with customers. Contract rates, for example, which became legal only in 1978, allowed such longer-term arrangements between shippers and carriers, thereby reducing investment risks of both parties. As one writer put it, they were a "license to innovate." Thus, railroads often agreed to improve service—which meant fewer wrecks—in return for guaranteed volume. On the Illinois Central Gulf, a 20-year contract required the company to upgrade track, in part with shipper-supplied funds. Deregulation also eased merger requirements, while getting rid of rate bureaus and legalizing contract rates raised the payoff to single-line service and thus to mergers.[43]

The reader may be familiar with the organizational difficulties that have resulted from airline mergers and the Penn Central debacle, but the railroad merger wave that began in the late 1970s mostly went smoothly and improved safety, for while unit trains and run-through service could—and did—involve interline transfers, multiple railroads made things more difficult. In economists' terms, they raised transaction costs. Consider: before the merger movement, if the Bangor & Aroostook wished to ship a carload of potatoes from Maine to Philadelphia, it went through the Maine Central, the B&M, the New Haven, and the Pennsylvania.

Coordinating a run-through among five carriers was difficult—especially since none of them except for the originating line had the slightest incentive to expedite the shipment—and so the end-to end mergers of the 1970s typically improved service. When the C&O and the Seaboard merged into CSX in 1981, it promptly instituted new run-through piggyback service from Philadelphia to Atlanta. The Union Pacific–Missouri Pacific–Western Pacific merger also resulted in a "substantial increase in the number of trains running through [Kansas City] . . . a major interchange point." The Santa Fe–Denver & Rio Grande marriage of 1987 had similar consequences, and so did later mergers. By 2003, the Norfolk Southern had alliances with the BNSF and the Union Pacific that offered four- or five-day coast-to-coast intermodal unit train service.[44]

While there seems little doubt that on balance the operational improvements and improved fiscal health that came from mergers reduced train accidents, the 1997 marriage between the Union Pacific and the Southern Pacific was a short-term exception, for the integration of the lines was anything but smooth. For the six years prior to the merger, the Union Pacific's collision rate stood at 67 percent of the national average; for the six years after 1997, it was 95 percent of the national average. Some of the increase involved spectacular and deadly main-line collisions that brought down a storm of criticism and publicity from newspapers, unions, and the FRA. An FRA investigation claimed that there had been a "fundamental breakdown in railroad operating procedures." In the *Union Pacific*, Maury Klein titled his chapter on the merger "The Nightmare." Not until 2003 did the UP's collision rate again begin to fall relative to that of other carriers.[45]

As noted, one of the ways that these improvements in marketing and organization benefited safety was by reducing the number of miles spent in switching, which was an accident-prone activity—the rate was 13.8 per million yard train miles in 1975, compared with 8.9 for other work. Moreover, while most yard accidents were low-speed "fender benders," there were occasional catastrophic hazmat wrecks such as that at Decatur, Illinois. Yard train miles declined from 29 percent of the total in 1975 to about 12 percent in 2015 and the overall accident rate—the weighted average of yard rates and rates outside yards—declined from 10.3 to about 2.5 per million train miles. Had the share of yard train miles remained constant at its 1975 level, the overall accident rate would have declined only to 4.34—a 58 percent decline instead of the actual 76 percent drop. In 1986, *Railway Age* wrote about "high productivity railroading." It quoted one railroader that "there just aren't any seven-railroad routings anymore." The journal explained the results: "Score one for the railroad merger movement. Score one for the long haul and the elimination of a lot of interchanges and intermediate switching," it crowed. And it might have added, score one for safety.[46]

Money for Maintenance

Better service also required, as Gus Aydelott of the Rio Grande had noted, much better maintenance, for dilapidated equipment and track not only led to derailments, they might require slow orders too, and these were disastrous to company efforts to compete. For this reason, reducing derailments was, as *Railway Age* put it in 1974, "The Problem Only Dollars Can Solve." The journal had a point, and far more important than the new track safety regulations was the sharp increase in spending on maintenance of track and equipment as well as capital expenditures on upgrading track that began about 1977. The reader will recall (Figure 2.3) that expenditures for both track and equipment declined from 1950 on, bottoming out in 1976, and then rose the following year. Unfortunately, changes in accounting procedures make the data for the period 1978–1982 not comparable with those of earlier or later years. However, maintenance expenditures and investment per track mile from 1983 to 2014 adjusted for price changes and productivity growth rose sharply. Nonfinancial data also reveal improved maintenance. New ties installed per mile of track doubled between the 1960s and the 1980s, and while tons of new rail did not, this was the outcome not of poverty as in previous years, but of much better rail steel and improved maintenance techniques (see below). All of which raises the question: where did the money come from?[47]

Based on GAO reports, I have calculated that from fiscal year 1970 to fiscal year 1985 about $10 billion of it (about two years' worth of maintenance of way spending in 1982 dollars) came in the form of federal loans, grants, and guarantees.[48] While this might seem like corporate welfare, one can also think of it as a return of capital that governments at all levels had been skimming off for a generation. Thus, the majority of the increase came from private sources. These were first, a reallocation of capital spending from equipment to track, as companies began to divest from the freight car business, and second, sharply improved cash flow. This, in turn, was due to deregulation, which gradually raised returns on investment, and the Reagan tax cuts of 1981, which by 1984 accounted for about 20 percent of the railroads' cash flow. The ICC's per diem rule of 1970 that raised the payoff to freight car ownership also seems to have attracted a flood of outside capital. As has been seen, deregulation also spurred maintenance spending by unleashing competition with other transport modes that in turn raised the payoff to an improved track structure. Third, accelerated track abandonment and—as Gallamore and Meyer point out—the end of passenger service staunched a cash drain and allowed concentration of funds on fewer miles. Finally, continuing productivity increases raised the payoff to investment and maintenance dollars.[49]

Many examples breathe life into these figures and help explain the sources of the turn-around. In 1975, after Conrail funding came through, that company began a massive rehabilitation program to make up for decades of neglect. By then Conrail had slow orders on 9,000 of its total of 48,000 miles of track. A survey showed it needed nearly 30,000 track miles of rail and the replacement of one-fifth of all its ties. By the end of 1975, it had 43 tie gangs at work and they had replaced 3 million ties. A year later with 65 tie gangs as well as a host of other specialized track crews the company had begun to reduce slow orders—on one piece of track allowable speed rose from 8 to 40 mph. By 1982, at the cost of about $1.75 billion, the company had installed 5,000 miles of continuously welded rail (CWR), resurfaced the entire line twice, and inserted 22 million ties. Derailments dropped from 1,520 in 1978 to 202 in 1981.[50]

While Conrail's rehabilitation grew out of political imperatives and funding, for other lines the pressures were economic. As they had in the late nineteenth century, the carriers began a massive effort to upgrade track and equipment. In 1977, the Illinois Central Gulf raised its track repair budget even as it was abandoning more than 1,000 miles of little-used line to concentrate on the heavy tonnage routes. Later it converted 500 miles of main line to single track, thereby concentrating maintenance and reinvesting the money received from scrap. The company increased spending on ties and CWR, and like many lines, began to focus on more fundamental problems by increasing ballast to keep the track structure "from pumping mud." In 1979, Robert Ahif, the company's chief of operations and maintenance, told *Railway Track & Structures*, that safe track was only a minimum requirement: "in addition to safety, a track must be economical for the use made of it." In 1980, the Canadian National reported installing twice as much ballast and twice as many ties as in 1975, and computerized a Total Inventory and Evaluation system to improve resource allocation. In 1983, the Seaboard too was pursuing a major program to upgrade its track or, as CEO Richard Spence put it, "get the mud out." The explanation, according to *Railway Age*, was that "under deregulation Seaboard will have to provide reliable, fast service to win customers." At Grand Trunk, a new management team in 1971 began to emphasize marketing and safety and upgrade physical plant. By 1981, it was calling itself "the good track road," its motive being to "preserve market share." Improved maintenance—due in part to the cash provided by the Reagan tax cuts of 1981—helped raise its on-time performance from 69 to 92 percent between 1973 and 1981. Even the Milwaukee began to upgrade. Having shrunk its track but not its maintenance budget by two-thirds, it was able to triple spending per mile to upgrade its main line.[51]

In early 1981, *Railway Age* reviewed the major carriers' proposals for track rehabilitation that year and it detected "a change of heart." Rail and tie renewals had been ticking up, and the writer thought he knew why: he found a "spirit of confidence . . . of cautious optimism." Businesses invest for the prospect of future returns; with the freedom to compete, for the first time in many years, the future for railroaders seemed to beckon. Preserving markets or winning customers required faster, more dependable schedules, and train accidents could have no place in this newly competitive world. Companies found that a culture of service also required a culture of safety: in short, market competition raised the payoff to safety.[52]

A Technological Revolution

In the period 1974–2008, railroad productivity increased 3.4 percent a year—more than twice as fast as in the previous quarter century and much faster than the rest of the American economy. What underlay the productivity boom was a nearly unparalleled gale of technological change, much of which derived from the computer/Internet revolution. This "creative destruction" not only raised productivity, it also drastically reduced rail employment, reshuffled the types of job and nature of work, and sharply reduced train accidents.[53]

Several aspects of this technological revolution stand out. First, some of the new technologies had origins dating back to the 1950s and 1960s and occasionally even earlier; what was new was their rate of adoption. Second, the level of research funding was by historical standards extraordinary. Third, the AAR and the federal government assumed much larger roles in railroad research. Fourth, railroad rivalries and intermodal competitive forces ensured that economics, as manifested in industrial engineering, played an increasing role in technological choices. Finally, computer and wireless technology along with increasingly sophisticated remote sensors enormously expanded information availability as they yielded low-cost monitoring of track, equipment, and their interactions.

Between 1970 and 1974, annual FRA R&D budgets averaged $32 million a year (although a substantial fraction of that went to high-speed rail) and $58 million in the following half decade. Moreover, as *Modern Railroads* noted approvingly, the research became increasingly practical and safety-oriented. As has been seen (Chapter 2), the AAR also made a major commitment to research about 1970, hiring William Harris as its director and beginning what would prove to be a sharp increase in funding. Its research budget rose from a mere $800,000 in 1971 to $2 million in 1973. By then, *Railway Age* was happily predicting an "explosion in track research." The AAR budget rose, that journal later reported, to $19 million in 1990 and $26.4 million in 1992.[54]

By the late 1960s, large carriers such as the New York Central, St. Louis, San Francisco, Santa Fe, and others were also increasingly involving industrial engineers in their decision-making. In 1978, the editor of *Railway Track & Structures* noted both the growing numbers of PhDs and industrial engineers. He observed that the latter group brought analytic economic techniques developed elsewhere to railroading, and concluded that "'horseback' decisions" were on the way out. This blending of economics with engineering analysis required the development of low-cost, sophisticated methods of data collection and analysis; collectively these brought a revolution of railroad maintenance.[55]

The importance of industrial engineering for safety was twofold. First, the ability to document the profitability of (say) improved hotbox detectors made it more likely that management would provide the funds to install them. While some railroaders thought—as the epigraph suggests—that economics contaminated engineering, others understood that the marriage of engineering with increasingly sophisticated economics could enhance the corporate status of technical people. When D. G. Ruegg, a vice president on the Santa Fe, spoke to the AREA in 1979, he reminded his audience of Arthur Mellen Wellington's dictum: engineering was "the art of doing that well with $1.00 what any bungler can do with $2.00 after a fashion." He pointed out that managers often overrode engineering safety factors as too conservative and too expensive. They were more likely to be respected, he argued, when backed by compelling economic analysis.[56]

Second, ranking and allocating funds for projects according to where they would yield the highest return increased the safety/output bang for the maintenance buck. A safety/output frontier from Chapter 1 (Figure 1.1) can illustrate the increased payoff to maintenance. Such techniques helped push inefficient companies closer to the frontier, allowing improvements in both efficiency and safety. This was hardly a novel idea; as Chapter 1 noted, companies had allocated block signals, steel rails, and steel passenger cars first to high-density lines. What was new was the range and precision of applications and the ability to generate and manipulate large amounts of information. As early as 1968, industrial engineers at the Penn Central were determining the optimal (least cost) spacing of hotbox detectors that balanced the costs of additional detectors against their savings from fewer hotboxes and accidents. Of course, such an analysis is only as good as its data, and the grouping of all loss and damage causes discussed above helped ensure that accident savings would not be underestimated. Conrail's industrial engineers also employed computerized cost-benefit analyses to rank projects and allocate funds by estimated return on investment in order to wring more payoff from its limited budget. With the increasing availability of low-cost computing power, industrial engineers applied these techniques to a widening range of potential projects, thereby enhancing the productivity of all forms of spending.[57]

The FRA's contribution to improved technology began with the tank car safety program noted above. Perhaps that organization's most important achievements were the creation in 1970 of the Volpe Transportation System Center, which would do research on all aspects of transportation matters, and the Transportation Technology Facility at Pueblo, Colorado.

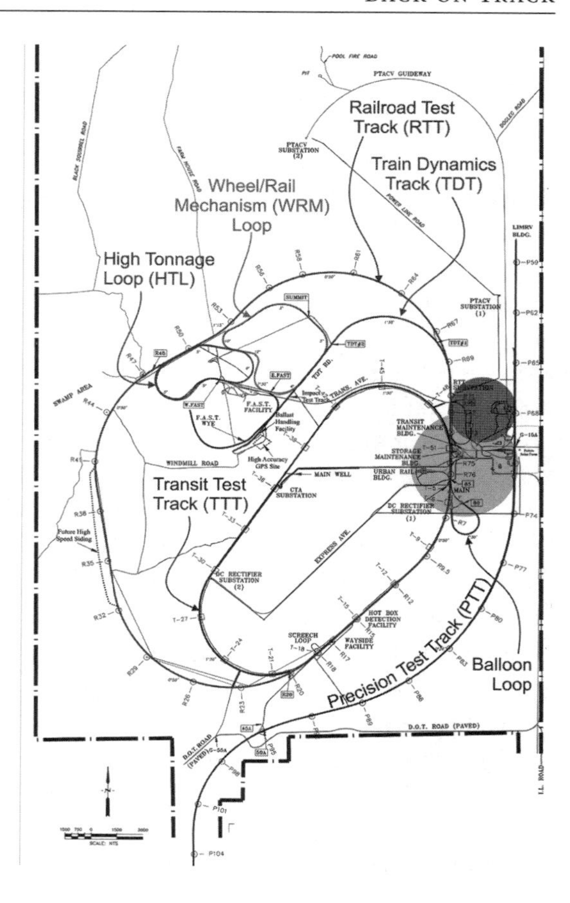

The Transportation Test Center (TTC), initially funded jointly by the FRA and the AAR, was the most important development in railroad research in a century. (Courtesy Luis I. Maal, On-Site Resident Engineer–Program Manager, US DOT–Federal Railroad Administration)

The FRA ran the test facility jointly with the AAR until the latter group took it over in 1982.[58] By 1975, the FAST (Facility for Accelerated Service Testing) 4.8 mile track was up and running. It included 22 sections of ties, ballast, fastenings, and so on. Running a continuous heavy train day and night accelerated "real-world" service testing of both track and equipment by a factor of 10. By 1981, a blizzard of valuable results had begun to emerge. Curved track had always generated problems, but the FAST results suggested the need for much more care. Engineers on both the Seaboard and the ICG reported that they had increased use of flange lubrication (that would reduce stress as trains navigated curved track) because of FAST experiments. The data also indicated both the need for more substantial track structure and how to achieve it. FAST also demonstrated the value of better rail (see below) and of improved tie hold-downs, such as screw spikes and clips.[59]

In 1988, FAST researchers began to study the effects of heavy axle loads (HALs) from 125-ton cars, thereby creating FAST/HAL. Although car weight had been steadily increasing from the earliest days of railroading, in the 1970s there had been considerable pushback from maintenance engineers against the newer heavier cars, and *Modern Railroads* repeatedly questioned their economics. Researchers discovered, to their surprise, that heavier loads resulted in the rail "dragging the ties through the ballast" but that more clips to prevent rail creep provided a quick fix. Although some lines such as the Florida East Coast had made extensive use of concrete ties as early as the 1960s, few others followed. Accelerated tests provided the experience necessary to create a "level playing field" (provide equivalent information) for concrete ties, and their use slowly began to spread, especially on curves. The studies also finally put the debate over larger cars to rest. They found that while maintenance costs went up much more rapidly with HALs, on balance larger cars were more economical. By 1990, the Burlington Northern had come to the same conclusion: employing a computer simulation developed by the AAR, its own studies showed about a 5 percent net cost reduction from increasing the payload of coal cars from 101 to 112 tons, and similar gains from use of large double-stack cars for container shipments. As had been the case for decades, bigger really was better.[60]

The third major FRA-AAR research initiative was the Track-Train Dynamics (TTD) program that ran from 1972 to 1986. It was the brainchild of William Harris, who when he became head of research at AAR, asked researchers from the Southern Pacific to develop a program. As Harris knew, several carriers, including the Southern, the Canadian National, and the Southern Pacific, had been researching track-train dynamics for several years and the latter company had even gone to Japan for experiments—as no test track then existed in the United States. An important goal was to puzzle out seemingly unexplainable derailments. Studies of dynamic braking by Electro Motive Diesel demonstrated that the newer, more powerful six-axle units generated so much braking force that they might easily jackknife a train. Other work established the relations between track-train dynamics and gauge widening or overturned rail.[61]

An important result of such efforts was a set of guidelines on train handling. Writing in 1985, Harris claimed that "almost all railroads apply the guidelines in training and operations." TTD also developed a number of mathematical computer models such as the Train Operations Simulator to help researchers—and locomotive drivers—understand and control train operation. As time-share computer costs came down, the program became economic, and in 1978 the Frisco used it "in major derailment analysis," where it helped pinpoint accident causes and led to modified train handling techniques. In 1974, TTD also published an accident investigation manual that became widely used. It investigated wheel failures and developed a model to analyze stress that manufacturers were required to employ in wheel design. TTD research also stimulated major innovations in freight cars and their trucks, and by the early 1980s new equipment was coming through that significantly reduced truck hunting and rock 'n' roll. While TTD officially disappeared in 1986, much of the work continued under the Vehicle Track Systems programs.[62]

A host of other developments also improved both the safety and the efficiency of train operations. While the new ABDW air brake of the 1970s was an improvement over the older ABW brake, it remained a poor way to stop a long train: on a 150-car train it took 13 seconds for the brakes to begin to apply and 70 seconds for them to attain full pressure. Electrically controlled pneumatic (ECP) brakes that would be instantaneous in operation had long been a goal, but the first serious testing dated from the 1990s as the cost and power of microchip technology improved. ECP brakes sharply improved safety: stopping distances were 50 percent shorter and, unlike the ABDW, they allowed gradual release. With many fewer moving parts they were also less failure-prone than older brakes and they generated less wear and tear on wheels and draft gear. Of course, as always, the railroads took some of the gains from better brakes in improved productivity, for the new brakes also allowed longer trains and higher speeds, and yielded fuel savings as well. *Railway Age* called ECP brakes "the most significant development in freight train braking since George Westinghouse." Yet the new equipment could not be mixed with the older brakes and its advantages would accrue to the carriers while the costs would fall on equipment owners, many of which were private. The FRA assisted development; because the new equipment was self-diagnosing, the agency waived some obsolete brake test requirements and funded a study that predicted high returns on investment. In 2007, the BNSF and the NS began to apply the new technology to unit coal trains where interoperability was not a problem. On one run the latter company found the new brakes saved 800–900 gallons of fuel a trip. ECP brakes have increased track capacity and average train speed while improving safety. *Railway Age* thought they "could do as much for railroad operations as . . . PTC [positive train control] mandated by law." ECP brakes are expensive, however, and there have been some problems with electrical interference and conductivity. Their use has spread slowly, but a 2015 FRA regulation—opposed as cost-ineffective by the AAR—mandates ECP brakes on trains carrying crude oil and similar hazards by 2023 (see below).[63]

In 1971, wheel manufacturers had begun to employ magnetic particle and ultrasonic inspection of all new wheels to comply with a new AAR rule. Wheel makers also developed heat-treated, curved plate wheels that could accommodate thermal expansion with lower stress and danger of fracture. Research also discovered that the traditional index of thermal stress used for wheel scrapping—discoloration that supposedly reflected overheating—was worthless, resulting in the removal of many useful wheels and failing to find others that had been dangerously overheated.

The net effect of these was a dramatic decline in derailments from bad wheels. An AAR survey done about 1985 showed that the low-stress, heat-treated, curved-plate designed wheels were one-sixteenth as likely to fail as older designs. In 1978, there had been 423 derailments from bad wheels, in 1990 there were 80, and by 2015 the number was down to 47.[64]

While the decline in yard work after 1980 reduced collisions and derailments throughout the whole period, yard accident rates only began to decline after about 2004 as companies introduced remote control technology that allowed a person on the ground to control a yard switch engine, thereby eliminating garbled radio communication or misinterpreted hand signals. New Zealand and Canada had employed the technology in the 1990s, reducing yard accident rates 40 percent or more. Labor opposition delayed its introduction in the United States but as the new approach spread, yard accident rates dropped sharply from about 21 to 12 per million locomotive miles between 2005 and 2015.[65]

The gradual substitution of CWR for jointed rail undoubtedly improved safety by reducing the number of failure-prone joints. However, it involved a risk-risk trade-off, for it might buckle during hot weather, causing a derailment. It had to be installed with an eye to the proper temperature and needed to be well spiked and ballasted. Rail will increase in length according the formula $\Delta L = \alpha L \Delta T$, where L is the length of the rail, T is temperature, and α is the coefficient of linear thermal expansion. If length is constrained, compressive forces result; a rise in temperature of 90 degrees Fahrenheit can induce 220 tons axial compressive force in the track—enough to generate a "sun kink."[66]

Indeed, there were 60 derailments from such causes in 1985, motivating research into track stability. Companies discovered that maintenance work was highly temperature sensitive: the Norfolk Southern developed the equation "disturbed track in hot weather + failure to follow instructions = buckled track." The AAR and the FRA developed Gage Restraint Measuring System (GRMS) vehicles that could measure the potential for gauge-widening as well as test the longitudinal stress in CWR. In the 1990s, the Union Pacific employed such information to modify both track-laying and train operation practices and successfully reduced derailments. As companies learned to use the new data they gradually came to employ it in maintenance planning. In 1995, CSX became the first carrier to use such equipment to evaluate tie renewals. More recently wayside rail stress monitors have arrived that can upload data to pinpoint danger spots. Better information and improved maintenance reduced sun kink derailments to 15 by 2014.[67]

Broken rails continue to be an expensive and dangerous problem, resulting in 182 accidents in 2015 (with 75 of them occurring in main track), down from 230 in 1990. Track evaluation technology, dating from Elmer Sperry's work on flaw detection in the 1920s, became progressively more sophisticated. In the 1970s, the Sperry Corporation began to employ road-rail equipment that raised productivity and to include ultrasound that could test within joint bars. It also offered flaw detection for new rail before installation. By the twenty-first century, detectors used overlapping technologies (ultrasonic, magnetic induction) and could find much smaller defects with fewer false alarms while operating at higher speeds. Economics also began to shape testing. Since rail flaws increased rapidly with tonnage, it made economic sense to increase inspection frequency with the age of the rail. Relying in part on software developed at the Volpe Center, about 1997, the BNSF began a long-running analysis to implement risk-based ultrasonic testing that varied with both the age and the expected use of the track. For example, passenger routes and hazmat routes were subject to more frequent examination, and the result was a sharp reduction in derailments. Yet detection remains an art as well as a science. About 1 in 200 rails that break in service (a break not found by a detector car) causes a derailment and one did on the Norfolk Southern on October 20, 2006. Surface defects masked a transverse fissure that eluded a Sperry car, and the resulting derailment dumped 23 cars of ethanol, which burned for 2 days.[68]

Other kinds of track evaluation equipment also appeared. As early as the mid-1960s, the B&M was using a mechanical inspector to evaluate track geometry and more sophisticated, faster cars soon appeared. In 1971, a track geometry car (TGC) on the Southern could measure gauge, cross-level, vertical displacement, super elevation (of the outside rail on a curve), and curvature. In the 1970s, the FRA began to employ such equipment in track inspection. By the 1980s, the C&O's TGC could computerize and rank a host of details, highlight critical defects, indicate maintenance priorities, and reveal places that violated FRA standards. By the twenty-first century, TGCs traveling at 90 mph were using laser scanners and onboard PCs to record virtually all aspects of track geometry and transmit the data by wireless to central computers and hand-held devices of inspectors in the field. Recently, the BNSF has applied to use drones for inspection of track and bridges in remote areas. New methods can now measure tie wear and plate cut, along with rail cant and corrosion, and machine vision is coming that will photograph and interpret images. The resulting data can be uploaded to an asset management database, and as prices have come down they are being increasingly widely used.[69]

A range of other new technologies began to improve track and roadbed in the 1970s. Companies employed geo-textiles and later geo-grids (plastic structures) to stabilize roadbeds and banks. In the 1980s, researchers began to investigate use of ground-penetrating radar to investigate roadbed. By the mid-1980s, the Osmose Company was offering in-place preservative treatment for old ties and wooden trestles. In the 1990s, the Burlington

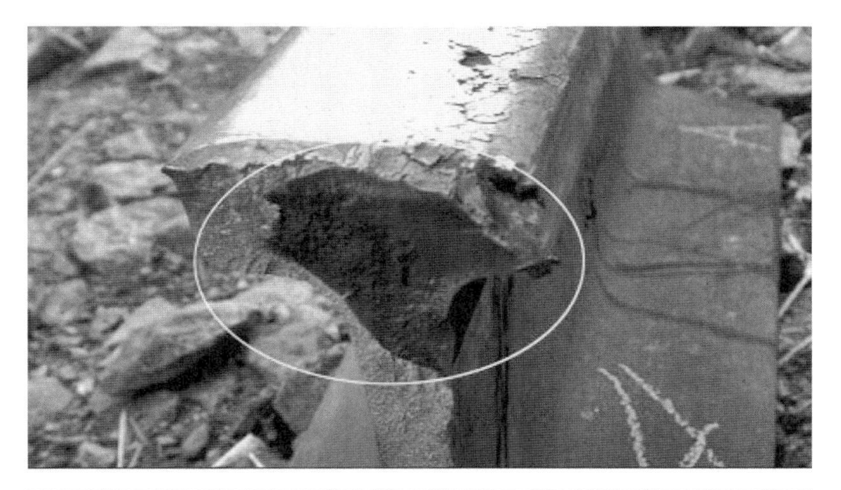

A recent ultrasonic inspection had failed to find a transverse fissure (*top*) that caused a Norfolk Southern train carrying 83 cars of ethanol to derail near New Brighton, Pennsylvania, on October 20, 2006 (*bottom*). (National Transportation Safety Board)

Northern experimented with sound waves to assess deterioration in wood ties. More recent efforts employ lasers and scatter X-ray imaging.[70]

In few areas of railroading were the changes greater than in rail manufacturing and maintenance. In the 1980s, research had demonstrated that crack growth often originated from small inclusions and as a result manufacturers developed clean steel—which yielded dramatically lower failure rates. FAST research also revealed the value of harder, heat-treated rail and companies found that it was cost-effective on the sharp curves of heavy traffic lines. A 1991 study showed that after 200 megatons of traffic, clean, heat-treated rails had a defect rate of 0.21/1,000 tons versus 1.48/1,000 tons of traditional rail. Of course, rail and wheel were part of a technological system, and in the 1980s railroad-FRA research showed that at high wheel loads metal fatigue rather than wear became the limiting factor on rail life. Studies also demonstrated the need for much more careful wheel maintenance as modestly out-of-round wheels on heavy cars resulted in dynamic impacts nearly twice the static wheel load, with potentially disastrous results for rail and track. Similarly, corrugated rail increased dynamic wheel loads.[71]

Rail maintenance provides a good example of the blending of economics, engineering, and computer analysis noted above. With rail costing $100,000 a mile (in 1996), companies searched for the best (most cost-effective) levels of maintenance. Railroads had long used grinding to remove rail corrugations and detail fractures; in the 1990s, they extended grinding to reshape the rail head to extend its life, employing data from test cars to determine the most economic grinding interval. Grinding strategies varied by type of equipment, speed, number of passes, and rail profile and accordingly benefited from much learning by doing. In the 1990s, the BNSF and the Canadian Pacific replaced corrective with proactive programs, thereby reducing detail fractures and increasing rail life. Companies employed data on defects and accident costs to determine optimal rail replacement intervals and when rail damage from out-of-round wheels made their replacement economic. By the mid-1990s, the CP computerized the data coming from its TGC to develop a track maintenance advisory system to determine least-cost intervals for rail grinding and renewal, tie renewal, surfacing, and ballasting. Continuing research at the Transportation Technology Center (TTC) and company experience led to vastly expanded use of rail lubrication, which extended rail life and, because it reduced lateral forces, prevented derailments. By the twenty-first century, newer top-of-rail, water-based polymer lubricants that promised fuel savings and safety made their debut.[72]

The outcome of these various technologies has been a much more scientific and economic approach to track maintenance. Moreover, their application has required complementary upgrading of employee training. Some companies have made formal arrangements with community colleges (Chapter 4) and all have extensive on-the-job training. Amtrak, for example, has systems of training and retraining, complete with exams, for track supervisors and inspectors that vary in strictness with FRA class of track and include courses in defect detection and field mentoring.[73]

The development of sensors, computer programs, and economic analysis also wrought a revolution in equipment maintenance. Trackside equipment monitoring devices (dragging equipment detectors, hotbox detectors) had begun to appear in the 1950s (Chapter 2). By the mid-1980s, Conrail and other companies were installing smart detectors; linked to microcomputers they could instantly measure minute temperature differences in wheel bearings, adjust for a host of other factors, and then provide voice readout of the results to the cab crew. Yet when roller bearings failed they could lead to overheating and catastrophic failure in less than a minute. Hence the need for on-board detectors, and the late 1980s brought the "smart bolt"—the cap of a roller bearing bolt that held a tiny radio transmitter that would send a signal when the bearing overheated.

In 1990, after the Burlington Northern had experienced 19 derailments that cost $3 million from journal burn-offs, it began installation of smart bolts on 10,000 cars.[74]

In the twenty-first century, use of these equipment-monitoring technologies has continued to spread, and they are being continually refined and employed in a widening range of applications. New acoustical detectors have arrived that can spot bearing problems before they become serious. There are, in addition, hot wheel detectors, broken or cracked wheel detectors, clearance detectors, wheel impact load detectors, truck hunting detectors, and shifted load and heavy car detectors. In addition, hazmat cars are being equipped with sniffers that can detect excess emission of chlorine or other dangerous gases. Companies are using an Internet-based Integrated Railway Remote Information Service that employs wayside devices with lasers to read passing wheels; it can identify tread defects, evaluate wheel profiles for preventive maintenance, and evaluate truck performance. Other systems can monitor everything about a train's performance from braking to engine oil, and smart cars can even send a text message describing their problem. In 2004, the AAR began an Advanced Technology Safety Initiative intended to develop a program of proactive maintenance. It takes readings from the wheel impact load detectors and sets a cutoff load of 90 kips (thousands of pounds of force) to determine wheel replacement. And just as machine vision technology has been employed to evaluate track and roadbed, by about 2010 companies had begun to employ it to provide much more rapid inspections and to evaluate and record every critical part of a train in motion.[75]

What If?

An implication of the above analysis and that in Chapter 2 is that had partial economic deregulation occurred earlier—say in the immediate post–World War II years—the course of safety might have been quite different. These years would have seen more rapid shedding of unprofitable lines and service, an earlier shift to end-to-end mergers, and faster diffusion of better technologies. Thus, these and no doubt many of the other changes of the 1980s would have occurred 30 years earlier and would have largely avoided the profitability collapse and deterioration of plant, equipment, and morale that led to the derailment explosion. The shift to much larger cars might still have caused a rise in derailments, but these would have been greatly reduced by better maintenance of track and roadbed and more money for research. Such a counterfactual suggests that there would have been no explosion in accidents after the mid-1950s.[76]

Regulatory Crosscurrents

The years after 1990 saw sharp changes in the approach to railroad safety regulation along with a large dose of "more of the same." In 1993, Jolene

Molitoris became head of the FRA. Seven years later as she left in 1999, *Railway Age* included her on a list of its "most respected and admired" railroaders of the twentieth century. While FRA regulation was never as adversarial as that of some other agencies, it invariably pitted unions against management. Molitoris had concluded that the existing approach to regulation was too conflict oriented, was too slow, tended to ignore labor (leading it to appeal to Congress), and was often unable to get at the root causes of accidents. Under her guidance the FRA institutionalized cooperative regulation, announcing at about the same time a policy of "zero tolerance" for accidents. This was a reflection of the Total Quality Management (TQM) ideas then sweeping through railroading (Chapter 4); happily enough, it also had the effect of blunting any criticisms that cooperative regulation was soft on safety. Molitoris also shifted the agency toward a more risk-based approach to enforcement. In fact, her new approaches had antecedents and were a response both to external criticisms and to the changing nature of railroad risks.[77]

Cooperative Regulation

Testifying before Congress in 1984, FRA Administrator John Riley revealed that the agency had begun to supplement its inspections with System Surveys, a descendant of the earlier targeting programs; in 1987, he reviewed the progress of safety and peered into the future. "We are now past the era in which we can generate dramatic improvements in safety performance . . . [by] focusing on track and maintenance," he observed. "If we are going to keep the accident rate moving downward, we need to develop mechanisms to deal with . . . human performance accidents." The figures bore out Riley's contention: when he spoke, human factors accounted for about 30 percent of all train accidents—most of which were collisions—compared with about 15 percent as late as 1978. Yet any effort to deal with accidents due to human error required the cooperation of workers and their unions, for the agency had no authority to compel them to do anything. Moreover, as we have seen, the old command and control approach to regulation had never worked well. Outside critics had noted the need to involve the railroads and others in rulemaking as early as the 1950s, and several GAO reports sharply criticized the process in the early 1990s. About that time, under Riley's successor Gilbert Carmichael, the agency also began to experiment in a small way with cooperative regulation.[78]

Riley's tenure marked the first significant attempts to deal with "human factor" accidents—rulemakings on drug and alcohol use and engine driver training in the late 1980s—but because their primary motivation was passenger safety, they are discussed in Chapter 5. Molitoris tried to push cooperative regulation by revamping the older System Safety Surveys. In 1995, the FRA renamed such reviews its Safety Assurance and Compliance Program (SACP). They now incorporated labor as well as company representatives and reflected the influence of the broader System Safety Movement, which originated in defense-space programs and emphasized both a holistic approach to the problem and the search for root causes of accidents. An SACP audit evaluated all aspects of a company's safety; in response, the affected railroad was to develop an action plan to address the deficiencies identified. Descriptions of these programs sound like they must have done much good, but as Figure 3.2 reveals, it is hard to find the evidence in train accident statistics.[79]

Early versions of the Federal Railroad Safety Act of 1970 had included a provision that would have required advisory committees composed of interested parties. While the requirement never became law, Molitoris developed Administrator's Roundtables that were eventually formalized as a Rail Safety Advisory Committee to work on safety issues. It was another effort to make regulation less contentious, for the committee included representatives of labor, management, the FRA, and other interested parties to identify safety priorities and propose solutions. "Only since Ms. Molitoris has come here to the FRA have we had a stake in the table," claimed James Brunkenhoefer of the United Transportation Union (UTU). Molitoris was the perfect facilitator; she was straightforward, was unassuming, and inspired trust. In an early Administrator's Roundtable meeting as participants identified themselves, she said simply, "I'm Jolene Molitoris and I work at the FRA." The Roundtable/committee might also have been an effort on the part of the agency to become less reactive and to take control of safety priorities away from the NTSB and Congress. The committee divided into a host of working groups, each tasked with the investigation of a separate topic. One of the topics was fatigue, which Molitoris later termed "the number one underling cause of human issue accidents." Whether by accident or design, a focus on fatigue also helped justify a requirement for PTC—a topic discussed in Chapter 5.[80]

Railroaders had long been concerned with dangers of fatigue—we acknowledge them when we say "asleep at the switch"—and the original Hours of Service Act of 1903 was an early response to the problem. Yet understanding of sleep and fatigue had progressed since that time and experts found the act very nearly irrelevant. A representative of the NTSB pointed out that the act allowed an engine driver to work 432 hours a month (30 consecutive 14-hour days) while airline pilots were limited to 100 hours. It also ignored a host of complexities involving circadian rhythms. The NTSB investigated several wrecks it thought were fatigue-related in the late 1980s. In one, a head-on collision of two Conrail freights near Thompsontown, Pennsylvania, on January 14, 1988, which killed four crewmembers, the board concluded that the engineman had been asleep and run a signal, probably because of fatigue from irregular work hours (Table A3.1). In 1990, it added fatigue management to its "most wanted list" of safety improvements, and about that time the GAO issued several reports on fatigue while the FRA also began to study crew scheduling.[81]

The carriers too were becoming concerned with fatigue. The AAR began to study the issue in 1992, joining with labor to create a Work/Rest Review Task Force that generated a large statistical base to study problems of fatigue. In 1999, a labor agreement established committees to investigate scheduling and participated with the FRA working group to study ways of coping with fatigue-caused accidents. One rather chilling discovery was that fatigue could essentially nullify the value of cab alerters. Recall that these are devices that sound an alarm and apply brakes if not acknowledged. A sufficiently fatigued driver could routinely acknowledge the alerter and still run a signal to collide with another train. By the late 1990s, an AAR-sponsored survey showed that most of the major carriers were experimenting with education and training programs to cope with fatigue. Conrail, the BNSF, the CN, and the CP had pilot programs that allowed naps on duty and all were trying to modify schedules to ensure greater predictability and allow at least 10 hours of rest. A survey update in 2007 again revealed much experimentation, with the Norfolk Southern emphasizing scheduling while the Union Pacific was developing a risk-based fatigue management program. The Rail Safety Improvement Act of 2008 included detailed rules governing allowable work schedules for train and signal employees. In addition, it required the carriers to develop fatigue management programs and provided the FRA with broad powers over work schedules. Most large carriers now allow napping on the job under specified circumstances and are trying to regularize employee work hours.[82]

Research into fatigue-caused accidents led to a study of close calls, for they were far more prevalent than were accidents and contained much useful information. One writer has likened their study to a vaccine that would protect against accidents and injuries. Other industries had also employed close calls as a way to identify accidents waiting to happen but a history of distrust between employees and managers, along with the 1908 Federal Employers Liability Act's (FELA) fault-based accident liability system, discouraged candid reporting by workers—another example of a principal-agent problem. In the early 1990s, the Burlington Northern had implemented an anonymous hotline that generated close call reports. About 2003, the FRA began to look at developing such a system—another market-augmenting action on its behalf. In 2007, after several pilot projects that showed promising results, it set up a Confidential Close Call Reporting System (C3RS) demonstration project that included the UP, the CP, Amtrak, and the New Jersey Transit.[83]

The close call program and other initiatives were part of a broader risk reduction strategy that was proactive and cooperative, but while the FRA under Molitoris stressed cooperation, it did not abandon regulation. The end-of-train (EOT) monitor was one of the increasing numbers of remote sensors and controls described in the previous section that appeared in the 1980s. Traditionally, monitoring the rear of the train, watching brake pressure, and—if necessary—applying the brakes were the jobs of the employees who rode the caboose. Hauling a caboose was expensive, however, and in a 1982 agreement, labor allowed its elimination by arbitration. Unions had always claimed that cabooses improved safety, but a comprehensive Railway Labor Conference study of 1984–1985 found them no more effective than EOT devices in reducing train accidents. Early versions of the devices could only monitor brake pressure, but by 1987 Canadian lines were employing a two-way device that allowed the engine driver to apply brakes at the rear of train. By 1987, the NTSB was urging their use in the United States. By 1996, it had investigated eight accidents it claimed such a device could have prevented. After two disastrous run-away wrecks down Cajon Pass in 1994 and 1996 (Table A3.1), the FRA mandated them, effective July 1997. The railroads beat the deadline and by then the device had already prevented at least one more run-away.[84]

Hazmat Problems Return

As Figure 3.2 reveals, collision and derailment rates rose slightly for about a decade after the mid-1990s and then continued their downward trend. The increase reflects a worsening of yard accident rates, which rose in the late 1990s before plunging after 2005. Main-line train accidents show an uninterrupted decline, reflecting a continuation of the improvements in technology and operations discussed above. Of course, no technology works perfectly; wheels contain hidden defects and rails break from undetected fractures. Human failures can still occur in an almost infinite number of ways. When these result in accidents to freights carrying hazardous cargo, they continue to drive public policies.

There has been no surge of hazmat accidents in the early twenty-first century; while such accidents have risen relative to train miles this has been offset by a declining fraction of cars releasing (Figure 3.1). Releases averaged about 24 a year in the period 2006–2015. Public anxieties have increased, however, and contributing to this was the attack on September 11, 2001. A search in *Newsday* for "railroads and hazardous" yields about 1,100 articles a year in the late 1990s; in 2001, the number was 1,800. Similarly, a search of "railroads and terrorism" yields 160–200 entries up through 2000 and nearly 1,400 in 2001. Some of these stories focused on possible sabotage of Amtrak or commuter lines, or hazardous grade crossings, but the terror attack on the twin towers also reawakened longstanding fears of hazardous materials being shipped through urban areas. In March 2004, the train bombings in Madrid, Spain, also kept concerns over railroad safety in the news. Not much happened, however, until in early 2005 a deadly spill brought matters to a head.

On January 6, 2005, an improperly set switch in "dark" (non-signaled) territory diverted a Norfolk Southern freight, Train 192, onto an industry track near Graniteville, South Carolina, where it collided with a parked train (Table A3.1). The wreck derailed three cars of chlorine and punctured one of them. The resulting cloud of gas killed 9 people, including the engine driver of Train 192, and caused the evacuation of about 5,400. The NTSB report pointed out that the car that leaked chlorine dated from 1993 and reflected the new, tougher standards for pressurized tanks. Yet it failed. The report also observed that the force on the cars depended on their speed and position in the train. In this, and in reports on several previous less publicized wrecks involving poisonous inhalation hazard (PIH) chemicals, the board recommended that the FRA investigate the forces acting on tank cars in wrecks. It also urged study of the strength of older cars and ways to improve the fracture resistance of new ones. Finally, it recommended reducing train speeds and placing cars toward the end of trains where impact forces would be less.

The wreck also seems to have coalesced what had been vague public worries about railroad hazmat shipments. A representative of the NTSB linked the wreck to fears of terrorism, telling victims' families that Graniteville was "your September 11th." Soon after the disaster, headlines began to appear such as "Dangerous Trains Pass through the Region" and "Keeping the Public Safe from Railroads' Cargoes." *The New York Times* headlined the wreck "Deadly Leak Underscores Concerns about Rail Safety," and several editorials decried "Deadly Trains" and "Trouble on the Tracks." Indeed, the specter of the robber barons appeared once again to haunt the *Times*'s editorialists. Apparently imagining that the railroads were indifferent to the cost of train wrecks, the paper denounced the "shoddiness of federal regulation and industry self-policing" and called for $50 million in fines as "the sort of toughness that public safety and security cry out for." Despite the carriers' routine failure to earn their cost of capital, *The Boston Globe* thoughtfully decried the railroads as a "profit crazed industry."[85]

Congress responded to these public worries, and in the Rail Safety Improvements Act of 2008 (P.L. 110-432) it largely mandated the NTSB requests. In 2009, the Pipeline and Hazardous Materials Safety Administration upgraded safety requirements for cars carrying PIH chemicals—mostly ammonia and chlorine—and it reduced train speeds to 50 mph for trains carrying such materials. The agency viewed these as interim standards until better cars could be developed.[86]

Hazmat accidents made the news again in 2013 after a runaway trainload of Bakken crude oil derailed and blew up in Lac-Mégantic, Canada, killing 47 and destroying much of the town. It involved a Class II carrier (the Montreal, Maine & Atlantic) and reflected a perfect storm of poor maintenance, dangerous tank cars, employee malfeasance, and, apparently, bad luck. Since then there have been at least five other spectacular derailments resulting in fires and evacuations. These wrecks reflected a confluence of events. First, there has been a mushrooming in shipments of crude by rail, which rose from fewer than 21,000 barrels in 2009 to about 1.1 million in 2014, although numbers have since declined. Second, crude oil from the Bakken formation is far more flammable than traditional crudes. Third, much of the oil travels in DOT-111 tank cars (which were the type involved in the Lac-Mégantic wreck). These cars lack head shields and insulation and the NTSB has warned of their dangers for years. Finally, the newer CPC-1232 cars with thicker, normalized steel have not held up as well as expected.[87]

Just as occurred in the 1970s, scary accidents resulted in panicked newspaper editors. In early 2014, *The New York Times* reported an "accident surge." Yet FRA data report 78 cars releasing hazmat of all sorts in 2013, up from 50 in 2012 but not that different from the 76 cars releasing in 2007. Once again, these disasters have provoked a strong and rapid response from both the carriers and the FRA. Following earlier research, the AAR has improved brakes on trains with 20 or more crude cars, implemented speed restrictions of 40–50 mph, increased inspections and hotbox detectors on crude routes, and instituted a Rail Corridor Risk Management System intended to move hazardous materials along lines where accidents will do the least damage. A recent FRA regulation tightens these improvements, mandating either ECP brakes or speed limits of 30 mph by 2021 and ECP brakes on all trains with 70 cars of crude by 2023, all of which the AAR argues would be expensive and cost-ineffective. The new rule also requires more careful classification of crude oil. Finally, a new car—the DOT-117—made with thermal protection, thicker, normalized steel, and full head shields has debuted, and rules also upgrade older equipment. At the behest of the rail unions, the FRA has also proposed requiring two-person crews on crude trains and perhaps other freights too—an idea that echoes the good old days of full crew laws and that prompted a disquisition from *Railway Age* on the etymology of the word "sabotage."[88]

The turnaround in freight train accidents after 1978 has been breathtaking, and the keys clearly lie in incentives, organizational innovations, resources, and technological and regulatory changes. Beginning in the 1960s, a series of interrelated revolutions swept through railroad safety. The advent of the FRA and the NTSB ended voluntarism and inaugurated a new regime, and with the Federal Railroad Safety Act of 1970 all aspects of railroad safety came within the federal purview. Despite these fresh departures, the dominant forces curtailing accidents lay in the market; in firm and industry structure; and in technological changes, some of which

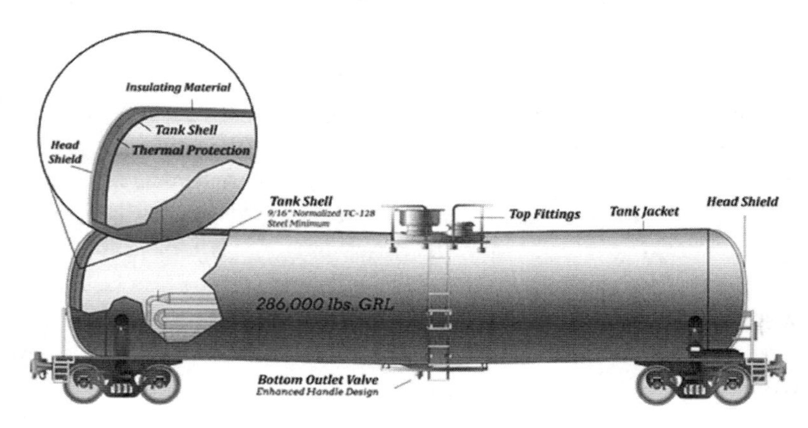

Intended to replace the DOT-111, the DOT-117 is part of a long line of improvements in tank car safety. It has head shields, a thicker shell, thermal insulation, rollover protection, and improved reclosable pressure-relief valves. (Federal Railroad Administration)

originated outside railroading. The carriers began to emphasize improved marketing, which required better service, much improved maintenance, and far fewer train accidents. Economic deregulation resulted in a wave of intermodal competition that reinforced the railroads' commitment to improve safety and service. Relaxing the economic shackles also helped provide both the resources for improved maintenance and the profit incentives to motivate it. In addition, deregulation nudged the carriers to focus on high-speed, long-haul, dedicated trains, thus reducing yard time and its attendant accidents. By implication, had economic deregulation occurred a generation earlier, the great rise in train accidents might never have occurred. Tax cuts and deregulation also raised the payoff to research, encouraging sharply increased support from the AAR. And so a gale of creative destruction has swept through railroading, improving both productivity and safety. New, cheaper remote sensors have dramatically expanded the information available on track, equipment, and human behavior. The availability of increasingly powerful computer analytics has also allowed researchers to blend engineering and economic analysis to implement risk-based techniques that improve the efficiency of maintenance.

The arrival of the NTSB and the FRA after 1966 shaped these outcomes in several ways. The former agency pioneered searching investigations into accident causes and has forced all parties to run a tighter ship. While the FRA's regulations seem largely to have mandated best practice, under Jolene Molitoris and after, the agency has performed a number of market-augmenting activities that have improved safety. Work of the cooperative

Safety Advisory Committee has helped integrate labor into safety decision-making and has eased the birth of potentially contentious improvements such as fatigue management. The FRA's system inspections show little impact, but they help spread information and reinforce company commitments. In the 1970s, the FRA initiated the Volpe Center and Transportation Test Facility and initiated federal funding for railroad research; as part of its risk reduction strategy it continues to support, publicize, and fund a host of proactive and cooperative safety initiatives, including fatigue management and close call reporting.

By 1980 or so, the basic institutions that brought freight train accidents under control were all in place. If the old system was voluntarism, the new approach might be termed "redundancy." The first line of defense remains private companies and the market. But they are backed up, far more than in the past, by public support and pressures. The outcome of these events has been a long decline in all types of train accident rates to low levels. Yet there has never been a time when the public thought railroads were "safe enough," and as recent developments suggest, public tolerance for hazmat accidents is low indeed.

4

"Our Goal Is Zero Accidents"

Work Safety in Modern Times, 1955–2015

An unsafe operation cannot provide superior service.

Jerry Davis, President, CSX, Congressional Hearings, 1989

This is the first safely program where somebody wanted to hear my views.

Gary Lansendorfer, Machinist, Conrail, Inside Track, *1989*

Throughout the first half of the twentieth century, railroad worker safety improved steadily, if unevenly (Chapter 1). Fatality rates that had averaged nearly one per million employee hours in the 1890s declined more than 80 percent by about 1955 and death no longer stared a worker in the face in quite the same way. It must have seemed inevitable to railroad workers of that day that these gains would continue, but the improvements in that past half century had been anything but inevitable. Rather, they had been the outcome of company investments in improving technology and endless campaigns by railroad safety departments. Railroad safety during these years rested on a three-legged stool. Workers supported company safety work because it benefited employees as well as others exposed to the dangers of railroading. Congress would leave safety largely up to the companies, as long as they delivered the goods. The third leg was that Interstate Commerce Commission (ICC) regulation would provide the necessary profitability to support safe railroading. As competition and regulation kicked out this third prop after World War II, profitability eroded, train accidents rose, and worker safety began to decline. Fatality rates rose after 1957, peaking in 1969, where they fluctuated until 1973 when they again began to decline.

This chapter contains four major sections. I begin with an analysis of the worsening of work safety in the years between about 1957 and 1973. Rising worker casualty rates reflected not only the decline in profitability, but also the faltering of railroad safety institutions during these years. As the following section demonstrates, safety again began to improve after 1973. The origins of this turnaround lay in the companies' increasing awareness that to compete they needed to improve marketing and customer service. This forced a focus not only on train accidents (Chapter 3), but on the need to reduce worker risks as well, for work accidents undermined efforts to improve service quality. Sharply rising compensation costs also made companies focus on improving worker safety. While these changes preceded economic deregulation, they could not have been successful without it. Although employment shifted toward relatively more dangerous jobs during these years, on balance there has been stunning improvement in worker safety after 1973. The third section demonstrates that the carriers' adoption of Total Quality Management (TQM) concepts reinforced these developments. The railroads also embraced behavior-based approaches to improve safety. The final section discusses federal policies to improve work safety and traces the continued improvements into the twenty-first century.

The Erosion of Work Safety

The course of worker fatality rates from 1947 to 1973 is depicted in Figure 4.1. As can be seen, they rose for about a decade after 1957 and in 1969 were 80 percent above their 1957 level, where they remained until 1973.[1]

Some of the deterioration reflected traffic changes. Passenger trains had always been safer to work with than freights, and thus the relative decline in passenger traffic shifted the composition of employment toward more dangerous freight work. More important, as Chapter 1 noted, safety work is subject to a Red Queen effect, in that "it takes all the running you can do, to keep in the same place." A representative of the Wisconsin Central put the matter somewhat less eloquently: "take your eye off the ball and it will bite you." Whatever the metaphor, it is clear that by the 1950s the carriers' old safety organizations had grown tired. This two-decade pause in safety progress resulted in large measure from declining profitability, which not only cut into maintenance and retarded the diffusion of new, potentially safer technologies, but also eroded morale. Safety work had always involved a commitment from both companies and employees, and as the carriers deteriorated it looked to workers like the bosses were welching on the bargain. Finally, as companies responded to declining profitability with other new, potentially dangerous technologies, their safety organizations failed to cope with the new risks. For decades, safety and profitability had come wrapped in a package, and as the finances of many carriers crumbled, safety faltered.[2]

The rot had been accumulating for a long time. In 1947, a Pennsylvania employee reported that "the ties are gone for 8 to 12 feet and no spikes for 18 to 20 feet. The joints are so bad that when two cars . . . pass they all but touch one another . . . Just on Xmas day one man was swiped off a car due to close clearance." In 1949, a local trainmen's union complained to the state regulatory board of poor maintenance in the Chicago Great Western's Minneapolis yard. There were close clearances, debris scattered about, and holes around the tracks. A 1962 survey of maintenance of way supervisors found that 21 of 40 thought deferred maintenance was "piling up." On some lines car inspectors were let go and the number of bad

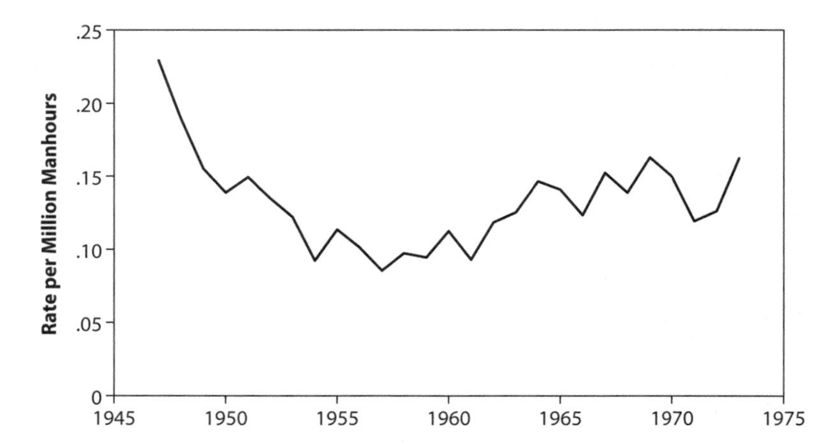

Figure 4.1. Fatality Rates, All Workers, 1947–1973. (ICC/FRA, *Accident Bulletin*, and author's calculations)

order cars skyrocketed. Workers reported that deferred maintenance led to cluttered yards and roadbeds that were choked with weeds and brush, causing treacherous footing and poor visibility. In 1978, a union representative wrote a superintendent on the Illinois Central complaining that "bushes and trees even raked the engine and caboose" so the men could not look out the windows to inspect the train.[3]

While financial constraints were surely the source of such poor maintenance, they also slowed the diffusion of new technologies that might have improved worker safety. Car retarders were electro-pneumatic devices employed to control the speed of cars in classification yards, and they ended the need for yardmen to ride freight cars. Retarders therefore saved labor and they eliminated many work injuries. In 1947, as one executive of the Pennsylvania explained, "I know of no place where we can get hold of our injuries to employees quicker than by the installation of retarders as they are, at the present time, the most valuable capital expenditure facility to decrease the number of employees exposed to hazardous occupations." In 1980, Ralph Strickland recalled the dangerous old days of yard work on the Seaboard before the advent of car retarders. "You had to be a monkey in order to get up there and grab them cars and slow them down. They don't do that now; they switch with air [pneumatic retarders] a whole lot now." Combined with the reduction in labor costs, the savings from reduced accidents encouraged investment in retarders. However, lack of funds limited their spread. In 1955, the American Railway Engineering Association (AREA) reported only 84 retarder yards in operation. Edwin Mansfield's study of new technologies, including not only car retarders but also such potentially safety-enhancing technologies as centralized traffic control, demonstrated that the carriers were slow to adopt even these profitable innovations—probably because the railroads had been unable to earn their cost of capital for decades.[4]

A statistical analysis for the period 1959–1965 that relates companies' casualties to net railway operating income finds a significant negative effect. Controlling for assets and employee hours, a 1 standard deviation increase in net income (which would have raised the carriers' return on assets from about 2.3 percent to 6.6 percent) would have reduced casualty rates about 8.8 percent. On some poverty-stricken carriers the rise in accidents was so sharp that it must have reflected an almost complete collapse of safety work.[5]

The erosion of profitability also contributed to and magnified the weakening of the carriers' safety institutions. Railroad safety work had always overemphasized exhortation and supervision, often featuring contests, cartoons, and ballyhoo, largely leaving engineering to others. And safety organizations had little say over such matters as train operations. When the Association of American Railroads (AAR) Safety Section proposed to employ a full-time safety engineer in 1948, with money tight the AAR refused to fund the position. Railroad safety organizations had typically included worker safety committees, but by the 1960s these seem to have largely lapsed. A survey by the Railroad Retirement Board in 1962 reported that only seven lines had retained such committees. Harry Brady, who had been a brakeman, conductor, and longtime safety committee man on the B&O, retired in the late 1970s. He told an interviewer: "they let it [safety] slip." On the Chicago Great Western in the 1960s, there had been "essentially no discipline . . . for many years," and the result was "many serious rules and safety violations." Another survey found that only 5 to 10 percent of carriers routinely involved the safety department in new purchases and construction. The result was sometimes the need to modify dangerous equipment on the job or by a later purchase. Moreover, modern research reveals how safety departments can sometimes evolve into pro forma organizations that go through the motions but have little impact on actual work practices. Thus, a 1979 survey of safety programs on 18 Class I carriers ranked them according to program content, procedures, equipment, training, and a host of other attributes. It found that the scores were uncorrelated with accident outcomes. Many were, apparently, paper programs only.[6]

Such failings had been around for a long time; the appearance of a flood of new track machines after World War II magnified their importance. The shift to a 40-hour week in 1949 raised the payoff to mechanization, and led by the Southern, companies purchased power adzes, tampers, wrenches, spike drivers, tie shears and saws, cranes, ballast cleaners and distributors, and more. This rapid transition to ever more mechanized track work affected worker safety in complex ways. The carriers shifted to smaller, more specialized track gangs, which should have improved supervision, and mechanization also reduced injuries from materials handling. Yet, in fact, its net impact was surely to worsen safety,

Table 4.1. Fatality Rates by Broad Occupational Group, Per 1,000 Employees, 1955–1957 and 1983–1985

Group	1955–1957	1968–1970	1983–1985
Executive, Professional, and Clerical	0.022	0.033	0.022
Maintenance of Way	0.266	0.376	0.264
Maintenance of Equipment	0.117	0.201	0.102
Transportation (Other Than Train, Engine, and Yard)	0.083	0.041	0.095
Transportation (Train and Engine)	0.483	0.512	0.272
Total	0.201	0.260	0.167

Sources: ICC/FRA, *Accident Bulletin*, and AAR, *Railroad Facts*.

for mechanization coincided with a sharp rise in maintenance worker fatalities (Table 4.1).[7]

How They Died

In the early 1970s, the Federal Railroad Administration (FRA) began to publish brief reports of employee fatalities from train and non-train incidents (i.e., all fatalities except those from train accidents). In 1975, near the peak for worker fatality rates, a total of 110 employees died on duty, only 17 of whom perished in train accidents. The remaining 93 met their fate in little accidents. Of course, most were blue-collar workers: of the 93, 32 were maintenance of way workers, 18 maintenance of equipment workers, and 30 train and engine workers. Many of the deaths reflected the dangers from moving trains: 24 were struck by trains or rolling stock, 6 were crushed by close clearances, 17 died servicing equipment, and 5 died falling from trains. Men were electrocuted; they fell from bridges or due to a shift in lading, and from a host of other causes.[8]

Many companies failed to integrate safety into purchasing and to develop proactive rules. Power adzes, for example, were initially without a guard and when the machine hit a spike it might throw the head with deadly force. In 1952, J. P. Hiltz, maintenance of way engineer on the Lackawanna, admitted that that the new equipment had caught his company's safety work "flat footed." Unions might have avoided this failure had they played a more prominent role in company safety organizations, for labor tended to be enamored with technological fixes for risks. E. L. Woods of the SP thought the problem was organizational, for mechanization had entirely upset traditional maintenance work. Section gangs became extra gangs, ranging over the entire line, using different equipment, working under different rules in unfamiliar places. The crux of the difficulty, he believed, was that company safety programs had failed to teach the old dogs new tricks. Thus, inadequate safety training and job planning contributed to the totals. On August 25, 1975, a Santa Fe freight struck a ballast tamping machine, killing the operator. The crew was working on an

The explosion of machinery use in track maintenance in the 1950s coincided with a deterioration of safety. Here compressed air spike drivers are being used by workers without hearing or eye protection. (*Railway Track and Structures*, September 1953, courtesy *Railway Age*)

interlocking that was a dead (un-signaled) zone even though it was in block signal territory. The foreman had not known the interlocking was a dead zone. Often such accidents reflected both equipment defects and poor safety procedures. On August 5, 1975, a tie-saw that the operator had failed to lock when he left the machine—a failure to follow rules—killed a machine operator on the Soo. Moreover, the machine was easy to engage when left in neutral—a design flaw. On December 2 that year, a loose wire—the result of both bad design and inadequate maintenance—energized a ballast tamper on the Penn Central, crushing a trackman.[9]

Sometimes the problem was simply a weak safety program that failed to develop rules for the new equipment. On June 3, 1975, a track foreman for the B&M was riding on a tie shear, which was slick from rain; he fell and was crushed to death by the machine. Company rules simply warned that riders should not be in "insecure" positions—which was little more than an admonition to be careful. Use of track inspection motor cars also provides examples of the dangers of mechanization in the absence of pro-active safety work. Many early cars carried no handholds and were front-wheel drive, making them prone to derailment. In addition, some also had no transmission and had to be jump-started. This was difficult on wet or icy rails, which might make it impossible to reach a setoff to avoid a train. Drivers of such cars were also understandably reluctant to stop before grade crossings, yet the cars were too light to trip crossing signals, and so many motor car-automobile accidents resulted. Motor cars and trains were also natural enemies; in a contest the train always won.[10]

William Doble was both a signalman on the New York Central and an official of the Signalmen's Union from the 1930s through the 1950s.

By the 1950s, safety work had become formulaic and it overemphasized worker negligence. Literature that blamed track car accidents on workers rather than lax company practices was unlikely to do much good. (http://www.jon-n-bevliles.net/RAILROAD/erie_safety_pamphlets.html)

Doble's correspondence contains many complaints that the company installed too few setoffs to allow a worker to remove the car when a train appeared. Such were the difficulties of trying to urge safety investments on impecunious carriers. In addition, track cars would not trip block signals and many companies' procedures for operating track cars remained hopelessly lax; train dispatchers sometimes gave car operators a train lineup but told them it could change at any time. Clearly, this was an important safety issue; yet it was beyond the authority of railroad safety departments. Doble's correspondence also reveals that some train dispatchers would not provide car operators with a written train lineup for fear of liability in case of an accident.[11]

Such sloppy procedures resulted in many deaths and loud complaints from the maintenance of way workers' union, which routinely featured articles with titles such as "More Blood Has Been Shed." For example, on the L&N, on December 4, 1951, the train dispatcher had informed a motor car crew that an extra locomotive had taken a siding. However, when the locomotive returned to the main track the dispatcher failed to alert the car crew, resulting in a head-on collision that killed two. Just how dangerous track cars were is reflected in the number of close calls they generated that did not result in an accident report. The Railroad Retirement Board investigated one carrier that had 40 track car accidents in 1958, only 3 of which were reportable because in the other instances the crew had jumped before the collision and no one was hurt. In 1954, at the request of the ICC, the AAR Safety Section belatedly began to look into the problem and it issued a code for proper use of motor cars. Procedures remained inadequate, however; track motor cars accounted for at least 94 fatalities in the decade ending in 1959 and the railroad brotherhoods made several unsuccessful efforts to induce Congress to legislate on the matter.[12]

These failings of company safety work might have been at least in part offset had either the ICC or the technical press performed well as outside critics, but they did not. In prior decades the railroad journals had been powerful voices for a host of improvements in equipment and practices. In the postwar years, however, concerns with regulation and crumbling finances largely drove worker safety from their pages. The ICC might also have been a much more powerful force for better safety. The commission investigated accidents and inspected freight cars, largely confining itself to counting defects, but it never publicized especially effective company safety organizations. Indeed, the commission admitted to Congress it knew nothing of such programs! It could have provided detailed studies of the benefits and dangers of new technologies, or pressed for better safety training, but its intervention in the track motor car issue was a rare exception. Marver Bernstein proposed that commissions often go through a lifecycle from youthful enthusiasm to gradual senility. In its youth the commission had been an independent force for worker safety, but by the 1950s in late middle age it had largely lapsed into "a passivity that border[ed] on apathy."[13]

The Great Turnaround

Worker fatality rates remained elevated between 1969 and 1973 before beginning a sharp plunge, falling about 60 percent in the 12 years from 1974 to 1985 (Figures 4.1 and 4.2). Moreover, as indicated by the data in Table 4.1, the improvement in safety was general, occurring in most broad occupational groups. After 1985, the decline was milder, with rates falling another 64 percent in the quarter century from 1985 to 2015. Injury rates, which are available on a consistent basis only from 1975 on, reveal similar safety gains. Thus, it was not deregulation that began the improvement in work safety. Nor was the cause some major change in technology such as the spread of unit trains, for as noted, the gains were general across most major employment categories. Neither were employment shifts a factor in better safety; indeed, as will be seen, they masked some of the gains. Finally, public policy played a comparatively minor role in shaping workplace safety before the late 1980s. In 1976, the FRA responded to a

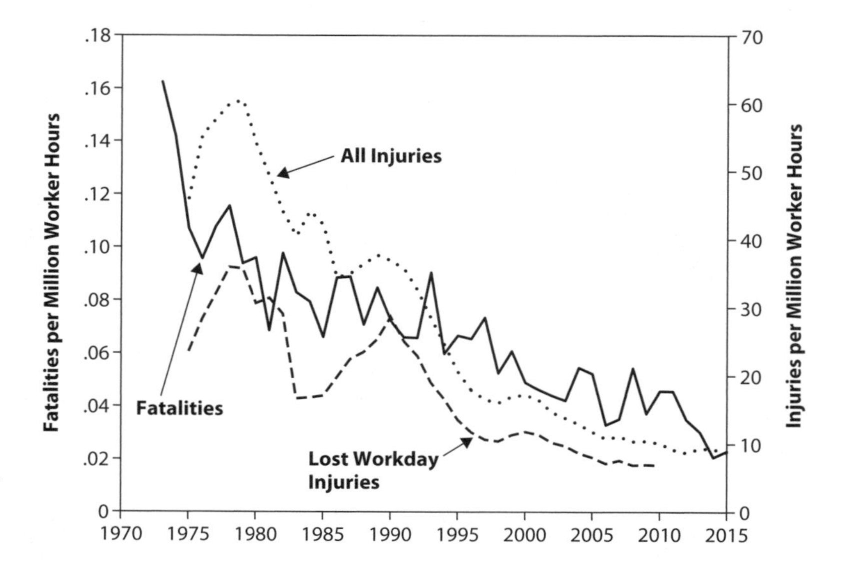

Figure 4.2. Fatality and Injury Rates, All Workers, 1973–2015. (FRA, *Accident Bulletin/Safety Statistics*, and author's calculations)

congressional mandate (P.L. 94-348) requiring "blue flag" protection for men working on rolling stock, to prevent inadvertent movement of equipment that could endanger workers. Some railroaders thought the law would likely worsen risks. It required, in addition to the warning flag, that workers also lock the switch to the track. This sometimes required long—dangerous—trips through the yard, and if the worker didn't lock the switch, someone who found it unlocked might well then ignore the flag.[14]

Yet if the origins of improved worker safety largely reflected the work of private enterprise, the motives leading to this improvement were partly political—as had been the motives leading to Safety First about 1912. As Chapter 3 detailed, the rise in derailments that had begun about 1957 resulted in FRA regulations in 1970 and congressional hearings nearly every year thereafter. These brought forward a seemingly endless flood of proposals for new laws and regulations, many of them backed by railway labor. In response, the AAR established its first director of safety in 1969. By the mid-1970s, the carriers were trying to shift the focus of concern away from train accidents. AAR studies demonstrated that train accidents yielded few employee casualties, while many workplace injuries—such as those from tripping and falling—were also unlikely to respond to new laws. When the railroads told Congress that "safety cannot be legislated," this was part of what they meant. The AAR also established a joint Railroad Safety Research Board with labor representatives to study safety. The carriers no doubt hoped the board might diminish union efforts to obtain make-work safety laws and demonstrate good faith to suspicious Congress personnel. The cooperative effort represented the beginnings of an important change from the carriers' older disinclination to involve unions in their safety work.[15]

EVER SEE A BRAKESNAKE?

A lot of people don't when they should, and they get bitten*. Getting off a car or engine and stepping on a brakeshoe often means a sprained ankle.

LOOK DOWN BEFORE YOU STEP DOWN.

Most worker injuries and fatalities resulted not from dramatic train accidents but rather from far more prosaic causes such as tripping over yard debris like this "brakesnake." (*Modern Railroads*, February 1975, courtesy *Railway Age*)

Economic Incentives and Worker Safety

Economic as well as political pressures were encouraging the carriers to redouble their efforts to improve worker safety, for injuries were rapidly becoming much more expensive. When a head-on collision on the Burlington Northern killed two engine drivers on June 14, 1984, a jury awarded their estates between $1.7 and $1.9 million each. The average cost for injuries settled under the Federal Employers Liability Act (FELA) rose from $1,699 to $2,155 between 1954 and 1963, and in the 1970s and 1980s claim costs exploded (Table 4.2). Such figures, moreover, do not include medical and litigation costs and on Amtrak, at least, these added another 28 percent to the cost of claims in the 1980s. If we adjust this increase for changes in freight rates, the result indicates that injuries that cost about 120,000 ton-miles of revenue in 1954 cost more than 164,000 in 1963 and about 300,000 in 1979. By 1991, it took a million ton-miles of revenue to fund a typical injury claim. Part of the reason for such high costs was that FELA lawsuits encouraged longer disabilities because these raised payoffs. In 2005, the median days lost for bruises

Table 4.2. Federal Employers' Liability Claims and Costs, 1954–2013

Year	Average Claim Cost (dollars)	Number of Claims Settled	Total Cost (millions of dollars)	Revenue Per Ton-Mile (dollars)	Ton-Mile Claim Costs[a]
1954	1,699	26,829	46	0.0142	119,648
1963	2,155	28,060	60	0.0131	164,503
1979	8,004	36,981	296	0.0260	307,846
1985	20,566	25,167	526	0.0304	676,513
1991	28,213	32,273	911	0.0259	1,089,305
2007–2013[b]	29,177	5,830	170	0.4848[c]	60,189[c]
				—	

Sources: "Injury Claims, a $100 Million Dollar Problem," RA 158 (February 15, 1965): 15; US 101st Cong., 1st Sess., House Committee on Energy and Commerce, Hearings, *Federal Employers' Liability Act* (Washington, 1989), 53; Transportation Research Board, Special Report 231, *Compensating Injured Railroad Workers Under the Federal Employers' Liability Act* (Washington, 1994), 74–75; Amtrak Office of Inspector General, *Injury Claims Trend Data for Fiscal Years 2010 Through 2013,* Report No. OIG-MAR-2014-008 (Washington, 2014).

[a]These are the number of ton-miles at $.0142 per ton-mile it took to defray a claim that cost $1,699 in 1954.

[b]Amtrak only.

[c]Per passenger mile.

was 4 for all industry but 40 for railroads; for sprains the numbers were 8 and 72, respectively.[16]

Yet direct compensation payments by no means exhausted the costs resulting from worker injuries and fatalities. As maintenance work became more capital intensive, it also required longer to set up. At the same time, rising traffic density made track time increasingly precious. Together these developments raised the opportunity cost of injuries that interrupted work. As Chapter 3 explained, beginning in the 1950s some far-seeing railroaders and their companies realized that the carriers needed to undertake dramatic changes in order to compete in the new intermodal world. In particular, they needed to make quantum improvements in service. The carriers responded, as has been noted, with unit trains, blocking, run-through trains, piggyback, and containerization, and they sharply upgraded employee training. Yet if rising costs from loss and damage encouraged a focus on train accidents (Chapter 3), improper handling was the largest single cause of damaged goods and the solution therefore led companies to cultivate more careful employee behavior. Damage to automobiles in transit was especially critical. In the early 1960s, the carriers were trying to claw back traffic in new cars and rough switching easily damaged them. The Pennsy tried more intense supervision to prevent over-speeding in yards. It also employed sex, hiring Georgia Malick whom it crowned "Miss Careful Handling." The company described her as "the best looking visual aid and safety symbol that ever graced a railroad yard" and distributed pins imploring workers to "Keep Georgia on Your Mind."[17]

While the success of "Miss Careful Handling" is unknown, the program reveals the carriers' dependence on worker motivation where monitoring is difficult. What needs to be emphasized is that companies were relearning that they could not motivate better service without addressing work safety. This was what Ralph Richards, the father of railroad Safety First, had understood as early as 1910: companies could not expect employee cooperation in achieving corporate goals if management seemed indifferent to the men's (and more recently women's) welfare. It is therefore not a coincidence that the men most closely associated with the railroads' rebirth—for example, Bill Brosnan at the Southern, Downing Jenks of the MoPac, L. Stanley Crane at Conrail, Richard Davidson of the MoPac and the UP, and John Kenefick and Mike Walsh of the UP, among others—were all dedicated to worker safety. Work accidents were by no means the only barrier to good employee relations. Mergers and declining employment were creating serious morale problems and the carriers responded with efforts to improve communication and respond to grievances, but none of this would have worked without better safety.[18]

Worker safety thus began to escape the compartmentalization that had bottled it up in safety departments. An early example comes from the Chicago & North Western. As Chapter 3 demonstrated, that company began a campaign against train accidents as early as 1966. In the process it began to see accidents as part of a much broader "quality control" problem. Its newly formed Accident and Loss Prevention Department included not only train accidents but also freight claims and personal injuries, and the company began to upgrade training and inspection. In the language of economists, such reorganizations were efforts to reduce the intra-company agency problems that prevented a full accounting of the costs of employee injuries and the benefits of safety.[19]

About the same time, the Soo tried to rejuvenate and broaden the focus of its safety program, which it too merged with freight damage and fire prevention. In 1967, when the Rio Grande set up its Transportation Management Center to improve service (Chapter 3), it did not neglect employee safety. *Railway Age* reported that the CEO read and responded to every injury report. R. C. Grayson, president of the Frisco, adopted a similarly holistic approach to safety when he spoke to the National Safety Council (NSC), linking personal injuries, freight loss and damage, and other losses that he claimed amounted to nearly half of railroads' net income. In the early 1970s, the C&O followed the North Western approach and placed all loss and casualty prevention in a single department, which enhanced its stature and apparently achieved results. It also developed imaginative safety training videos and integrated casualty prevention into its maintenance of way work planning.[20]

Other companies were also rediscovering Richards's approach. In 1968, when the Santa Fe woke up its worker safety program, it found that "derailments and freight loss and damage are showing trends that almost exactly match the personal injury improvement." As at other companies, training, rules examination, and worker motivation constituted an important part of the program. The best division or shop received at the end of the year a "safety train," instead of a trophy, while a slippage in the safety record resulted in an HO Scale gauge freight car marked "bad order." CEO John Reed later explained that "we . . . found that safety and good performance went hand in hand. The same has been true of our efforts in . . . quality control." The company emphasized "total management commitment" to safety. Its safety rules met or exceeded Occupational Safety and Health Administration (OSHA) standards and it developed its own detailed accident investigation form. It had a large safety organization; to sharpen the safety message, the company began a focus on particular kinds of safety problems—for example, eye protection—and cut such injuries sharply.[21]

In 1972, *Modern Railroads* featured the Missouri Pacific as its "railroad of the year." Under the leadership of Downing Jenks, that company had upgraded its plant, become a leader in containerization and computerization, and instituted one of the industry's best training programs. It emphasized communication with two groups: customers and employees. That company still had active safety committees that included both workers and supervisors, along with union representatives, and *Railway Age* described its shop safety as "excellent." It tried to encourage compliance with rules and its work led to conferences with employees who had accidents. That company also developed training videos on yard safety and instituted surprise checking. As one of its vice presidents informed Congress in 1976, "safe operations are simply a matter of good business."[22]

Sid Showalter, who joined the Southern Pacific as a brakeman in 1951, was asked "How did you learn your job?" He responded: "On the job. Very little training." This began to change in the late 1960s as companies inaugurated many more formal employee training programs. These reflected both internal dynamics and outside pressures. Rising accident and injury costs, and the increasing use of sophisticated locomotives, computers, and maintenance of way machines, all raised the payoff to training, while FRA safety regulations and informal prodding also nudged companies to improve instruction. Moreover, as workers undertook greater responsibilities, the costs of accidents rose commensurately. The Southern Pacific and the Santa Fe pioneered in the use of locomotive simulators in the late 1960s, and the companies thought the program improved both service and safety. Training programs included not only engine drivers and other blue-collar workers, but office personnel and mid-level supervisors as well. About 1965, the Canadian National began a massive training program that emphasized both customer relations and safety, and within a decade the Southern, the MoPac, the SP, the North Western, the Santa Fe, the BN, and most other large carriers had followed.[23]

"Training to improve safety is getting special emphasis these days," the *Railway Age* noted in 1974. The Rio Grande and the Santa Fe emphasized employee training in their safety programs and both examined workers on safety rules every year. For managers, the new training programs invariably included marketing and sales, as well as safety. *Modern Railroads* saw the connection. "Higher Capacity Railroads Need Higher Capacity People," the journal argued, claiming that "railroads can no longer afford high levels of employee injury or employee-caused loss and damage." Moreover, after 1980, as accidents resulting from bad track and defective equipment receded, those resulting from human factors became relatively more important, underlining the need to improve training. In 1982, the National Transportation Safety Board (NTSB) informed Congress that its investigations found "a much lower [train] accident rate for those railroads that have and use formal structured training methods and programs." These programs have continued. In 1988, the Burlington Northern began a training program in conjunction with a community college, and by the mid-1990s the Union Pacific and CSX had made similar arrangements. These programs included courses for road masters, instruction in track welding, and reading blue prints and similar work, along with much on-the-job instruction.[24]

The deregulation of rates and service that began with the "Four-R Act" of 1976 and culminated with the Staggers Act in 1980 benefited work safety because it sharply reinforced the carriers' commitment to better service and marketing. UP President John Kenefick nicely summarized the outcome of the new regime: "railroads are market driven," he concluded. As Chapter 3 demonstrated, companies responded to competitive forces with blocking and run-through trains. These not only reduced train accidents in yards but other sources of employee dangers as well. For example, in 1978, 37 of the 122 employees killed that year died in switching and train operation, most of which occurred in yards. By 1984, the number of such fatalities had declined to 12.[25]

Vandalism, Service, and Worker Safety

Another example of the intertwining of worker safety and customer service comes from the companies' efforts to cope with vandalism that rose in the 1960s. Railroads had suffered from vandalism and other related crime for as long as there had been railroads. The nineteenth century had seen Indian raids and outlaw holdups—memorialized in a thousand westerns—and both of these sometimes took the form of a deliberate derailment of the train. Deliberate wrecks also occurred during labor disputes and when companies tried to clear their tracks of trespassers. Small boys regularly took pot shots at the logo on freight cars and pilferage was a chronic problem. In the 1960s, vandalism rose sharply, reflecting in part the nationwide increase in crime rates. While it was never a major source of worker risk, vandalism demoralized employees who felt under assault and caused property damage—new cars were an especially tempting target—and customer irritation the carriers could ill afford. Vandalism was therefore a threat to the carriers' emerging efforts to improve service quality and safety.

Much of the problem involved property damage. In 1962, the L&N found damage to new cars arriving in five different terminals from sticks, stones, and bullets. Companies also suffered from theft, while repair of broken windows in locomotives and cars cost the tottering Penn Central more than $2 million in 1970. Yet property was not the only target and the problems worsened through the 1970s as nationwide crime rates took off. "It seems to be a reflection of the times," one AAR spokesman dourly remarked. In 1971, an Amtrak engineman reported that vandals hung a grocery cart from a bridge in North Philadelphia, tearing the pantograph off the locomotive. That company also reported five train accidents from vandalism from 1971 to 1975. On February 19, 1975, a 13-year-old boy dropped a rock through the dome on one Amtrak car, hitting a passenger on the head and killing him. On the B&M in 1978, vandals pushed a shopping cart off an overpass that hit a Buddliner cab, giving the driver a concussion.[26] That same year vandals tampered with both a switch and a signal light, sending a Santa Fe freight onto a siding where it derailed, injuring two of the crew. When gunfire hit a conductor riding in a caboose, a jury awarded him $180,000 because the company had failed to install bulletproof glass. A teenager threw a half full beer bottle at the engine of a Grand Trunk train, smashing the windshield and killing the engine driver.[27]

Labor troubles were probably the cause of switch tampering on the Florida East Coast that resulted in a derailment in February 1979. That company also reported ties, pallets, and barrels placed on tracks and sometimes camouflaged; a crossing signal shorted out; a crossing gate removed; a damaged switch lock. Locomotives made especially attractive targets—for rocks, beer bottles, concrete blocks dropped from overpasses, and .22 rifle shots. "A bullet zinged through the window right by my head and ricocheted out the back door," one brakeman on the SP told the *Los Angeles Times*. As Table 4.3 indicates, in 1978 there were 757 incidents of trains being fired on. Train crews started naming some especially dangerous stretches of track: in Philadelphia it was the "combat zone"; in Detroit, the "Ho Chi Minh Trail."[28]

Companies developed a wide array of responses. In 1960, the New York Central increased police patrols and equipped them with portable radios.

Table 4.3. Types and Targets of Vandalism, 1978–1979

Category	Incidents, 1978	Incidents, 1979
Multilevel Cars	29,878	22,488
Track and Signals	5,823	5,788
Switch Tampering	1,718	1,446
Track Obstructions	7,695	7,329
Rock Throwing	9,009	8,262
Shooting	757	257
Other	8,043	8,123
Total	62,923	54,213

Source: "Vandalism: Time to Crack Down," RA 181 (December 29, 1980): 55.

About the same time, the L&N appealed to workers through its employees' magazine to be more vigilant. By the 1970s, companies were shipping automobiles in freight cars with protective siding and beginning to install protective glass in passenger cars. Companies also began to protect engine drivers with safety glass. In the early 1970s, several Chicago roads combined to hire a helicopter to monitor that city's yards and the Southeast Pennsylvania Transit Authority (SEPTA) and the New York Metropolitan Transit Authority (MTA) tried the same approach. Property damage and injuries both declined. After vandals threw rocks at the *Super Chief*, the Santa Fe hired guard dogs to patrol yards and installed a motion picture camera on its route; it later produced anti-vandalism posters.[29]

Needless to say, vandalism was a major concern of the Brotherhood of Locomotive Engineers (BLE). When the Long Island offered a $1,000 reward for information on vandals, the *Locomotive Engineer* enthusiastically reported "Long Island Railroad Sets an Example." The union had wanted a federal law prohibiting vandalism but had to settle for a regulation instead. In 1979, after considerable prodding by that organization, the FRA finally required safety glass for locomotives, cabooses, and passenger cars for carriers under its jurisdiction. The new rule required glass that would withstand the impact of a 24-pound concrete block at 30 mph and a 40-grain .22 bullet traveling at 960 feet per second. The cost, *Railway Age* estimated, was about $75 million, but the program had "the enthusiastic support of management and labor alike."[30]

Neither company rules nor the FRA regulations immediately ended vandalism, but along with the national decline in crime accidents after 1980, they helped contain the problem. The number of incidents probably peaked in the late 1970s. In 1978, there had been 172 train accidents from vandalism and 179 in 1979. Train incident assaults in those years had been 289 and 311, respectively. By 1990, the numbers had dropped to 55 train accidents and 250 injuries from assault. Loss and damage expenses peaked at nearly 2 percent of revenue about 1979; by 1990, they were down to less than 0.4 percent.

Workforce Composition and Employee Safety

The improvement in worker safety that began after 1973 occurred as the carriers were merging, spinning off pieces, and shedding workers at an unprecedented rate. These changes in structure and employment resulted in important shifts in job composition. Nearly everyone has heard of the long-running battles between the carriers and their unions that finally resulted in the demise of the fireman in the 1980s, and so one might suppose that technological change was reducing the share of dangerous jobs. In fact, just the reverse was occurring because the employment shares of train and engine workers and maintenance of way workers (the two most dangerous categories) increased between the late 1960s and the early 1980s, modestly raising fatality rates, and the process has continued. Computerization allowed the carriers to make drastic cuts in their white-collar labor force, while rising traffic and the need for better track maintenance increased the employment shares of maintenance of way and train and engine workers on Class I lines from 44 percent in 1983–1985, to 57 percent in 1996, and to 63 percent in 2013–2014. While data limitations preclude calculation of fatality rates for subcategories of workers after 1996, if employment had not shifted to more dangerous jobs between the mid-1980s and 1994–1996, fatality rates would have been about 8 percent lower (0.128 versus 0.139 per million employee-hours).

With economic deregulation after 1976, the railroads also began a process of vertical disintegration—that is, as competitive pressures increased they found they could contract more cheaply for work they had once done in-house. With track maintenance becoming ever-more specialized and capital intensive even large carriers began to hire outside firms for some work. Factory-made track panels also appeared. In addition, as the carriers disinvested in freight cars—carrier-owned cars declined about 40 percent between 1980 and 1989—railroad equipment maintenance fell and some lines contracted out work on their own cars and locomotives. Companies also purchased white-collar services such as engineering and computer programming. Even derailment investigations are now available through the market from consultants such as Rail Sciences. These changes in employment patterns shifted workers and their injuries "off the books." Contractors were subject to the same rules that governed the railroads, and so it is unlikely that such work was more risky, but it is impossible to know whether outside contracting shifted employment composition in ways that raised or lowered railroads' injury and fatality rates.[31]

Finally, as Class I carriers peeled off excess capacity under the pressures of deregulation, some track and some workers reappeared in new, smaller railroads, and these lines were typically relatively risky. For example, in 1990 injury rates were 53 per million employee hours on Class III railroads versus 34 per million employee hours on Class I carriers. In

1981, Class I carriers accounted for 91 percent of all railroad employee hours, but the proportion shrank to 85 percent in 1985 and 80 percent in 1996, where it appears to have stabilized. In the latter year the employee injury rate for all railroads combined was about 18 per million hours; had the 1980 Class I share of employment obtained in that year, the rate would have been 2 points (about 11 percent) lower. In all, a reasonable summary of the effects of these complex employment changes is that, on balance, from the 1960s on they probably raised casualty rates.[32]

"Our Goal Is Zero Accidents"

By 1985, worker fatality rates had fallen 60 percent from their peak in 1969 to their lowest level ever. Several years earlier *Railway Age* had surveyed this decline, focusing on three lines that had improved safety in the face of difficult circumstances. These were the Santa Fe and Rio Grande discussed above, and even the wheezy old Boston & Maine. Under energetic new management, that line, which had only recently emerged from bankruptcy, was able to cut its injury rate from more than 40 per million man hours in 1980 to less than 6 in 1982. All three success stories, and others too, the *Age* concluded, reflected no dramatic new technologies and strategies but rather the effective application of ideas that date from the earliest days of the Safety First Movement. On all these lines managers made their commitment to safety in ways employees found believable—by deploying resources, investigating accidents, designing and rigidly enforcing safety rules, and involving employees. In modern terminology, they had developed an effective safety culture. What the *Age* failed to stress, however, was the market forces that were encouraging such commitments. As the carriers struggled to succeed in an increasingly competitive transportation market, worker injuries, which disrupted service and eroded morale, were a form of inefficiency that was becoming increasingly intolerable.[33]

Total Quality Management Comes to Railroading

These efforts to become competitive made the carriers receptive in the 1980s to the concepts of TQM, which derived from the work of W. Edwards Deming, Philip Crosby, and Joseph Juran. A number of lines had hired Deming as a statistical consultant by this time, and they were undoubtedly familiar with his views on quality control.[34] The core ideas stress the need to put customers first; to emphasize quality—and the costs of poor quality—in all aspects of the organization; to involve and empower employees; to strive for continuous improvement; and to communicate and work across organizational boundaries. In addition to quality, Deming's ideas about statistical process control also applied to safety. Crosby popularized the idea of "zero defects," while another branch of the movement focused on achieving "six sigma" quality, which is a process that is 99.99 percent defect free. Accordingly, since the bane of quality is failure in some form, TQM stresses prevention and the need to determine root causes of problems.[35]

As has been noted above, many carriers had been trying to improve service quality since the 1960s, and *Modern Railroads* had trumpeted that goal for decades. Corporate emphasis on training and the need to improve labor relations also ensured that TQM ideas about employee involvement fell on receptive ears. Thus, while TQM might seem to be just one more in a seemingly endless string of management fads, it reinforced and validated existing company efforts to cut costs and improve customer service. As many writers grasped, safety was an integral part of the process. "Everything safety is about relates to the absolutes of quality management" was the way Crosby put it. And when Jolene Molitoris arrived as administrator of the FRA in 1993, she brought the approach to that organization too. "Our goal is zero accidents," she informed the bridge and building engineers in 1994. "How could we have any other goal?"[36]

Safety integrated easily with TQM in part because the search for root causes promised to be a fruitful way to analyze accidents and it meshed with ideas of the System Safety Movement that emphasized identification and control of hazards and work planning. The Southern Pacific, for example, had been applying PERT (Program Evaluation and Review Technique) methods to safety since the 1960s. These had been borrowed from defense contractors and stressed detailed job planning to prevent both inefficiencies and accidents. As injury and fatality rates declined, companies began to experience problems from what one writer has termed the "unrocked boat." That is, with injuries rare, it becomes increasingly difficult to maintain a high level of safety commitment. By the mid-1980s, commentators observed that the injured were increasingly among the older workers with years on the job; systematic planning was a way to prevent complacency.[37]

To defend against complacency, safety work also included multiple defenses and so accidents also increasingly reflected the interaction of a series of highly improbable events. In the language favored by those trained to think of event trees, there were often many "PCFs"—possible contributing factors. The "Swiss cheese model" provides a homier analogy: the cheese slices are multiple layers of defense, but if the holes all line up—a rare occurrence—you get an accident. An example comes from a death that occurred on the Alaska Railroad on the morning of February 24, 1984. A truck driver with 14 years of experience took a forklift of 4,000 pounds capacity to transfer lumber from a loading dock. The lift had no overhead guard, although other guarded lifts were available. The ground was uneven because of snow and ice, and he picked up 2 stacked 2,300-pound pallets of lumber that were covered with ice. When he raised the package, the top pallet slipped backward and because the lift had no guard, it fell on him, killing him. Consider the PCFs: he chose the wrong lift; he then overloaded it; the ground was uneven; the pallets were ice-covered.[38]

Four days later, on February 28, 1984, the holes in the cheese again lined up on the MoPac when a ballast-tamper and a rail motor car operated by a track crew collided, killing the track foreman. The rules were faulty because the dispatcher issued overlapping "track and time" to both crews without notifying the tamper operator of the track crew's whereabouts. The track crew had a radio but failed to use it, and both vehicles were going too fast. Preventing accidents such as these required hazard identification via a search for root causes, and then systematic planning of work and an almost religious attention to communication, rules, and detail, all of which fit easily within the confines of TQM.

TQM also meshed well with the new behaviorist thinking about safety that emerged in the late 1970s. The "Three E's" of railroad safety work had always been exhortation and enforcement, with a side dressing of engineering (Chapter 1). For behaviorists, they were empowerment, ergonomics, and evaluation. Such ideas, which derived from the work of the psychologist B. F. Skinner, emphasized the involvement of both employees and management as a safety team, the value of positive feedback, and peer-to-peer counseling to develop a safety culture. Thus, psychology joined economics (Chapter 3) as the social sciences began to play a greater role in safety work. The railroads became interested in behaviorism in the late 1970s, and in 1982 the AAR, along with the FRA, conducted a large-scale study on the Burlington Northern, Duluth, Missabe & Iron Range (which had employed Deming as a statistician in the early 1970s), the Illinois Central, and the Southern. The findings confirmed the effectiveness of positive reinforcement and the idea spread to other carriers. For a railroad trying to enhance its competitiveness and encourage quality improvements, positive reinforcement encouraged not only better safety but also the idea that workers were a valued part of the team.[39]

CSX was an early adopter of TQM, implementing some of these ideas as early as 1982. As one writer put it, "quality is something employees from track to boardroom are giving . . . [special] attention." In the 1970s, the company had begun to integrate its various loss and damage functions in a Casualty Prevention Department (see above). It now included train accident prevention and developed cause-finding seminars that instructed representatives from all relevant departments to find the root causes of accidents. To facilitate the process, the company began detailed accident reporting and computerized data analysis.[40]

At the Union Pacific, better safety and behaviorism preceded TQM. That company had emphasized safety since the time of Edward Harriman, but the Red Queen effect appeared in the 1970s and injury rates rose. In 1981, the company decided its approach needed a shake-up and hired DuPont Safety Management Services to provide an evaluation.

DuPont had a long tradition that stressed commitment by top management. As one of the company's executives claimed: "when you have convinced your people that you are truly concerned for their safety, then you will gain the support you must have to overcome your management problems." DuPont also emphasized a behavioral approach to safety: it underlined the importance of overlapping safety audits at each management level (i.e., a supervisor visits her subordinate and the two of them perform the audit), as well as two-way communication, continuous training, and the need to integrate safety across departments. DuPont's safety consultants once arrived late to a meeting with UP managers due to their inability to find a taxi with seatbelts. "It was at that point I began to realize how serious DuPont was about safety," a UP executive recalled. The consultants found a program that had gotten tired. They discovered instances of poor housekeeping and lack of fall protection and found workers who had failed to wear eye protection. They restructured the program and also reoriented the company away from reacting to an accident toward prevention. From 1981 to 1983, the UP's injury rate fell 30 percent; company managers thought that the safety gains also increased productivity.[41]

TQM came to the UP in 1986 when Mike Walsh arrived as the company's CEO. He was an outsider, from Cummins, where he had learned about TQM. While its implementation ruffled many an executive feather at the UP, in fact, it simply pressed managers for further improvements in quality, customer service, employee involvement, and safety. The company began to survey customers about the quality of its services and discovered that "you guys are among the worst." It investigated failure costs across departments, discovering, predictably, that derailments and worker injuries were far more expensive than had been thought, and began employing root cause analysis to reduce them. The company set a goal of zero defects for its locomotive suppliers, and "all injuries are preventable" became the company mantra.[42]

Not all companies were able to apply TQM methods to safety successfully. In the early 1990s, the Florida East Coast applied the new approach to resuscitate a floundering safety program. Older safety work had been rules- and discipline-based; one company representative concluded that it had led to "poor attitude[s] and boredom," rising injury rates, and underreporting. The company began its "Fast Track to Quality" by involving employee safety committees in decision-making. The newer program emphasized training, safety orientation, and accident follow-up. Yet despite high hopes, worker injury rates rose after 1990 and remained elevated for the remainder of the decade, although whether the program failed or simply encouraged more injury reports is unclear.[43]

At least some railroad suppliers were independently discovering TQM. In 1980, Portec, which made rail friction control systems, wayside data collection products, and track components, learned of TQM from Philip Crosby, the father of "zero defects," and then hired his brother David to become the company's director of quality assurance. In the mid-1980s, the AAR also showed the influence of these new ideas when it implemented its Quality Assurance Program for certain freight car parts. The program focused on journal bearings, wheels, axles, and other safety-critical parts. Suppliers were required to adopt the program and were subject to audit. Failure costs in such parts were high and no doubt rising since freight cars were becoming larger and trains longer. A 100-car freight ran on 800 wheel bearings, and the failure of any one might cause a very expensive wreck. While zero defects might be impossible, TQM ideas emphasized that quality was a special imperative in such vital parts.[44]

While TQM reinforced an already strong safety program on the Union Pacific, at Norfolk Southern the experience of strengthening the safety program seems to have led that company to TQM. Formed by a merger of the Southern and the Norfolk & Western in 1982, Norfolk Southern combined two organizations with strong safety traditions. About 1987, however, management decided the program needed rejuvenation—as one supervisor concluded, "NS decided safety was 'good business' "—and brought in DuPont safety consultants who exposed the company to the ideas of behaviorism and TQM. Apparently the safety shakeup was successful—injury and fatality rates fell sharply thereafter and the company then adopted a TQM approach based on the experience of 3M. As CEO Arnold McKinnon told Congress in 1991, his company's adoption of TQM "really grew out of our success with the safety program." By 2007, NS had had DuPont back five times.[45]

Conrail managed to maintain a strong safety program—routinely discussed in its company magazine *Inside Track*—even as it shed workers by the thousands. As early as 1980, the company was using behaviorist concepts such as positive reinforcement, employee recognition luncheons, and awards for safe work. It urged use of protective equipment by nominating and publicizing workers whose eyes had been saved by safety glasses, which was part of its "Wise Owls" program. Local committees were active and management followed suggestions, which "built up morale." Under CEO L. Stanley Crane, in 1982 the company began a "quality circles" program that involved labor-management teams and was intended to improve communication, safety, and service. Conrail also hired DuPont as a safety consultant in 1985 and it made the usual recommendations. Probably in response to DuPont's stress on behaviorism, in 1988 the company revised its program to reemphasize teamwork and involve blue-collar employees in injury analysis and investigation and in counseling for workers with safety problems. "This is the first safety program where somebody wanted to hear my views," one employee said with some surprise.[46]

When James Hagen became Conrail CEO in 1989, he meshed the program with other TQM ideas he had learned from his previous work at CSX. The company instituted a Conrail Certified Quality Supplier Program that required suppliers to develop their own programs. Hagen also saw the connection between quality and safety. "You can take out a lot of . . . costs of bad quality, things like running through switches. Once you add them up they amount to a lot of money. Then there's the personal injury side—that's a total waste," he informed *Railway Age* in 1994. That year the company created a Risk Management Department and charged it with reducing the cost of risks, broadly conceived. About two years later, Conrail hired Aubrey Daniels International, a safety consulting firm specializing in behaviorism.[47]

Like the Norfolk Southern, Amtrak contracted with DuPont for a return engagement, in 1989 importing a Safety Training Observational Program (STOP) that focused on teaching supervisors to analyze unsafe work practices. It also incorporated the newer, nonconfrontational, more cooperative approaches to employee safety, emphasizing teaching rather than discipline. The Chicago & North Western had begun to implement TQM in 1985; like Conrail it also employed quality circles, but the program seems to have petered out after 1988. The Southern Pacific started about 1990. By that time the Norfolk Southern's president claimed that 80 percent of the Class I railroads had embraced some form of TQM.[48]

The influence of TQM and behaviorism have continued into the twenty-first century. Researchers and executives have begun to talk about corporate safety cultures, which is shorthand for how well or poorly these various ideas are implemented. As numerous articles in the trade press attest, the emphasis on employee training and communication remains strong. The BNSF increasingly involved its unions in safety work and it stressed training and counseling to encourage compliance. That company also developed a "closed loop safety process," which derived from TQM and systematically focused on root causes of risks and developed action plans for their reduction. In 2009, it created "Listening Posts" composed of teams of managers that traveled to divisions to hear workers and root out safety problems. Its company magazine, *Railway*, has a "Focus on Safety" section in each edition, and it routinely stresses peer-to-peer communication and emphasizes positive reinforcement. The Norfolk Southern's *BizNs* magazine has an annual safety issue, emphasizing the need to create a culture of safety. In 2008, when Joe Boardman moved to Amtrak from being head of the FRA, he instituted a behavior-based "Safe-2-Safer" program that emphasizes training, teamwork, and peer-to-peer counseling. It seems, however, to have ignored discipline and as a result injury rates have risen.[49]

In 2003, the UP was calling the framework for its approach to safety the "total quality management system." The company's chief engineer emphasized safety, quality, and productivity in its maintenance of way equipment, with "safety being the most important." In 2001, *Progressive Railroading* reported that the UP, the BNSF, CSX, and the NS were all importing "six sigma" quality control processes from GE. The need for improved service quality led to scheduled railroading—an old idea—receiving new emphasis on CSX and the CN, with the latter carrier scheduling not only train times but also dock-to-dock times for individual shipments. In 2002, the CP instituted a "5-alive" program that emphasized blue flag protection, use of derails, and applying lockout-tagout, among other things. Workers received their own individual locks to ensure that crucial safety equipment was not removed and violators of the rules—including supervisors—were fired. Recently, the Norfolk Southern has employed Aubrey Daniels International to implement a behavior-based program.[50]

These new management ideas underlined and reinforced earlier efforts to improve marketing and quality and they made safety integral to all operations. Moreover, the goal of continuous improvement and the competitive forces driving it ensured that safety work would be less likely to atrophy, as had occurred during the 1950s and 1960s. In short, the Red Queen effect would be less likely to reappear.

Machinery and Ergonomics

Maintenance of way was among the most dangerous of railroad jobs (Table 4.1). As noted above, mechanization during the 1950s and 1960s had coincided with rising fatality rates. If anything, the blizzard of new, special-purpose machines increased in the 1970s and 1980s. There were production, spot, and multipurpose ballast tampers, along with machines that could pull spikes, remove ties, spike new rail, distribute rail anchors, align track, unload and distribute ties, space ties, distribute ballast, cut brush, straighten joints, and lay continuously welded rail (CWR). Yet after the early 1970s, even as mechanization proceeded, and in contrast to earlier years, the new emphasis on safety discussed above sharply reduced maintenance of way injuries and fatalities.

In maintenance work as elsewhere, the rising costs of work accidents and the need for better service helped motivate improvements in safety. In 1986, E. B. Burwell, president of the Southern, addressed the bridge and building engineers. He stressed the importance of high-quality maintenance in the new world of competition: "we cannot compete if the railroad is covered up with a bunch of slow orders," he informed his audience. A decade later, an executive from the Wisconsin Central reemphasized the importance of maintenance to marketing. "Service is our business," he explained, "it's moving freight to meet customer expectations . . . to gain a profit." Many other speakers made the same point. Because injuries reduced track time, such concerns translated into a need to improve safety.

Nor could companies expect high-quality maintenance work in an unsafe work environment. Thus, maintenance of way engineers' concern with safety stemmed from the system-wide focus on service quality. Work planning and hazard analysis spread accordingly.[51]

Work accidents were also expensive, and because injuries were far more prevalent than fatalities they were far more costly. Strains and sprains were leading causes of injuries. In 1987—a typical year—out of 16,113 lost workday injuries, 33 percent involved sprains and strains to the torso—that is, back injuries—and they were especially prevalent in maintenance work. Such injuries, one maintenance of way supervisor noted, were the "number one area for claim payouts"—what he called "green poultice"—because settlements sometimes ranged between $250,000 and $350,000. Human factors, often interacting with machinery, were the major causes of fatalities and injuries, and the behavioral approach to safety provided a way to attack these and other kinds of accidents. It focused on the interactions between worker behavior and the (management-created) work environment—ergonomics—and in the late 1980s companies began to apply these ideas to redesign jobs and equipment.[52]

This was a much more subtle way of thinking about equipment safety. Previously, the focus had been on ensuring that machines were not actively dangerous to life and limb. Ergonomics also investigated ease of use and other less dramatic human-machine interactions. Such ideas were not new—in 1971, Humble Oil announced the Blivet, which was a 50-pound container of rail lubricant. Moving it yielded fewer personal injuries than had the traditional 400-pound barrel. Other suppliers also occasionally stressed the ergonomic advantages of their products. Still, a major push waited until the mid-1980s. In 1987, *Railway Age* reported that the Safety Research Division of the AAR Research and Test Department had drawn on the National Institutes of Occupational Safety and Health and private-sector research to computerize a biomechanical model that they used to generate ergonomic guidelines for job and machine design and worker training. Rising injury costs were the apparent motivation.[53]

Spurred by TQM ideas, researchers also began to look for the root causes of maintenance of way injuries—which included poor training and supervision, or improper job or equipment design. Wendell Pyles, an SP engineer, explained how TQM could help companies focus on ergonomic problems and their solutions. He told of an engineering team that began by identifying slips, trips, and falls as the major source of maintenance of way injuries and then found the root causes for each—for example, slips and falls occurred most often from getting on and off roadway machines—which led to the redesign of decks, steps, and handholds. Finally, companies began to develop quality teams that included workers to assist in machine design and use.[54]

By the 1990s, the need to reduce injuries and fatalities had made equipment safety a major concern, not only in maintenance of way work, but in yards and repair facilities as well, and a range of new, more user-friendly equipment began to come through. The Unit Rail Anchor Company patented an E-Z wrench that caused less back strain to use when applying rail anchors. Early AAR research into switches—which were often difficult to throw and had been a prolific source of back injuries—resulted in new maintenance procedures and better switches with a 17:1 mechanical advantage ratio, which was 2 to 3 times better than that of older models. Some companies reported 25 to 33 percent reductions in injuries from this source. Similarly, investigation of ergonomic problems at a UP repair center in Palestine, Texas, resulted in increased use of winches for lifting and redesign of dollies and work tables. The new designs cut low back injuries from 70 percent of the total to zero. Hydraulic jacks replaced mechanical designs; they were lighter and eliminated the danger of kickback when the load was released. New Fairmont tie-replacing machines could hold the tie plate in place so a worker did not risk back strain in doing so. Other maintenance of way equipment began to come through with negative pressure cabs, engines that were more accessible for repair, safety steps and catwalks, easier operator access, safety glass, better suspension, and ergonomically designed seats. Cranes appeared with computers that could monitor for unsafe loads. Tie adzing had always been among the most dangerous jobs, but with new equipment "the operator can remove his metal leggings, full face protection, respirator and coveralls," one supervisor enthused. Classification yards were also improved; the UP's Jerry Davis hump yard employed flatter grades to reduce the dangers of runaways.[55]

The Union Pacific, the Norfolk Southern, and other major carriers have continued to work with suppliers to improve equipment safety. Design of switch stands has become more ergonomic, while companies are marketing machines with motion alarms and that provide better visibility and require less maintenance. Suppliers have also improved personal protective equipment. Newer hearing protection can attenuate noise yet still allow vital personal communication. Reflective clothing has become common while better shoes provide improved slip and fall protection.[56]

Wireless communication has also resulted in several important new safety technologies in the twenty-first century. Remote control—of yard locomotives, ballast dump cars, yard switches, track tampers, blue flag and derail placement, and a number of other activities—not only raises productivity, it also sharply reduces exposure to risk. Remote control of locomotives stirred up both the United Transportation Workers and the Brotherhood of Locomotive Engineers and Trainmen (BLET). The new technology threatened job losses, which motivated the BLET especially to try to ban them. An FRA study discovered that remote control of yard locomotives reduced work injuries by 57 percent.[57]

Wireless communication is also reducing the risks of track work. In 2001, the BNSF began to deploy a Hi Rail Limits Compliance System that employs GPS technology to keep track inspection equipment from straying out of safe territory. More recent innovations in wireless technology have developed systems intended to protect track workers. Its origins stem from a Massachusetts Bay Transportation Authority (MBTA) fatality in 2003 when one of its track workers was hit by a CSX freight, leading the former company to join with Protran to develop a portable warning device. In 2007, two more track worker fatalities motivated Bombardier, which was then operating MBTA, to develop Smarttrain, a similar technology. Both devices include a train-mounted unit along with trackside sensors and personal monitors that alert train crew to the presence of road workers, and vice versa. Their use has spread to include the MBTA, the Metropolitan Atlanta Rapid Transit Authority (MARTA), SEPTA, the Cleveland Regional Transit Authority, the DC Metro, and the Maryland Transit Authority, among others. These new technologies are not substitutes for traditional safety work; rather, they are a backup for when it fails.[58]

How important were all of these new ideas and programs? Safety began to improve on CSX, NS and UP about 1980. The later decisions to hire DuPont and implement TQM contributed to these gains but they were an effect of this prior commitment, as well as a cause of future improvements. The brief rise in casualties on the UP after 1996 reflects in part the difficulties that company experienced in merging with the SP, but injury rates at other carriers also rose, suggesting business cycle influences were part of the UP's problem.

Harassment, Intimidation, and Injury Underreporting

There have always been incentives to underreport injuries. In the 1920s, as the railroads inaugurated safety contests and vied for Harriman Awards, the ICC had warned that one result was to suppress casualty reporting. David Surratt, who began as a fireman on the Southern in 1913, told of losing a finger to a coal chute and never going to the hospital and probably never reporting it. There have also been instances of overreporting: when James Larson joined the Chicago Great Western he learned that injuries always increased the first day of fishing and hunting season. After the company moved the claim agent's office out of the yard and into downtown Des Moines, making it less convenient, the injury rate fell in half, Larson claimed.[59]

The problem of underreporting surfaced again in the 1980s. The competitive pressures that were motivating companies to improve safety also generated perverse incentives within corporate hierarchies—an agency problem again. As top managers squeezed middle- and lower-ranked bosses and foremen to reduce injuries, they, in turn, sometimes pressured workers not to report them. When Harry Brady was asked if the problem was B&O managers pressuring workers to slight safety in order to save money he said, "No. I think it is just local officials that do these things."[60]

Initial complaints to Congress involved alleged harassment of "whistle-blowers"—individuals who had reported safety violations to the FRA. The agency, it turned out, routinely informed the railroad of the informers' names. In the 1990s, however, the carriers' unions began to report that companies were harassing and attempting to intimidate employees to suppress reports of injuries too. A good example comes from a 1997 complaint filed by the BLE against Conrail. The incident involved a back injury to an engineman who fell when a stair railing broke. He informed the superintendent that while he didn't need medical attention, he wanted to report the injury. Conrail then had a "safety shares" program that awarded company stock to the division with the lowest injury rate. The supervisor responded, "I'm not trying to coerce or intimidate you into not reporting this but I just want to let you know your co-workers will be out their $450.00 in safety shares. How do you want to handle this?" On the BNSF, for a time the procedures involved points (demerits) with more going for reportable than non-reportable accidents, which surely encouraged underreporting. About the same time, the FRA was performing a Safety Assurance and Compliance Program (SACP) audit of CSX during which it found that 27 of 62 injured employees interviewed (44 percent), thought the company used intimidation. In one case a supervisor got a doctor to revise his recommendation for restricted work activity to make an accident non-reportable.[61]

Top executives swore that such incidents were unrepresentative and were not company policy. They brandished anti-harassment policy statements, noted that they had instituted confidential hot lines for reporting such violations, and, in the case of the BNSF, had fired two supervisors for harassment. Still, a steady stream of complaints continued in both United Transportation Union (UTU) and BLET publications, and in 2007 the problem resulted in lengthy congressional hearings entirely on the topic of harassment. Yet unions were sometimes complicit: the FRA reported mysteriously finding few sprains and strains at one facility CSX had threatened to close unless injury rates were reduced, and union representatives were among the "most concerned about injury reporting." The idea that injuries were underreported was also useful; it allowed the BLET to dismiss the FRA's findings that remote control of locomotives reduced yard injuries.[62]

There is every reason to believe the protestations of company executives, for harassment was an outcome not of policy but of competitive pressures, rising accident costs, and the new approach to safety. It was, as noted, a principal-agent problem. Unions thought safety competitions were part of the difficulty. The Harriman Award had been instituted by the widow of Edward Harriman in 1913 to reward the lowest injury rates, but to labor these were a fraud and an invitation to intimidation. In 1997, all the rail unions picketed the ceremony, terming it the "Harassment Award."[63]

The intention behind the TQM and behavior-based approaches to safety that spread during the 1990s was to create a safety culture in which workers took responsibility for their own and others' safety. Injured workers or those who endangered themselves and others received more supervision and "counseling," This new approach could affect injury reporting in several different ways. As noted above, Amtrak's injury rate rose after it instituted the "Safe-2-Safer" program in 2008 and TQM might have increased injuries reported on the Florida East Coast described above.

Yet the new safety procedures might simultaneously encourage harassment. Counseling, for example, need not always be gentle, and as the Conrail example demonstrated, much of it involved peer as well as supervisor pressure. Matthew Rose was COO at the BNSF; the company's goal, he said, was to make safety a "way of life." He testified of the need to prevent harassment and intimidation and to change the company's culture. The company had shifted to "performance accountability" from an emphasis on discipline. "And we're doing this all at a personal level," he added. Probably one motive for its "Listening Posts" was to encourage more candid reporting. Yet in 2010, OSHA found that the BNSF had harassed a UTU member and ordered it to pay him $95,000. As noted, safety has always required villains and even gentle counseling and peer pressure inevitably stigmatized the injured. Thomas Prendergast, president of the Long Island, understood the problem. "There's been a tremendous amount of pressure within the organization to find out why accidents happen . . . [and to] lean on people to say they shouldn't happen." To a manager "leaning on" people might imply holding a worker strictly responsible for a safety lapse, but the same "leaning" could easily feel like harassment to its recipient.[64]

Have such pressures led to large-scale underreporting of injuries? In the late 1990s, the UTU claimed that half of all injuries went underreported and the UTU *News* regularly featured claims of harassment. Yet neither an FRA regulation of 1997 (intended to tighten company internal controls) nor the Rail Safety Improvement Act of 2008 (which increased penalties for interference with workers' efforts to obtain medical care) generated a jump in reported injuries. Either the harassment had little impact on underreporting before the new policies, or they failed in their objective. The decision of the AAR in 2012 to drop the Harriman Award suggests that the problem continues, as it might have been responsible for some of the pressures that led to harassment.[65]

Federal Policies and Worker Risks

Until the mid-1980s, the FRA and Congress were preoccupied with train accidents—especially those that might involve hazardous substances—and they focused most sharply on those that resulted from defective track and equipment. Since even at their peak in 1978 train accidents accounted for only about one-fifth of employee fatalities and less than 1 percent of injuries, this emphasis ignored most of the problems of worker safety.[66]

Worker Safety and Human Error

The regulatory focus began to change in the 1980s. As train accidents receded, the FRA became more concerned with worker safety and attention also shifted to include human factors as causes of all types of accidents. Similarly, as employers' liability costs continued their steady rise, companies increasingly focused on human factors. Alcohol and drug programs provide one example. The requirements of every railroad in the country had long included Rule G, which forbids alcohol consumption or intoxication on the job, but enforcement had always been a problem. In 1994, Charles Monin, a member of the BLE, recalled the old days: "when I came on the railroad in 1964 the culture was that you took the top off [the bottle] and threw it away . . . because you were going to finish what you started."[67]

In response to these problems, the Union Pacific and its unions inaugurated Operation Red Block (named because a block signal that shows red requires an absolute stop) about 1984. Reflecting the influence of behaviorism, the program employed peer pressure to emphasize prevention rather than punishment. It also instituted a "bypass" program that allowed an employee to enter a substance abuse program without running afoul of Rule G. Federal efforts to curb alcohol and drug use and those of other companies will be discussed in more detail in Chapter 5 because their primary motivation was to prevent passenger and hazmat disasters. Yet the benefits of such programs included reductions in worker injuries from sources other than train accidents. The FRA study of non-train incidents in 1975 noted above attributed 9 of the 93 fatalities to alcohol—including one where the blood alcohol level of the deceased was 0.26 percent. In the early 1980s, several men killed on the job had blood alcohol levels of 0.36 percent and higher. A 1978 FRA study of seven large carriers estimated that alcohol and drug use cost them $108 million in 1978 due to property damage, absenteeism, and injuries. Medical and safety officers on these carriers admitted they had no hard data on injuries that resulted from alcohol and drug use, but they all concluded that the number was substantial, ranging from 10 to 15 percent to as much as 50 percent of the total. One officer claimed that four of the six fatalities that had occurred on his road during the previous seven years had been alcohol related.[68]

The federal law of 1989 (Chapter 5) required drug and alcohol testing in only certain safety-critical jobs (engine driver, train dispatcher), but it also provided a rationale for companies to engage in broader testing. In 1993, about 0.7 percent of workers tested positive in random drug and alcohol tests. By 2010, some of the large freight railroads gave annual tests to half their hourly workers and found positive results for only 0.3 percent. Of course, some programs were more successful than others; Amtrak's inspector general pointed out that its positive tests averaged about 40 percent higher than the industry average. Still, it seems clear that some—perhaps a large—fraction of the improvement in worker safety during these years resulted from the employment of a more sober labor force. On many lines such as the UP and CSX, the drug and alcohol programs were developed with strong union support, deemphasizing discipline and emphasizing counseling and rehabilitation, thereby reinforcing the carriers' new approach toward labor relations.[69]

In 1993, when Jolene Molitoris became FRA administrator, ideas from the TQM Movement also started to shape policy. She stressed the need for better "customer service"; in an effort to speed up the regulatory process and make it less contentious, the FRA began to emphasize participatory rulemaking. As noted in Chapter 3, the Rail Safety Advisory Committee included representatives from the carriers, their unions, and other interested parties. This more cooperative approach appealed to labor, which had felt excluded from the regulatory process, and it dovetailed with the carriers' new-found interest in incorporating labor into safety programs. The SACP, begun in 1995 also demonstrated that the FRA's focus was widening to include more than just train accidents. Its aim was to diffuse information, provide informed criticism, and encourage cooperation, and it was therefore another market-augmenting activity. FRA representatives would meet with both management and labor to assess the company's safety culture. The process also included a detailed safety audit, similar to that which DuPont or another outsider might provide. The goal was to focus on root causes with the end being zero accidents. In 1995–1996, the FRA undertook a review of many company safety programs. CSX, the NS, and the UP, all deeply committed to safety, had only marginal improvements in safety.[70]

In the twenty-first century, the FRA has continued to focus on human factor accidents, and to emphasize cooperation and behaviorism. A Switching Operations Fatality Analysis (SOFA) working group that included representatives of unions, management, and the FRA began in 1998 and built on the cooperation that had begun with Operation Red Block. One union representative wrote the secretary of transportation that in his 39 years of railroading "serving on this committee has been the most productive and rewarding assignment of my career." The group issued a report in 1999 that included five major recommendations that the Class I carriers had begun to implement by 2000.[71]

In cooperation with individual carriers and their unions, the FRA has also developed and tested a number of experimental approaches to reduce "human factor" accidents. For example, in 2002 as part of its Clear Signal for Action Program, which involves behavior-based safety and continuous improvement, Amtrak's Chicago terminal employed peer-to-peer communication and continuous feedback to reduce employee injuries. Between 2005 and 2008, the FRA and the Union Pacific tested such an approach in the company's San Antonio transportation department. A review found that the program had cut derailments about 80 percent over two years. The Confidential Close Call Reporting System (C3RS) begun in 2007 (Chapter 3) was primarily intended to reduce train accidents, but it contributed to broader safety goals as well. It also reflected the complaints of employee harassment and intimidation discussed above. In 2009, a group similar to SOFA, the Fatality Analysis of Maintenance of Way Employees and Signalmen (FAMES) committee, began to look at the continuing risks of maintenance work. While none of these various groups and investigations leaves a dramatic footprint in the fatality data, collectively they undoubtedly did—and do—much good, functioning somewhat like the old AAR Safety Section meetings in developing and spreading information and expertise and pinpointing weaknesses, but with the important difference that the modern approaches include labor.[72]

Under Molitoris and her successors, there have also been a number of regulations intended to improve worker safety. In the Rail Safety Act of 1988 Congress mandated regulations governing safety of bridge workers, and in 1992 an FRA rule arrived requiring training and fall protection (49 CFR 214). The most important of such regulations was the Roadway Worker Protection rule of 1996 (49 CFR 214), which became effective in early 1997. The rule was the result of negotiations by the Rail Safety Advisory Committee: it mandated that companies develop a safety program for roadway workers to be approved by the FRA. There were requirements for training, coordination of work with trains, and much else. Again, it is hard to find the imprint of the new regulation on maintenance workers' injury rate statistics presented in Figure 4.3, which show no sharp improvement after 1997 and indeed seem largely to mimic overall safety trends. The most likely reason is that most of the new regulations made at best marginal changes in the status quo, and that is hardly surprising. To imagine otherwise would be to believe that either the FRA knew more than the carriers, which is unlikely, or that the carriers had little incentive to provide worker safety, which as detailed above, was patently not true.[73]

More recently, the NTSB has also begun to address accidents to maintenance of way workers. Such concerns were rare in the 1970s, but have

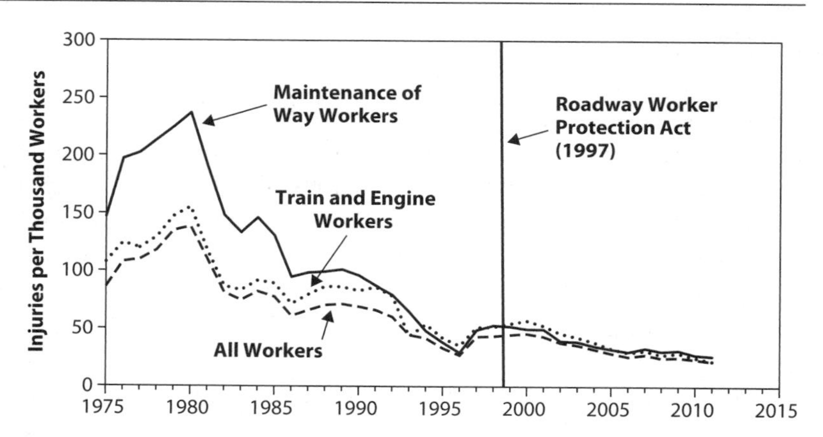

Figure 4.3. Maintenance of Way Worker and All Worker Injury Rates, 1975–2011. (FRA, *Accident Bulletin/ Safety Statistics*, and AAR, *Railroad Facts*, various years)

become more common as train accidents have declined. A 2013 review of 11 fatalities to roadway workers found that these tragedies typically occurred at companies with weak safety cultures—which it spelled out as inadequate hazard identification and mitigation, and failures in peer-to-peer support—all of which it laid to the feet of top management, and it stressed the need for redundancy in safety systems. These might include, for example, not only the usual allocation of track rights to protect roadside workers, but also use of backup electronic devices such as those of Bombardier and Protran described above. The FRA has responded with a safety advisory that also stresses the need for redundancy.[74]

Locomotive Safety

In the late 1990s, the FRA began to study locomotive crashworthiness. Needless to say, this was a topic dear to the hearts of engine drivers and for decades the BLE had been a powerful force pushing for safer locomotives. Federal regulations have governed steam locomotive safety since 1910 and more than steam locomotives were soon included. In 1954, pressed by the union, the ICC promulgated the first requirement for design and inspection of MU (multiple unit) locomotives. By 1970, a host of regulations covered lights, handholds, footboards, safety tread, and similar items affecting worker personal safety on all types of locomotives. The BLE campaigned for cab safety regulation throughout the 1970s; its journal reprinted NTSB accident reports that called for safer cabs, and it complained of inadequate federal inspections. The carriers and manufacturers developed the "clean cab" design about 1972; it was an effort to remove obvious dangers such as sharp edges. As noted above, the BLE also midwifed protective window glazing, which arrived in 1979.[75]

Although the FRA funded various studies of locomotive crashworthiness throughout the 1970s and 1980s, little seems to have come of them. In 1990, after consultation with representatives of labor and the FRA, the

AAR promulgated AAR-S580—its own standards for locomotive crashworthiness, which specified, among other things, requirements for anti-climbers and the yield strength of collision posts. There were also requirements governing fuel tanks, means of egress, interior configuration, and the strength of the short hood. In 1992, Congress passed the Railroad Safety Enforcement and Review Act (P.L. 102-365), which directed the FRA to study locomotive crashworthiness, leading it to fund a blizzard of studies on that topic. In 1998, the agency formed a working group under the Rail Safety Advisory Committee that resulted in a proposed rule in 2004. Passenger car standards adopted in 1999 (Chapter 5) covered the cabs of "cab cars" and MU operation trains. Regulations for traditional locomotives arrived in 2006. For locomotives constructed after 2009, these mandated the AAR S-580 requirements or, alternatively, any other configuration that would withstand certain collision forces. In 2010, Amtrak ordered its first crashworthy locomotives. Here again, the regulation essentially codified industry practice and it is too soon to know how well the new equipment will perform. However, based on studies of collisions before the implementation of positive train control (PTC) (Chapter 5), the agency claimed that it would prevent two to three fatalities a year.[76]

The focus on locomotive cab safety also reflects a broadening concern with the well-being of engine drivers. As fatalities have declined and injuries have become more costly, there has been an increasing attention to locomotive and cab redesign to prevent more minor injuries, such as back strain, and to reduce noise levels. In addition, men and women who run locomotives confront not only physical risks on the job but sometimes emotional stress too. The BLE had campaigned against accidents at grade crossings and to trespassers for years (Chapter 6). Grade crossing collisions often wrecked the train, but even if the train crew came through unscathed, the horror of hitting and killing automobile passengers or trespassers sometimes stayed with them the rest of their lives. The sound was "like hitting a pumpkin," recalled Tom Donovan, an engineman on the New Jersey Transit, when his train killed a trespasser.[77]

Some engine drivers became unable to function and left the job after such accidents. Others experienced panic attacks and other emotional difficulties. Gradually the carriers and their unions began to develop "critical incident" programs to help individuals cope with the trauma. Donovan helped found such a program on the New Jersey Transit. By 2008, when Congress required the FRA to develop regulations requiring such programs, most large carriers had already established one. These typically defined "critical incidents" broadly, although their motivation was clearly grade crossing and trespassing accidents, and they provided a variety of counseling and time off from work to help in recovery.[78]

How They Die

In the twenty-first century, the FRA has again begun to publish details of worker fatalities and the NTSB has continued to probe such incidents. In 2006, there were 14 railroad employees killed on the job (and 7 contract employees). Of the railroad workers, seven were train and engine employees and of these, six died while switching cars; there were also three each from the maintenance of way and maintenance of equipment departments; one police officer was murdered. These tragedies reflected the interaction of a whole series of improbable events—what the FRA called possible contributory factors, or PCFs, as the reader will recall. Most of these reflected human mistakes, which in turn reflected lapses or inadequacies in corporate safety cultures.

The deceased were nearly all experienced, sober men who had recently undergone safety training. Consider the death of a brakeman during switching in the Union Pacific yard at Watsonville, California, on October 13, 2006. He fell under a car he had been riding, in violation of company rules. Yet the company knew that employees regularly ignored such rules, and it neither emphasized the problem in safety training nor stepped up enforcement. The locomotive that was kicking the cars was also traveling too fast. Alternatively, reflect on the many blunders that culminated in the death of a 55-year-old BNSF car inspector with 25 years' experience who was crushed on February 7, 2006, by a car he was working on. The investigation revealed that he had violated rules by crawling under a car supported only by a jack, and that the job foreman had failed to stop him. There also had been inadequate job briefing. Also, a longer cutting torch would have allowed him to work outside the car.[79]

Such tragedies are a reminder that safety cultures are imperfect and that the Red Queen effect, although diminished, has not entirely disappeared. Yet as Figure 4.2 reveals, they have become far less common than they were only a generation ago. Between 1980 and 2015, worker fatality rates declined an average of 3.3 percent a year. The following exercise suggests what these gains mean for the men and women who deliver America's rail passengers and freight every day. Eighteen workers lost their lives on the job in 2015: had the injury and fatality rates of the 1970s continued to prevail, there would have been an additional 37 fatalities that year, in addition to 21,049 more injuries, including 1,786 extra fractures and 89 more amputations.[80]

This tale of eroding safety in a major industry followed by its renaissance contains a number of lessons. One is surely the contingency of workers' safety. It is tempting to look at the long-term decline in work accidents in railroading or elsewhere as inevitable—the result of abstractions such as progress or technology. The carriers' two decades after 1957 provide a cautionary tale that such is not the case. Safety was contingent in part on the availability of resources—for deteriorating profitability kicked a major prop from the work safety movement. The story also reminds us that organizational structures can become dysfunctional; by the 1950s, safety work had become narrowly conceived and intra-company agency problems impeded an accurate assessment of its costs and benefits.

The turnaround in worker safety that began in the 1970s reflected political pressures and companies' desires to shift the gaze away from train accidents. Such pressures also encouraged them to welcome labor into the safety tent. The decline in worker injuries and fatalities also coincided with a regime change; voluntarism, which had left most decisions about worker safety to the carriers, was scrapped and the new approach has seen increasingly detailed congressional and FRA intervention in the work environment. The FRA's sponsorship of innovative programs and its efforts as an honest broker have encouraged labor-management safety partnerships and have augmented private safety markets. The importance of FRA regulations is less clear; it is difficult to see their impact in the injury and fatality statistics because they essentially mandated best practice at a time when market incentives were sharply raising the payoff to improving safety.

Thus, the story also reveals the centrality of private market incentives, organizations, and institutions, and those who deplore market forces and profits might reflect on these realities. The sharp rise in the cost of worker injuries was a powerful spur to improve safety, but injury costs were not the only way that market forces propelled better safety. In the 1970s, as the railroads strived to improve service and cut costs to survive in an increasingly competitive transportation market, they increasingly realized that worker safety was integral to their core mission, relearning a lesson that Ralph Richards had discovered nearly a century earlier. Therefore, the turnaround did not originate with deregulation, but the continued decline in casualties was surely contingent on the carriers' economic renaissance. Worker commitment to provide high-quality railroading required company commitment to provide a high-quality work environment and that meant unceasing efforts to improve work safety. These ideas encouraged a new emphasis on behaviorism and they meshed easily with the TQM Movement. Thus, if the railroads were like an old dog trying to relearn an old lesson, mastering it has required some new tricks—a dedication to quality and service unheard of a generation earlier and a far more complex understanding of human behavior and organizational commitment.

5

Passenger Safety in Modern Times, 1955–2015

Federal safety regulations for passenger cars are less comprehensive than regulations for locomotives and freight cars.

GAO, Amtrak Safety, *1993*

He pulled on the window handle and nothing happened . . . He stated that removing the emergency window took about three minutes.

NTSB, Collision and Derailment . . . Near Silver Spring, Maryland, *1966*

The year 1995 marked a minor milestone in American railroad history: it was the first time since 1846 that American carriers had gone a full year without killing at least one passenger. Alas, the triumph was short lived, for the following year saw 12 passengers killed, 9 of them on trains, of whom 8 died in a fiery head-on collision between Amtrak Train 29, bound from Washington to Chicago, and a MARC commuter train that also killed 3 trainmen near Silver Spring, Maryland. Still, if 1995 was not typical, neither was 1996. From the end of World War II down to about 1980, railroad passenger fatality rates from train accidents declined steadily, if unevenly. Yet even in bad years like 1996, a significant minority of passenger fatalities resulted not from train accidents but rather from a host of what I have termed "little accidents," while year in and year out a majority of passenger injuries resulted from such undramatic causes. The press and public policy largely ignored these little accidents but they were a major force shaping the safety of railroad passenger travel. Overall, the safety of train travelers improved steadily until the early 1980s and, thereafter, not so much.

The passengers whose safety is the topic of this chapter are those who ride long-distance trains—the private carriers until 1971; thereafter, Amtrak and the Alaska Railroad—and commuters on lines connected to the rail network. These are the lines that the Federal Railroad Administration (FRA) regulates; they exclude the independent light rail and rapid transit systems such as the Bay Area Rapid Transit (BART) in San Francisco, the Metro (Metropolitan Area Transit Authority) in Washington, the Dallas Area Rapid Transit (DART) in Dallas, and others not connected to the rail network. However, because these systems have much in common with commuter rail, they make an occasional appearance here and many of the developments discussed in this chapter apply to them as well.

The chapter begins with a brief review of the course of passenger safety. The following section focuses on train accidents and tries to explain their diminishing importance as a source of passenger mortality. I then turn to the roles of technology and public policy as they have shaped these trends. The following section depicts the increasing focus on accident survivability and tentatively assesses its success. The final section of the chapter looks at passenger risks from causes other than train accidents.

Getting There in One Piece: The Course of Safety

The Interstate Commerce Commission (ICC) divided the sources of injuries and fatalities for passengers into three broad cause classes: train accidents, train service accidents, and non-train accidents. Train accidents, the reader will recall (Chapter 2), involve the movement of on-track equipment and are defined by a monetary threshold, while train service accidents (train incidents) involve movement of on-track equipment where damages do not meet the dollar threshold. The FRA has modified this terminology, changing "train service accidents" to "train incidents." Casualties not involving the movement of on-track equipment were the final category. Until 1957, passengers injured in stations or otherwise lawfully on railroad property were termed "travelers not on trains," and added to the totals of passengers. In that year the ICC defined a passenger more narrowly as "a person who is on, boarding, or alighting from a railroad car for the purpose of travel, without participating in its operation." This did away with the category of non-train accidents, for individuals injured in stations and elsewhere, but who are not boarding or leaving a train, are called "other non-trespassers," even if they are ticketed passengers. The commission's motives for this redefinition are unknown, but it was not a trivial change; in the years 1950–1956, fatalities and injuries to travelers not on trains amounted to about 8 and 24 percent, respectively, of total passenger casualties.[1]

Down to about 1970, passenger safety was almost entirely unregulated at the federal level. The reasons for this were that the rail unions had provided the political push for most safety legislation and most of them worked with freight equipment. The threat of regulation was always present, however. Indeed, the spate of passenger wrecks during World War II had motivated the ICC to mandate an expansion of automatic train control—which seems to have had modest effects (Chapter 1). Hence, the decline in both train accidents and other sources of risk was largely the outcome of private enterprise—a regime I have termed "voluntarism." Beginning about 1970, public policies played a more important role. The National Transportation Safety Board (NTSB) arrived as a public scold in 1967. Amtrak federalized long-distance passenger traffic after 1971, and commuter lines were increasingly taken over by public authorities, although the actual train operation was (and is) often left to the private carriers or Amtrak or other contractors. Until the late 1980s, the FRA paid little attention to passenger safety; thereafter, partly in response to NTSB and congressional worries, it played an increasingly important role.

For a long time public interest in railroad safety had largely focused on the dangers to passengers from train accidents, while the courts and juries held the carriers absolutely liable for passenger safety. Railroad executives had responded to these incentives and gradually passenger travel became very safe indeed. As World War II ended, the carriers entered the postwar years with high hopes of reviving their passenger business and safety was to be one of its selling points. In 1946, an Association of American Railroads (AAR) study group reported cautious optimism. The group acknowledged that the carriers could not match airlines for speed, but that rail travel had other advantages and it stressed "safety, dependability and comfort." Five years later, in 1951, *Railway Age* reported, "Train Travel is SAFE Travel," and quoted AAR President William T. Faricy that "safety is the biggest talking point we have." Indeed, the article contained statistics for 1949 that showed a death rate for passengers on trains of 0.8 per billion passenger miles compared with 13 for airlines and 20 for automobiles. Despite such happy numbers, however, Faricy lamented that the press had "widely publicized" several "unfortunate" train accidents that he thought had greatly exaggerated public perceptions of railroad risks. A series of ads by Pullman in the 1950s claimed that railroad travel was "above all—safe," implying, perhaps, that air travel was not. These stirred up *Aviation Week*, which termed the ads "vicious." In 1956, the New York Central surveyed public attitudes toward train travel and found the image of rail patrons was that of "drab relics of the McKinley era" and that emphasizing safety and dependability simply reinforced that stereotype. Nor did other carriers stress safety in their advertisements. The C&O thought that the difference between the railroads and other common carriers was small—which was truer in 1956 than it had been in 1949—and that automobile accidents "always happen to the other guy." Thus, selling safety proved to be a one-way street: while train accidents stirred up the press and the public, safety proved hard to market.[2]

Yet while passenger safety failed to sell, it continued to improve, as Figure 5.1 depicts. In fact, because both fatalities and injuries had fallen to such low levels by this time, the data are highly sensitive to random fluctuations; to remove such influences, the figures show five-year moving averages centered on the middle year, and as can be seen even these smoothed data are subject to eruptions. The bulge in fatality rates in the 1970s demonstrates how sensitive the data are to individual accidents. It was the result of just two wrecks—an Amtrak derailment near Salem, Illinois, that occurred while the train was traveling at 90 mph on June 10, 1971, that killed 11, and a horrendous rear collision on the Chicago commuter

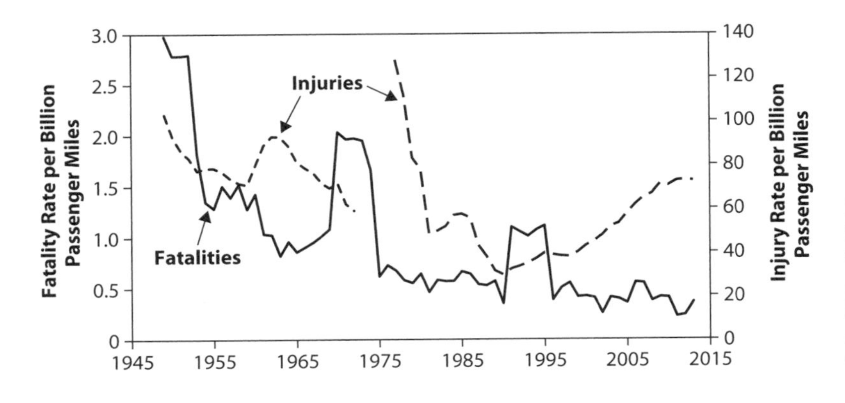

Figure 5.1. Passenger Injury and Fatality Rates, All Causes, Five-Year Moving Averages, 1947–2015. (ICC/FRA, *Accident Bulletin/Safety Statistics*, and author's calculations)

lines of the Illinois Central Gulf in 1972 that killed 45 (Table A3.1). There is a discontinuity in the injury data because the FRA changed the definition of injuries in 1975 to include medical cases as well as those incapacitated. In addition, although this chapter relies on injury data at times, the author believes that changing cultural norms and perhaps legal considerations shape these data sufficiently that one should treat inferences based on them with special care. One other conclusion emerges from Figure 5.1: the long-term improvement in passenger safety largely ended in the mid-1980s. Indeed, a statistical analysis confirms that there has been no downward trend in the passenger fatality rate since 1980.[3]

Train Accidents and Passenger Safety

In this section, the focus is on fatalities and injuries from train accidents. The figures for train incidents (what I have called "little accidents") will be reviewed in that section below. For the postwar years, train accident injury and fatality rates tell a story of safety progress down to the 1980s (with some spectacular blips), apparently followed by stagnation. Injury rates from all causes rise sharply after about 1990 (Figure 5.1), while those from train accidents do not, revealing that it was little accidents that drove the increase during these years.

The chance of a passenger getting killed or injured in a train accident depends on the likelihood of the wreck and—should a wreck occur—the chance of being hurt. That is, the number of fatalities from accidents per passenger mile equals the number of accidents per passenger mile times fatalities per accident. Fatalities per accident in turn depend on the fatality rate for passengers in the accident times the number of people in trains experiencing accidents. More succinctly:

$$\frac{\text{Fatalities}}{\text{Passenger Miles}} = \frac{\text{Accidents}}{\text{Passenger Miles}} * \frac{\text{Fatalites}}{\text{Accidents}} \quad \text{and}$$

$$\frac{\text{Fatalities}}{\text{Accidents}} = \frac{\text{Fatalities}}{\text{Passengers}} * \frac{\text{Passengers}}{\text{Accidents}}$$

Table 5.1 presents overlapping decadal average fatality rates (fatalities/passenger miles) from collisions and derailments—which accounted for most passenger mortality in train accidents—along with train collision or derailment rates (accidents/passenger miles) and fatalities per person on trains in accidents (fatalities/passengers). As can be seen, the average fatality rate from collisions and derailments fell about 79 and 84 percent, respectively, between 1947–1955 and 2001–2010 (Table 5.1, columns 1 and 4). Over the same period, passenger trains in collision per passenger mile fell nearly 91 percent, while trains in a derailment fell about 84 percent (columns 2 and 5). Moreover, at least since the mid-1970s, the odds of dying if you were in a collision worsened a bit (column 3), while survival chances show little change for passengers in derailments (column 6). The most likely explanation for these results is that rising train speeds have mostly offset improvements in survivability and emergency procedures. Injuries per passenger in a collision or derailment (not shown) follow a course similar to that of fatalities.[4] While the great variability of casualties in train accidents makes even these conclusions imprecise, what they depict is probably correct: the decline in casualties from train accidents since World War II has mostly been the result of fewer accidents, not diminishing accident lethality.

Moreover, the average number of passengers per train in an accident (passengers/accidents) has risen about 25 percent since the late 1980s. Given accident lethality, this rise leaves death rates unaffected, for it raises both passenger miles and fatalities in the same proportion, but it does increase the likelihood that any given accident will have a large death toll.

Table 5.1. Causes of Passenger Fatalities from Collisions and Derailments, 1947–2015, Overlapping 10-Year Averages

	Collisions			Derailments		
	Fatality Rate[a]	Trains in Collisions[a]	Fatalities to Passengers in Collisions[b]	Fatality Rate[a]	Trains in Derailments[a]	Fatalities to Passengers in Derailments[b]
1947–1955	0.73	3.06	1.82	0.61	9.24	0.67
1951–1960	0.20	2.14	0.99	0.75	5.72	1.37
1956–1965	0.09	1.82	0.49	0.59	4.88	1.37
1961–1970	0.04	1.88	0.18	0.29	6.03	0.46
1966–1975	0.45	1.60	2.30	0.27	6.53	0.34
1971–1980	0.50	1.72	1.87	0.22	5.85	0.24
1976–1985	0.04	1.39	0.16	0.13	4.25	0.19
1981–1990	0.13	0.99	0.82	0.07	2.43	0.16
1986–1995	0.18	1.14	0.98	0.32	2.18	0.89
1991–2000	0.57	4.25	0.79	1.35	9.25	0.87
1996–2005	0.08	0.56	0.84	0.13	1.83	0.37
2001–2010	0.16	0.28	3.14	0.10	1.46	0.39
2006–2015	0.17	0.23	3.76	0.06	2.16	0.14

Source: Author's calculations from ICC/FRA *Accident Bulletin/Safety Statistics,* various years.

[a]Per billion passenger miles.

[b]Per thousand passengers on trains in collisions or derailments, respectively.

If people especially fear disasters, then an increasing number of people on trains may raise popular perceptions of risk.

Technology, Public Policy, and Passenger Train Accidents

While it has been the decline in collisions and derailments rather than their lethality that has improved passenger safety, these gains have fluctuated from year to year—a fact that is masked by the moving averages in Table 5.1 that are intended to reveal trends. Indeed, a graph of annual collisions and derailments per passenger mile (not shown) would bear a family resemblance to the depictions of all collisions and derailments presented in Chapters 2 and 3 that are dominated by freights. There is good reason for this, for many of the same forces that drove freight train accidents also shaped passenger train safety. These—to repeat—have been safety incentives, and changes in technology, organizations, resources, and regulations. Thus, like freights, passenger train derailments declined down to the mid-1950s, for passenger safety also benefited from the spread of creosoted ties and use of rail flaw detectors and continuously welded, heavier, controlled-cool rails. Again, from the mid-1950s through the 1970s, the rise in passenger derailments roughly mimics that of freight trains. Passenger equipment traveled over the same deteriorating track as did freight cars and suffered from diminished expenditures for maintenance—as anyone old enough to remember riding the trains of the 1960s will recall. In 1967, when the Pennsylvania Railroad employed an old inner tube to hold a coupler cut lever in place, *Railroad Trainman* headlined that "Some PRR Trains are Held Together by 'Rubber Bands.'" In 1974, the FRA found sections of Penn Central main-line track in Indiana used by Amtrak that failed to contain one sound tie every 8 feet. Not surprisingly then, many wrecks resulted from poor maintenance.[5]

Two derailments on the Baltimore & Ohio and Louisville & Nashville railroads in 1967 and 1968 capture some of these problems. On February 24, 1967, B&O first-class passenger Train 31, consisting of 5 cars and traveling about 43 mph derailed on a 6 degree curve near Toll Gate, West Virginia, killing 2 passengers and 2 crew and injuring 48 passengers and 2 crew. It was one of the 129 passenger train accidents the FRA recorded that year. An investigation revealed a host of bad track conditions. The rail was poorly spiked, was badly worn, and had turned over; the gauge was too wide and the super-elevation on the curve was far below American Railway Engineering Association (AREA)-approved practice for such speeds and curves.[6]

The following year bad track also felled a passenger train on the L&N. On January 29, 1968, first-class passenger Train 704 consisting of 15 cars—of which several were box-express—derailed near Casky, Kentucky, injuring 25 people. The subsequent investigation found 60 percent of the ties deteriorated within the vicinity of the accident, and 31 percent of the spikes did not hold. Cross level measurements varied as much as 1 3/8 inches and the gauge was 1/2 inch too wide. The box-express cars rode on freight car trucks which—unlike those of passenger equipment—had no stabilizers and were therefore prone to rock 'n' roll. One of them—owned by the Pennsylvania Railroad—had worn side bearings and had apparently rocked off the rails.[7]

As Chapter 3 noted, sometimes wrecks resulted from use of supposedly safer technology—an example of a risk-risk tradeoff. While continuously welded rail (CWR) reduced passenger as well as freight derailments on balance, it yielded "sun kinks" that also caused accidents—especially in weak track—as the Penn Central found out on June 28, 1969. The temperature was about 98 degrees Fahrenheit; the rail had been installed at a temperature of 70 degrees Fahrenheit and compressive forces had built up. Recent maintenance near Glenn Dale, Maryland, had done a poor job of ballasting and tamping—demonstrating that maintenance itself is dangerous. There was a slight irregularity in the track that buckled when the trucks of the *Silver Star* passed over it going about 80 mph. The train derailed, sending 144 passengers to the hospital. Yet if CWR contributed to this accident, it also helped protect the train. Investigators noted that the unbroken rail guided the derailed cars along the roadbed, preventing them from colliding with a massive H beam that supported the catenary.[8]

Such wrecks continued to mar Amtrak's record. In 1990, a sun kink caused an 8-car derailment near Batavia, Iowa, on Burlington Northern track while the train was going about 79 mph, injuring 86 people. Track maintenance some months earlier had probably welded the rail at too low a temperature, leading to excessive compression so that it buckled due to the train's dynamic loading. On April 6, 2004, improper maintenance again resulted in buckled track that derailed 9 cars and a locomotive of the *City of New Orleans*, near Flora, Mississippi, killing 1 and injuring 36 passengers and 11 others.[9]

Yet the rise in passenger train derailments was far more modest than that of freights, for passenger and freight equipment led very different lives. As Chapter 2 noted, some of the rise in freight train derailments resulted from the spread of large freight cars that had a tendency to rock 'n' roll, which was less likely with passenger equipment—although not unknown when passenger and freight equipment were mixed, as the L&N wreck above reveals. Passenger equipment was also much more likely than freight cars to sport roller bearings that reduced axle failures and maintenance was much better. Coaches were lighter weight than earlier generation equipment and rode on heat-treated, low-carbon wheels. In addition, passenger trains were much shorter, and they were increasingly likely to use tight-lock couplers, which had arrived in the 1930s. Hence, there was no parallel to freight derailments that resulted from running in the slack. Moreover, tight-lock couplers ensured that cars in a collision were less likely to telescope; in a derailment they tended to stay together, often upright, dissipating energy in the roadbed and preventing one from ramming the side of another. Passenger car trucks were also chained to the car (unlike freights, which simply sat on the trucks) and this too helped reduce telescoping. Passenger trains had always had better brakes than did freights, and with the arrival of higher-speed trains in the 1930s, Westinghouse developed electro-pneumatic brakes. In the 1970s, improved versions appeared for Metroliners and commuter trains, and they were equipped with composition shoes that shortened stopping distances and reduced heat buildup. By the 1990s, Amtrak specified dual disc and wheel brakes. Finally, a fruitful source of derailments (and collisions) for freights was yard work, but passenger cars received little switching.

The decline in derailments after the late 1970s was even more precipitous than the rise. There had been 90 derailments in 1976; there were 22 in 1983. Since then the numbers have fluctuated in the twenties and thirties even as train miles, passengers transported, and passenger miles have risen. Some of the gains might have reflected the FRA's track safety standards (Chapter 3), which were extended to dedicated commuter track in 1984. As Chapter 3 also detailed, federal money helped rebuild Conrail and several other freight railroads that transported Amtrak trains after 1971.[10]

The reduction in economic regulation after 1976 that led the freight railroads to upgrade road and track also benefited Amtrak and computer lines, as did the gales of technological progress that were sweeping through all the carriers (Chapter 3). To take but one example, in response to continuing problems of track buckling discussed above, all the carriers developed elaborate company rules governing installation and maintenance of CWR (Chapter 3). By the twenty-first century, the BNSF was developing a computer model derived from work at the John A. Volpe National Transportation Systems Center that integrated economic and engineering analysis to focus maintenance on areas with high predicted risks.[11]

Congress also poured about $2.5 billion into the Northeast Corridor Improvement Project that resulted in some 440 miles of new CWR and 338 miles of concrete ties. Beginning with the Urban Mass Transportation Act of 1964, Congress also began to shore up commuter rail, allocating about $937 million from 1965 to 1976 and another $2.2 billion by 1983. Thereafter, funding for capital improvements increased sharply, reaching $709 million in 1984 and nearly $1.3 billion in 1986, while state and municipal expenditures added to these totals.[12]

The new funding resulted in major upgrades to a host of commuter lines. In Chicago, the regional transit authority began a capital improvement program of $100 million a year from 1983 to 1987, the largest share of which—$158 million—went to track and structures. The Long Island, which carried more passengers than Amtrak and in 1983 had been "creaking with old age," began a five-year $1.1 billion capital expenditure plan that year that included about $230 million for better track and new equipment. In 1978, the Government Accountability Office (GAO) had issued a report that was sharply critical of the safety of Conrail's commuter lines running out of New York, noting that failure to maintain track had resulted in several derailments. By 1983, New York's Metro North had taken over these lines, and it began a five-year $861 million upgrade in 1984 that included $85 million for track and structures including CWR, among other improvements. Boston's Massachusetts Bay Transit Authority (MBTA), the New Jersey Transit Authority, and Philadelphia's Southeastern Pennsylvania Transportation Authority (SEPTA) also began major upgrades at this time. Since the early 1990s, annual capital expenditures for commuter rail have more than doubled in nominal terms, yet adjusted for inflation and passenger miles they show virtually no increase.[13]

These improvements in track and equipment are reflected in the reduction in derailments and declining passenger fatality rates in Table 4.1. Ironically, however, the upgrading in track might have yielded offsetting effects that tended to increase accident lethality. An FRA survey of Amtrak routes revealed that the fraction of miles designated as Class 5 track (maximum speed 90 mph) had risen from 5 to 39 percent between 1980 and 1984, while the share of Class 4 track (maximum speed 80 mph) had declined from 82 to 51 percent. Better track meant fewer wrecks but with higher speed they were more deadly.[14]

The decline in collisions and collision fatality rates was more uneven. Every year 80 to 90 percent of all collisions reflected some form of human error due to the inability of carriers to monitor engine drivers to ensure their fidelity to the rules. In the early days of the Timetable and Train Order system, train dispatchers might make mistakes or engine drivers forget or misinterpret their orders, or simply fail to stop in time. By the late nineteenth century, block signals had begun to supersede the train older system; in the 1920s, cab signals appeared that mimicked those on the right of way. In the 1920s, centralized traffic control appeared. By 1982, it governed about one-fifth of all tracks—and no doubt far more passenger miles—compared with about 5 percent in 1950. In the 1940s, the ICC had mandated block signals where passenger trains traveled over 60 mph and automatic train control where speeds exceeded 79 mph. By 1965, train control covered about 10,000 miles of passenger road. By 1977, Amtrak's entire owned track had some form of block signal as, in all likelihood, did the track of the freight roads over which it operated.[15]

The expansion of federal and state spending on commuter lines also generated many improvements on already comparatively well-signaled tracks. In the 1970s and 1980s, the Urban Mass Transit Act largess discussed above improved signaling on the Long Island, the Metro North, the

Chicago Transit Authority, the MBTA, and other commuter roads. The improvement included such changes as blocking for higher speed and two-way traffic in each of two tracks to expand capacity.

Better training of engine drivers also contributed to declining collision rates. By the 1960s, such training had become much more formal than in the days of steam. The L&N chose candidates for apprentice engine drivers by competitive examination and then ran them through a program it administered jointly with the Brotherhood of Locomotive Engineers (BLE). While these carriers were predominantly freight roads, they often ran the early commuter lines and they initially supplied engine drivers to Amtrak and the newer municipal commuter lines. By the late 1960s, the Santa Fe was sending its apprentice enginemen—and one woman—through a "college" that employed a locomotive simulator, and such equipment soon spread to other roads. By the late 1980s, according to the AAR, typical programs screened engine driver candidates for visual, hearing, respiratory, and cardiovascular problems and then provided classroom training using a computer-driven train dynamics analyzer. Candidates attended air brake and train handling classes; they had to pass exams on company operating and safety rules and demonstrate familiarity with signal systems. Prior to certification, they were required to take trips over the territory for which they would be qualified, and they then operated trains under the supervision of a road foreman. After certification they underwent periodic retraining and retesting. Whatever the sources of "human factor" accidents in passenger trains might have been, inadequate training of engine drivers does not seem to have been among them.[16]

Disasters and Regulation

After about 1970, the NTSB began to focus on collisions, for block signals were not foolproof, as a head-on collision on the Penn Central in its waning days demonstrated. On August 20, 1969, two Penn Central multiple unit (MU) commuter trains collided near Darien, Connecticut, demolishing the front section of the 2 lead cars, killing 3 crew and 1 passenger, and injuring 1 crew member and 40 passengers. The line was single track, protected by manual block signals. The dispatcher had issued an order for one train to take a siding, but apparently through some form of miscommunication the engineman failed to do so. The NTSB's report was thorough as usual. The agency found no evidence of inadequate crew training, but it urged the FRA to reopen a study of automatic train control—which it would continue to recommend for the following three decades—and to investigate ways to improve cab car safety.[17]

The NTSB and the FRA had different missions: the exclusive focus of the former agency was safety, while the FRA needed to balance costs and benefits. Automatic train stops or speed control were expensive, and with passenger traffic unprofitable neither the carriers nor their regulators had any stomach for a further expansion. And so, despite continued NTSB prodding (see below), train control would largely remain a dead letter until the twenty-first century.

One of the worst rail disasters of the post–World War II decades occurred on the Illinois Central Gulf commuter lines running into Chicago on October 30, 1972, and it again motivated the NTSB to recommend study of automatic train control. During the morning rush hour, Train 416 overran the Twenty-seventh Street station and backed up. Train 720, which was following closely, could not stop and telescoped the rear car, killing 45 passengers and injuring 320. The collision occurred on track protected by automatic block signals, but when Train 416 backed up, it reentered the block after Train 720 had already passed its approach signal. The engineman of Train 720 claimed he was going about 40 mph—the train had no speed recorders—and saw Train 416 when it was only about four to six car lengths away. As usual the investigation revealed a host of causes that contributed to the disaster. For example, Train 720 was probably going 50–60 mph, not 40 mph, but the speedometer did not work; a walkway obstructed visibility of the signal protecting the station; and, according to company rules, Train 416 should have had flag protection while backing. The NTSB report not only recommended a study of train control, it again noted that car design was inadequate and urged the FRA and the Federal Transit Administration (FTA) to begin full-scale crash tests. The FRA ignored both suggestions; however, in 1973, it promulgated rules mandating rear visibility markers on commuter equipment.[18]

Far more consequential was the NTSB's campaign to require federal regulations governing drug and alcohol use. While drugs and liquor resulted in both freight and passenger wrecks, it was a specular wreck in 1987 involving Amtrak that finally yielded regulation. As Chapter 4 pointed out, every railroad in the country had long forbidden alcohol consumption or intoxication on the job, but enforcement was difficult. In one famous example, in 1963 the entire crew of a Southern freight was apparently drunk and partying with a prostitute as the train took a curve too rapidly and ran into the ditch. Yet an unofficial code prevented employees from informing on each other and officials sometimes looked the other way too. Not surprisingly, therefore, before the 1970s available company and ICC accident reports linked drug or alcohol use to few train accidents—another example of the ways in which social customs can influence accident statistics. The NTSB first called attention to the problem in a report on the rear collision of two SP freight trains near Indio, California, on June 25, 1973. The engineman of SP's Extra 8992 West, who died in the wreck, had a blood alcohol level of 0.16 percent. That wreck also provided an early demonstration that the alerter was no panacea, for the locomotive was so equipped and it also had an over-speed device (a type of governor), but the engineman had disabled both of them. His average speed had been 94 mph, substantially above company rules, and he entered the yard near Indio traveling about 60 mph when he hit a standing train, killing the front brakeman in addition to himself, and doing about $1.6 million in damage.[19]

The NTSB recommended stricter enforcement by the SP and that the FRA issue an equivalent of Rule G, governing use of drugs and alcohol by employees. While the FRA seems to have done nothing, that accident and a later one also on the SP proved a wake-up call to that company, and in 1980 it proposed random drug tests. The matter became embroiled in a labor dispute, however, and the National Labor Relations Board disallowed the program. Similarly, a court disallowed the BN's use of sniffer dogs to find drugs and alcohol. By about this time nearly every major carrier had some sort of program to combat drug and alcohol use. The Union Pacific had also become concerned with such problems, and as Chapter 4 noted it inaugurated Operation Red Block, which became the model for cooperative union-management anti–substance abuse programs and other cooperative safety ventures as well.[20]

In 1982–1983, several other accidents occurred in which alcohol might have played a role. One on June 23, 1982, near Gibson, California, involved a fire on an Amtrak train run by a crew from the Southern Pacific, one of whom was inebriated. A hazmat derailment on September 28 that year was even more unsettling; it resulted when a drunken engineman on the Illinois Central Gulf—authorities estimated his blood alcohol level was 0.19 percent—turned over the throttle to his girlfriend while he took a nap. Witnesses claimed he consumed "14–16 ounces of 86 proof Bourbon liquor during the 5 1/2 hours preceding the accident." Unsurprisingly, right after the wreck officials described him as "befuddled." The pair derailed 43 cars of a 100-car train containing corrosives, poison, and various flammables near Livingston, Louisiana. The pile-up caught fire, resulting in a BLEVE that caused the evacuation of 3,000 townsfolk and did $14 million in damage.[21]

These wrecks led to another call by the NTSB for federal drug and alcohol tests as did a third accident on a Seaboard freight in 1983. Sometime that year the FRA seems to have concluded that it needed to do something, but realizing the contentiousness of the issues involved, it moved slowly. The carriers' unions favored prevention programs such as that on the UP, but adamantly opposed testing except after an accident. The BLE, which invariably favored stricter safety regulations on the carriers, alternatively claimed that alcohol and drug use was not a problem but that if it was no federal or state law could solve it.[22]

In 1983, the FRA issued an Advanced Notice of Proposed Rule Making—the regulatory equivalent to sticking your toe in the water—and held hearings. Finally, in 1985 it issued a Rule G and required pre-employment screening for drug and alcohol use and testing for "reasonable

The Amtrak-Conrail collision in Chase, Maryland, on January 4, 1987, that killed 16 people briefly made drug and alcohol use by railroad crew a national issue as this image from the *Chicago Tribune* on January 18, 1987, reveals. It also led to more stringent federal regulations governing employee testing for drug and alcohol use. (© John MacNelly Ed: 1987 MacNelly, Inc. Distributed by King Features Syndicate, Inc.)

grounds" and after accidents. Labor sued, and in December that year a district court stayed the law until February 1987, when the 9th Circuit Court invalidated it.[23]

What finally tipped the scales for regulation was a dreadful passenger wreck near Chase, Maryland (Table A3.1). On January 4, 1987, about a month before the 9th Circuit Court issued its opinion, Conrail Train ENS-121 left the yard in Baltimore heading north on track 1, while at roughly the same time Amtrak Train 94 left Baltimore's Penn Station, heading north on track 2. Near Chase, Maryland, the engineman of ENS-121 ignored both the approach and home block signals and the signal in his cab, which were set against him, and entered a crossover to track 2, where Train 94 traveling 120–125 mph struck him from behind, killing its engineman and 15 passengers and injuring 174 people. Subsequent investigation revealed that the engineman of Train ENS-121 had been smoking marijuana; he had also disabled the alerter, partially disabled the cab signal, and muted the whistle that would sound in the cab should he fail to acknowledge a change in the signal.[24]

The accident provided a powerful rationale for drug and alcohol testing when the US Supreme Court upheld the FRA's regulation. Indeed, the Court's ruling was so broad as to support random testing as well. This might have been one of the FRA's most consequential regulations. In a 1979 survey of seven large carriers it found that 23 percent of operating personnel were "problem drinkers." The FRA also found that between 1972 and 1983 alcohol or drugs had been involved in at least 21 significant wrecks, or about 1.75 per year, along with a number of incidents that did not meet the dollar threshold to qualify as a train wreck. Moreover, this was surely the tip of the iceberg for neither the NTSB nor the FRA

Table 5.2. Railroad Drug/Alcohol Testing Programs, 1984–1987

Railroad	Type of Testing	Authority	Time Period	Number of Tests	Positives	Percent Positive
Southern Pacific	Mandatory/ Reasonable Cause	Railroad	1/1/84 to 12/7/87	4,806	373	7.60
Union Pacific	Mandatory/ Reasonable Cause	FRA	1/1/86 to 12/31/87	637	60	9.40
CSX	Mandatory/ Reasonable Cause	FRA	1/1/87 to 12/31/87	113	4	3.5
	Medical	Railroad	8/1/87 to 12/31/87	216	11	5.1
Florida East Coast	Mandatory/ Reasonable Cause	FRA	1/1/87 to 12/31/87	20	0	0
Conrail	Mandatory	FRA	1/1/86 to 12/31/86	69	5	7.2
	Medical	Railroad	3/15/86 to 12/31/87	13,929	581	4.2
Denver & Rio Grande	Medical	Railroad	1/1/87 to12/31/87	664	18	2.70
Central Vermont	Mandatory/ Reasonable Cause	FRA	1/1/87 to 6/1/88	1	0	0
Illinois Central	Mandatory	FRA	1/1/87 to 12/31/87	20	1	5
	Medical	Railroad	1/1/87 to 12/31/87	258	10	4.9
Burlington Northern	Mandatory/	FRA	1/1/87 to 12/31/87	3	3	100
	Reasonable Cause	Railroad	1/1/87 to 12/31/87	1,718	131	7.6
All	Excludes Medical	FRA and Railroad	All Periods	7,387	577	7.80

Source: NTSB, *Alcohol/Drug Use and Its Impact on Railroad Safety* (SS 88/04, Washington, 1988), Table 2.

investigated more than a small fraction of all train accidents. Writing in 1984, the journal *Modern Railroads* termed the FRA data on alcohol-related accidents "wildly inaccurate." An NTSB study for 1984–1987 (Table 5.2) found alcohol or drug use in 7.8 percent of all tests that were mandatory or for reasonable cause, which is to say, 7.8 percent of the 2,359 collisions and derailments that year, or about 184 wrecks. By 1993, the result had declined to 2 percent while random testing of 42,000 employees showed a 0.7 percent positive test rate. By the 1990s, drug- and alcohol-related wrecks had become scarce and it seems reasonable to conclude that the federal regulations and company programs prevented far more than one or two accidents a year.[25]

More recently, company efforts to monitor employees for drug and alcohol use (and more generally for their fidelity to operating practices) have sparked labor controversies. A number of carriers have installed video cameras in cabs—some facing outward, some facing inward—to monitor crew activity and provide evidence for accident investigation. In 2008, a horrendous collision on Metrolink motivated that company to install inward-facing cameras. On September 12 that year, a head-on collision between a Metrolink commuter and a Union Pacific freight near Chatsworth,

This image from the Chatsworth disaster of September 12, 2008, gives an idea of the tremendous forces involved in railroad collisions. Chatsworth spurred regulations governing the use of personal electronic devices by train crews and the congressional mandate for positive train control. (National Transportation Safety Board, docket DCA08MR009, document 248)

California, killed the commuter engine driver and 24 passengers and ultimately cost the company $200 million to settle passengers' claims (Table A3.1). The Metrolink engine driver had been texting and missed a signal, leading the FRA to issue an emergency (later permanent) ban on the use of cell phones and similar devices. Amtrak has also begun to install inward-facing cameras after the derailment near Philadelphia in 2015. While unions have fought these rules, the courts have upheld them and they seem likely to spread.[26]

The drug and alcohol focus also symbolized the shift in federal policy discussed in Chapter 3 to include not only "things"—track and equipment—but also human behavior. The accident at Chase reinforced this shift. Thus, it finally moved the FRA to mandate automatic train control on all locomotives in the Northeast Corridor (not just those on passenger trains).[27] It also led to investigations that revealed widespread tampering with safety devices on the part of train crew—the FRA found 108 instances during the first 6 months of 1987. The Rail Safety Improvement Act of 1988 reflected these concerns. It prohibited tampering with safety devices and mandated event recorders.[28]

Chase also led to congressionally mandated oversight of railroad training programs in the same law. As has been noted, the Class I railroads had sophisticated training programs, which they detailed in congressional testimony; moreover, the problem at Chase was not an untrained engineman but one who was under the influence and violated rules. Testifying at the hearings on the Chase accident, FRA Administrator John Riley observed, "we have not found weak training for an engineer to be at the heart of the accidents we have investigated." The carriers did not oppose mandated training but thought it irrelevant. It provides a good example of congressional micromanagement of rail safety and might have been a poor use of the FRA's resources. Moreover, while the regulations required an eye examination, that test proved not to be foolproof. On February 9, 1996, a head-on collision between two New Jersey Transit trains near Secaucus, New Jersey, that killed 1 passenger and injured 155 resulted because an engineman ran a red signal. A later investigation revealed that the initial test was probably improperly administered, for he was color blind.[29]

The next major passenger wreck was Amtrak's *Sunset Limited* on September 22, 1993, which derailed on a bridge near Mobile, Alabama, that a tugboat had hit. Forty-two passengers and 5 crewmembers drowned and 103 were injured. As usual, the NTSB report was thorough and thoughtful. It raised questions about emergency preparedness at both Amtrak and the Federal Emergency Management Agency (FEMA). It also proposed that the Department of Transportation (DOT) undertake an assessment of the risks from marine collision to every railroad and highway bridge in the country. In congressional hearings labor pushed for federal bridge standards, but the FRA apparently surveyed bridges and decided to do nothing at the time.[30]

The bridge episode provides an example of the drawbacks of congressional micromanaging and regulating by disaster. After a previous bridge accident in 1979, the NTSB had proposed that the FRA investigate a regulation that would connect bridges to track circuits so that misalignment would set block signals to danger. In 1981, the agency reported that such a rule would cover 85,000 bridges at a cost of $10,000 each for a capital cost of $850 million with annual maintenance costs of $85 million. Assuming a 10 percent return on investment and for simplicity that the devices last forever, annual capital and maintenance costs would be $170 million (= $85 million + 0.10 * $850 million). Chapter 1 noted that regulations can generate offsetting behavior. If they cost too much, they can also worsen safety by diverting resources from more valuable uses. Economists have estimated that a rise in U.S incomes of $15 million will yield one less statistical death. Thus, $850 million invested elsewhere at 10 percent might avoid (170/15) = 11 fatalities a year. There were no fatalities from bridge accidents between 1980 and 1992. For readers who find such abstract arguments unconvincing, consider the following. Nearly everything railroads do involves safety and so $170 million a year sunk into bridges implies less frequent track inspections, fewer hotbox detectors, slower acquisition of electronically controlled pneumatic (ECP) brakes, and so on.[31]

As Chapter 3 discussed, the FRA also tried to develop a more proactive approach to safety with its System Safety Program that looked at all

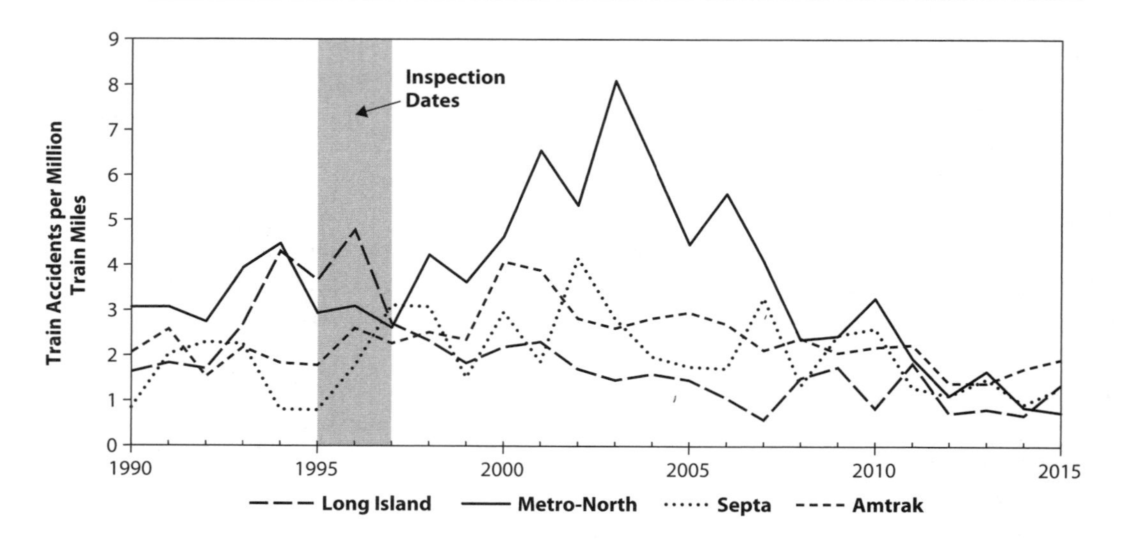

Figure 5.2. Passenger Train Accidents Before and After FRA System Safety Inspections, 1990–2015. (GAO, *Federal Railroad Administration's New Approach to Railroad Safety*, GAO-RECD-97-142 [Washington, 1997], Appendix II)

aspects of safety on a railroad by railroad basis. In 1984 and 1985, it undertook such assessments of Amtrak and the SEPTA. While there is inadequate evidence to assess its impact on the commuter line, Amtrak's fatality rate seems to have been unaffected by the program—perhaps because the FRA largely gave it a clean bill of health. As noted, in the 1990s under Jolene Molitoris the FRA shifted these assessments to a more cooperative stance and it also included representatives of labor. These inspections evaluated passenger as well as freight roads, apparently with about the same effect. Figure 5.2 charts train accidents for four major passenger lines before and after their System Safety Reviews of 1995–1996. As can be seen, only on the Long Island is there any discernable change in the accident trajectory. In 2005, in response to criticism from the DOT's inspector general, the FRA developed yet another risk-based National Inspection Plan; its impact to date is not encouraging (see below).[32]

More recently, a series of dramatic accidents have moved the NTSB and the FRA to undertake major company safety assessments of the Washington Metro Area Transit Authority (DC Transit) and New York's Metro North. In the former case the precipitating accident was a rear collision just outside the Fort Totten Station on June 22, 2009. Train 112, traveling about 36 mph, struck standing Train 211, telescoping the first car of the striking train, killing 9 and injuring 52 people. Nineteenth-century railroads operating under the old train order system had littered the landscape with rear collisions, but this one occurred on a line with automatic train control. The investigation revealed that electrical interference had prevented the system from recognizing the presence of Train 211. Worse, a similar failure had occurred in 2005, nearly causing an accident. Engineers had developed a test to determine track circuit integrity—and found that the problem had existed since 1988—but DC Transit had not routinely employed the test. The investigation was typically thorough: it delved into previous collisions and derailments, discovered that the state regulatory apparatus was largely toothless, and pointed out that in 2006 it had urged retirement of the cars involved in the accident because they were insufficiently crashworthy. The report centered on organizational ineffectiveness, concluding that DC Transit's safety culture was weak and pointing to shortcomings in internal communications, hazard recognition, risk assessment, and implementation of corrective action. A year after this dressing down, the company brought in a new general manager. In 2013, *Railway Track & Structures* reported that DC Transit was finally on the right track, but more recent problems—including several electrical fires—suggest that its celebration was premature.[33]

Four years later, a series of train accidents and workplace fatalities on New York's Metro North resulted in even more scathing safety assessments from the NTSB and the FRA. Between May 2013 and March 2014, the company experienced 4 train accidents in addition to 2 worker fatalities that did not involve train accidents, resulting in 6 killed and 126 injured. One of the six accidents—the derailment and subsequent collision with another train that occurred near Bridgeport, Connecticut, on May 17, 2013—reveals the company's problems. The cause was two broken joint bars and the break was sufficiently large to be visible to the naked eye. The joint had been broken seven weeks earlier, but the repair had been made incorrectly; the breaks resulted from inadequate ballast that was also easily visible. A track inspection two days before the wreck found and reported the bad ballast, but it had resulted in neither repairs nor slow orders. The break resulted in a derailment that another commuter train then hit, injuring 65. That no one died probably reflected the value of the company's new energy-absorbing Kawasaki M8 cars (see below). The other accidents reflected similarly slapdash procedures. The NTSB detailed a host of organizational failings, concluding that Metro North's safety management plans had degenerated "into nothing more than complex systems of paper"; it also criticized the FRA, suggesting that its National Inspection Plan "may not be effective." For its part, the FRA observed that the company demonstrated an "overemphasis of on-time performance . . . an ineffective safety department and poor safety culture." As with DC Transit, the assessment brought changes that included a close-call reporting system, better inspections, and train-monitoring systems.[34]

Positive Train Control

As Chapter 1 briefly noted, the idea of automatically stopping or slowing a train in the face of danger had been a gleam in the eye of safety advocates well before World War I. By the 1920s, inventors had developed a workable system of train stops controlled by track circuits. But since train stops promised no economic benefits save better safety, which the carriers thought was available more cheaply by other means, they were uninterested. In 1922, the ICC had mandated automatic train control on at least one passenger division and by then speed control was also becoming available. The benefits from the ICC order are illusive; probably the railroads were correct in claiming that there were better safety investments for the money. A second order came in 1947, requiring train control where passenger train speeds exceeded 79 mph. This extended the system by about 2,400 miles of road and probably reduced speeds elsewhere. By 1965, after the two ICC orders, train stops and speed control covered only about 10,000 road miles. But to the newly formed NTSB, they were the holy grail of collision prevention.

As has been noted, the NTSB had urged development of some form of automatic train control after the Penn Central wreck at Darian in 1969, and it repeated the request in the Illinois Central Gulf commuter collision of 1972. Finally, after the disastrous wreck near Chase, Maryland, in 1987 that killed 16, the FRA mandated train control on locomotives in the Northeast Corridor—about 20 years after the NTSB had first proposed it in the Darien accident. In 1990, the NTSB advanced train control to its list of "most wanted" safety programs.

By this time, however, a new, communications-based, approach, termed "positive train control" (PTC) was superseding the old technology whereby track circuits controlled trains. This was a marriage of the increasingly sophisticated remote sensors discussed in Chapter 3 with radio controls, computers, and GPS or some other location-finding technology. It origins lie in what were called "advanced train control" systems that emerged during the 1980s. These systems came with varying levels of sophistication. The simplest version, which the CP installed as early as 1985, was a computer-aided manual block signal system. The package included two-way voice communication between the train and the dispatcher and a computer that would check for any route conflicts and display the dispatcher's instructions. Fifteen railroads, including the CP, the UP, the CN, and the BN, were also experimenting with more advanced models that could monitor location, speed, and fuel consumption as well as a host of other safety and efficiency-related train characteristics. The primary goal of these systems was not safety but better productivity, for a precise knowledge of location and train characteristics (e.g., stopping distance), along with appropriate computer software might schedule much more precise meets, thereby eliminating much of what *Railway Age* termed the "hurry up and wait" of railroading and expanding track capacity.[35]

By the 1990s, the new term for advanced train control was "communication-based train control," of which PTC was a subset. The simplest systems (called "positive train separation") would simply overlay computer controls on existing signals to stop or slow trains under certain conditions. Thus, in territory governed by block signals the track circuits would still govern signals but these via some form of wireless technology now linked them to the locomotive, which would not be allowed to pass stop signals or violate speed control. In "dark" (non-signaled) territory, wireless technology would link switches to the locomotive, thus avoiding collisions or derailments from improper switch alignment. More advanced systems were being developed that would replace existing signals and track circuits. These combined GPS location technology with computer controls to essentially surround each train in a "bubble." Each train's bubble would comprise its safe stopping distance, while computerized controls would prevent any other train from intruding in it. With track characteristics programmed in the computer's memory, it could also enforce speed limits and communicate with track crews. At grade crossings such systems could calculate train speed and trip the signal to allow a constant warning time. The bubble systems would also increase track capacity because track circuits were designed for the longest stopping distance of any train using the track, thereby wasting space for all the others. Yet these more advanced systems also had two potential flaws: because they replaced track circuits, they could not catch broken rails and they might not be fail safe.

Amtrak had inherited an older track circuit–based form of automatic train control when it took over the track between Washington and New Haven from the Penn Central. This system included both speed control and cab signals. About 1991, in response to the disaster at Chase, Maryland (see above), and with prodding from the FRA and the NTSB, Amtrak began installing an Advanced Civil Speed Enforcement System (ACSES). This was an upgrade designed to deal with the newer, high-speed trains; it employed transponders to enforce conformity to the block system and to require compliance with speed restrictions. It was operational on May 12, 2015, when an Amtrak train took a curve near Philadelphia at 106 mph, although the posted speed was 50 mph. While the company could have programmed speed control into its system, apparently no one had imagined that it was necessary on that curve. The wreck killed 8 passengers and injured 200; as of this writing, the NTSB has concluded that the engine driver lost situational awareness, which is bureaucrat-speak for saying he took the curve too fast.[36]

In 1992, Congress directed the FRA to report on PTC, which it did two years later, but by that time events had given the topic a nudge. One event was a low-speed head-on collision between two trains on the Northern Indiana Commuter line near Gary, Indiana, on January 18, 1993 (Table A3.1). On that date Train 7, a two-car commuter heading east from Chicago to South Bend, ran a signal, coming to a stop and blocking a crossover where westbound Train 12 struck it. The wreck killed 7 passengers

The high-speed Amtrak derailment near Philadelphia on May 12, 2015, that killed 15 people could have been prevented by positive train control (PTC). It also reveals how casualties can be contingent on the behavior of cars during the derailment. The first car struck several catenary supports and was completely destroyed, while the locomotive and the latter two cars were largely unharmed. (National Transportation Safety Board)

and injured 95 and moved the NTSB to call yet again for train control and for sturdier cars (see below). A freight wreck the same year reinforced the message. In early 1994, the FRA began discussions with the AAR and individual carriers to begin pilot studies of PTC systems for which it provided financial assistance with the AAR chipping in an additional $20 million. In 1997, it turned the topic over to a working group of its Railway Safety Advisory Committee. The pilot projects included one in Illinois and another in Michigan, an Amtrak-NEC venture, a joint Norfolk Southern-CSX test, and another joint venture of the UP and the BNSF in the Seattle-Portland corridor. The Illinois project was conducted on UP track that involved Amtrak, the FRA, the AAR, and the Illinois Department of Transportation and was the most sophisticated of the group, as it employed a bubble to control trains.[37]

By about 2002–2004, these experiments had generated four firm conclusions. First, several forms of communications-based train control were technically feasible. Accordingly, in addition to Amtrak's ACSES, a number of stand-alone commuter rail systems began installation, including the New York City Transit, the New Jersey Transit, the SEPTA, and the BART. The latter system divided the right of way into control zones and linked trains by radio; it increased capacity in the trans-bay tube from 16 to 30 trains an hour. About this time Alaska railroad, with a $12 million cash assist from the FRA, also began to implement an overlay system. A second conclusion is that the freight railroads were mostly interested in the simplest and cheapest systems. CSX and the BNSF both began to implement versions of PTC that overlaid radio and computer controls on existing signal and traffic control systems. Third, the requirements for interoperability and other technical complexities of the advanced systems were nightmarish. For example, PTC and braking systems needed to be integrated, yet braking distances might be highly variable. Thus, a system designed to stop trains under the worst circumstances might reduce track capacity. In addition, the freight railroads were leery of the dangers from derailments or break-in-twos that might result from automating braking of mile-long freights. Fourth, the high costs relative to benefits prohibited widespread adoption of the most complex systems that did away with wayside signals altogether. Indeed, a 1999 study by the Railroad Safety Advisory Committee found PTC flunking a cost-benefit test, while later work found that a bare bones PTC system would cost about $1.2 billion with benefits of $500 million. Fully implemented, the system would cost $7.8 billion and prevent 40–60 accidents, 7 fatalities, and 55 injuries a year.[38]

Research and modest efforts to implement various train control systems continued until the 2008 Metrolink disaster (see above) moved Congress to act. The wreck was the perfect case for train control, for—the reader will recall—the Metrolink engine driver had missed a signal because he had been texting. The Graniteville, South Carolina, hazmat disaster of 2005, in which an incorrectly lined switch had led to a wreck and a chlorine leak that killed nine people (Chapter 3) provided a further spur to action. Sometimes it is argued that when a means to prevent serious accidents exists, there is a moral imperative to use it. That logic had impelled Congress to mandate air brakes and semi-automatic couplers for freight trains in 1890 and it finally carried the day for PTC. In 2007, a representative of the NTSB told Congress: "it will be a device that will . . . save lives." That, it seems, was good enough, for the board finally got its wish. In 2008, Congress passed the Railroad Safety Improvement Act, requiring the carriers to implement PTC on about 60,000 miles of track—main lines that carry passengers or certain hazardous inhalation substances (e.g., chlorine, anhydrous ammonia)—by 2015.[39]

Congressional action speeded up company efforts; by about 2010, several suppliers had developed systems that were modular—companies could choose to implement them in pieces—and interoperable. To assist the process, in early 2010 the FRA published rules governing the requirements for safety and interoperability. Yet congressional action also focused companies on meeting a short-term deadline. As a result, all the approaches involved an overlay on existing signal systems—none were of the bubble type discussed above. Even these "simple" systems have generated nightmarish and unanticipated problems involving the inability to obtain the necessary radio spectrum as well as the need to satisfy the Historic Preservation Act when siting communication towers. "Oh what a mess," *Trains* concluded in 2014. Because of the problems of merging braking distances and train control noted above, *Railway Age* thought that these systems would reduce track capacity. Thus, as that journal editorialized, “PTC is about safety, period,” and its supposed business benefits were moonshine.[40]

The difficulty with mandating a device simply because “it will save lives” is that there are a number of means to that end. Thus, the question arises: is PTC a wise use of resources? It will likely prevent many of the 20 to 30 main-line collisions a year and some derailments from excessive speed, but an overlay system cannot prevent wrecks due to signal failures. It will avoid a few grade crossing wrecks where a vehicle is stranded on the tracks—although crossing systems are available with the same capabilities (Chapter 6). The FRA recently estimated that initial installation cost would be about $5.4 billion and the present value of total costs over 20 years about $9.4 billion. An AAR study estimated these to be about 11 times the present value of benefits over 20 years. This conclusion seems plausible: from 2001 to 2015, all railroad collisions killed an average of 4.3 people a year and injured 89. Some of these occurred in yards and PTC would not have prevented them; moreover, these numbers are likely to decline due to other safety improvements.[41]

Previously it was pointed out that if regulations cost too much they can worsen safety by diverting resources from more valuable uses. If, as economists claim, a rise in US incomes of $15 million will yield 1 less statistical death, investment of train control’s $9.4 billion elsewhere at a yield of 10 percent might have a payoff of ($940 million/$15 million = 63) fewer fatalities a year. Moreover, in a litigious society like modern America’s it is hard to imagine that the carriers lack incentives to avoid injuries, fatalities, and property damage. To repeat an earlier argument, nearly everything railroads do involves safety. As the editor of *Railway Track & Structures* pointed out, the $9.4 billion present value of costs amounted to about a year of maintenance—“no new crossties, no bridge rehab, no track renewal . . . just PTC.” If PTC were a good safety investment for the freight railroads, they would likely be installing it voluntarily.[42]

A more modest alternative might have been to require PTC on high-density passenger lines only. The Northeast Corridor is already largely covered as are some commuter lines. Relying on FRA figures, I estimate that additional covering for commuter lines only would have an initial cost of about $800 million while maintenance would amount to about 15 percent of the total, or $120 million a year. Again, assuming a 10 percent return on investment and that the investment lasts forever, annual costs are ($120 million + 0.10 * $800 million) = $200 million a year. These lines typically account for 40 percent of the fatalities and 36 percent of the injuries from train accidents. Thus, a focus on commuter railroads would probably achieve 36 to 40 percent of the safety benefits of the existing rule at 15 percent of the initial capital costs. Even this might not be a good safety investment as annual costs would be about $200 million a year to avoid perhaps 2 fatalities ($0.4 * 5.3 = 2$) and 41 injuries ($0.36 * 113 = 41$) a year.[43]

Surviving a Wreck—or Not

In the bad old days of the nineteenth century, passengers bought travel insurance and discussed the best spot in the train to sit in case of a wreck. These were the years when wooden cars telescoped during a collision and then perhaps caught fire from the car stove. "The safest [car] is probably *the last car but one* in a train of more than two cars [italics in original]," the *American Railroad Journal* advised its readers in 1853. A century later, those days were long gone. Or were they? In 1974, a reporter for the *Washington Post*, planning a train trip and alarmed at the spectacular rise in accidents, cautioned his wife—perhaps in jest, perhaps not—"we'll pick a spot toward the rear of the train (not the last car) with protected windows."[44]

The survivability of a wreck depends first on the structural integrity of the car. The safety of the interior design also matters—whether there are sharp edges, the flammability of materials, the stability of seats and luggage. Finally, the availability of emergency exits and the training of the crew might also be crucial, as well as their ability to communicate with each other and with passengers during emergencies. Experience is a harsh taskmaster, but as companies learned from accidents, by World War II American passenger and commuter cars had become much safer.

The deadly car stove had departed in the 1890s, while by about World War I the carriers were buying steel cars (Chapter 1). By the 1920s, Pullmans had the truck chained to the car frame, which impeded telescoping. A decade later, cars were routinely built with windows of safety glass, and about that time the substitution of high-strength steel and aluminum alloys in place of simple carbon steel increased a car's ratio of strength to weight. In 1939, the AAR had established passenger car standards that remained in force during the postwar years. These required center sills to withstand an end load of 800,000 pounds and collision posts to have shear strength of 600,000 pounds where they joined the sill. These were minimum standards, but competition ensured that most equipment was in fact much stronger than this: in 1946, a representative of Budd proudly claimed that the company designed its cars to resist two million pounds compression at the center sill. The Central of Georgia cars could withstand 1.8 million pounds at the sill and the collision posts were twice as strong as required. By 1950, the *Railway Age* concluded: "experiences with modern cars in accidents suggest an amazing reduction in the number and severity of personal injuries when the integrity of the structure is maintained." The caveat about maintaining the integrity of the structure was well taken, for in the postwar years, as noted above, such changes had little impact on a passenger's odds of walking away from a derailment, while in collisions improvements were apparently largely offset by higher speeds (Table 5.1).[45]

One of the NTSB's first reports was an horrendous crossing collision that occurred between a Boston & Maine diesel rail car (a Buddliner of the sort the author remembers well—if not fondly) and a tractor trailer loaded with 8,300 gallons of fuel oil at 12:10 a.m. on December 28, 1966, in Everett, Massachusetts. The truck had lost its air brakes, and a "safety" feature had then locked the brakes just as the rig sat on the crossing. Traveling about 55 mph, the rail car was unable to stop even with an emergency application of the brakes and it struck the truck, rupturing the fuel tanks and spraying fuel over the front of the rail car, which caught fire. The calamity killed 11 passengers and 2 crew members. Autopsies revealed that only 1 had died from the collision; burns and smoke inhalation killed the other 12 because the car was a trap with no emergency exits. Moreover, the doors opened inward and were jammed closed by panicked passengers. Adding to the terror, the lights also went out and there was no emergency lighting. Rescue equipment that might have allowed passengers to break the windows was in an inconspicuous spot and apparently overlooked. The NTSB recommended the FRA require emergency lighting and exits, but it seems not to have responded.[46]

Thereafter, the NTSB regularly discussed passenger car construction in its reports. Two 1969 wrecks also raised questions about car design. The destruction of the two lead cars in the Darien wreck noted above was one where the board observed that the car had offered little protection to either the engine driver or the passengers. While the sill and collision posts met AAR specifications, the closing speed at the time of collision was 60 mph, which overwhelmed the safety features and telescoped one car. Another was the derailment due to buckled track near Glenn Dale, Maryland, on June 18, 1969, discussed above. In that accident, seats swiveled and luggage and passengers were strewn about. In both cases the board urged the FRA to investigate the safety of passenger car design. In another crossing accident, this one in which a truck plowed into an Amtrak train in Collinsville, Oklahoma, in April 1971, passengers were thrown through the windows while others were injured by sharp objects and flying baggage. The NTSB urged the FRA to study window design. Six of the 11 passengers killed in a 1971 derailment near Salem, Illinois, had been thrown through the windows of the coach and were crushed by overturned cars. In the 1972 collision of Illinois Central Gulf commuter trains, the NTSB pointed out poorly welded and designed collision posts of one car, while the trucks of the train's rear car had been unsecured, allowing more telescoping than would have otherwise occurred. In the rear collision between a Conrail commuter and an Amtrak passenger train near Seabrook, Maryland, on June 9, 1978, the board found that the

Amtrak seats had defective locking mechanisms while the commuter train had many unpadded metal surfaces that yielded injuries. The emergency door operating instruction and mechanism were also locked up. In a 1982 fire on an Amtrak train near Gibson, California, the board discovered that there was no coherent evacuation plan; passengers were "left to their own devices," and the emergency exits were hard to find and hard to use.[47]

The NTSB was speaking to deaf ears at the FRA, which did little. The result was a curious hole in railroad safety rules and regulations. There were no federal regulations governing construction of passenger equipment except for MU cars, which had been covered in 1954 (Chapter 4), while the AAR never updated its standards of 1939 because after 1971 its members no longer hauled passengers except as contractors. The FRA had funded some studies of passenger equipment crashworthiness in the late 1970s, and its annual reports to Congress described ambitious efforts to make passenger cars safer, but nothing came of them. In June 1982, the agency held hearings on the safety of passenger equipment—at which Amtrak testified in opposition to regulation—and in early 1984 it submitted a report entitled "Railroad Passenger Equipment Safety" that sang the praises of the passenger lines. The agency claimed that since the DOT was Amtrak's banker "we can accomplish these safety objectives more directly and efficiently by direct involvement in the company's policy-making process than by instituting single carrier regulatory proceedings that could take years to complete."[48]

Such a position ignored both the dangers of commuter roads and the need for considerable engineering research before any regulations would be feasible. By the 1980s, the NTSB had given up on the FRA and dealt directly with Amtrak, and while the company opposed formal regulation, it responded to the board's prodding, improving emergency training, and after the 1987 accident at Chase, Maryland, it acted on the recommendations to improve seat locks and luggage restraints and other aspects of interior design safety. Nearly 30 years earlier, the automobile industry had begun to design cars with safer interiors that included collapsible steering wheels, seatbelts, better door locks, and padded dashes. Not all of these are appropriate for rail cars, but they were widely reported and what would become the DOT's National Highway Traffic Safety Administration was an advocate.[49] Thus, the failure for nearly three decades of managers and regulators to transfer the technologies between industries and indeed "across the hall" seems remarkable indeed.[50]

On January 18, 1993, the collision between two commuter trains near Gary, Indiana, that killed seven passengers finally seems to have moved the FRA to act. The two trains were MU commuter cars; for regulatory purposes, these were locomotives and included collision posts. In addition, the contractor had installed corner posts with shear strength of 150,000 pounds. Yet the wreck was an off-center collision and the posts were far too weak; the accident tore through a side post and pealed as much as 27 feet of the left side of both lead cars like a banana. The NTSB concluded that "the use of collision energy absorption structures in the corner post assemblies . . . may have prevented . . . serious injuries," and it called on the FRA to develop collision energy–absorbing structures for such cars. Later the same year, the bridge accident near Mobile, Alabama, that drowned 42 passengers (see above) also revealed the need for much better emergency evacuation procedures.[51]

Finally, about a year later, in September 1994, the FRA formed a working group consisting of interested parties and announced that it would begin rulemaking for passenger car safety standards. An additional push came in 1993: the new Clinton administration had proposed an ambitious plan to spend $1.3 billion on high-speed rail, and higher speeds demanded safer equipment. Congress reinforced agency backbone with the legal requirement to develop standards that were included in the Federal Railroad Safety Authorization Act of 1994.[52]

Two major accidents in 1996 kept the heat on. One was the Secaucus wreck noted above; while the cars in that accident both had corner and collision posts, it was yet another off-center collision and they also failed; moreover, crew training for such emergencies seems to have been deficient. A second wreck was the head-on collision between Amtrak Train 29 and MARC commuter Train 286, near Silver Spring, Maryland, on February 16, 1996, that killed the 3 trainmen on the MARC as well as 8 passengers and injured a total of 26 passengers and crew on both trains. As usual, the wreck was rich with lessons. This too was an off-center collision that wiped out a corner post. The wreck also demonstrated that the emergency window exits on the MARC train were difficult and time-consuming to open; the "quick release mechanisms for the exterior doors . . . [were] located in a secured [locked] cabinet some distance from the doors," while fire had destroyed the emergency door mechanism and some interior materials failed later smoke or flammability tests. This prompted the FRA to issue an emergency order requiring, among other things, testing of windows to ensure their operability.[53]

In 1998 and 1999—more than 30 years after the NTSB first began to call attention to such matters—the FRA issued two regulations governing passenger car emergency preparedness and equipment safety. These required passenger lines to have an emergency plan, to train certain personnel, to have emergency equipment on board (fire extinguishers, pry bars), and to mark emergency windows. There were also requirements governing flammability, the strength of seats and luggage racks, and the inspection and testing requirements for brakes and wheels. Cars needed to have manual overrides for exterior doors; new equipment was to have two additional side doors. At about the same time, the American Public Transit Association (APTA) inaugurated a Passenger Rail Equipment

Safety Standards (PRESS) task force that would develop a set of recommended practices for lines unregulated by the FRA such as rapid transit systems. It adopted most of the FRA requirements and in some cases strengthened them.[54]

Managing Crash Energy

Neither better emergency exits nor safer interiors will make much difference if a wreck demolishes the passenger car. As noted, collisions on nineteenth-century railroads often featured telescoping—where one wooden car rammed through another, largely absorbing the collision energy. "A gentler way of stopping a train could not well be [imagined]," observed the *Railroad Gazette*'s distinguished editor, Arthur Mellen Wellington, in 1886. Sometimes passengers elsewhere in the train did not even know there had been a wreck—but when a car telescoped, the results were catastrophic for its occupants.[55]

There are two alternative ways to try to prevent such crushing. The older technique—enshrined in the AAR's 1939 regulations—was to make each car with a strong, rigid frame. Modern FRA regulations for new equipment intended to travel at 125 mph or less (Tier I equipment) follow these 1939 AAR rules, although the requirements for new cab cars are stronger. The difficulty with this strategy, as George Bibel points out in *Train Wreck: The Forensics of Rail Disasters*, is that train accidents routinely involve millions of foot-pounds of energy. Moreover, since kinetic energy rises with the square of velocity, cars strong enough to withstand the forces generated in even a medium-speed wreck would be uneconomic to haul. The alternative is to design cars that will absorb collision energy like a giant shock absorber—what is now called crash energy management (CEM)—so that the force of the collision is dissipated by each car in the train. As Chapter 1 detailed, in 1913 both Pullman and Barney & Smith car builders had developed cars with ends designed to absorb energy. Nevertheless, while *Railway Mechanical Engineer* was advocating such designs as late as the 1930s, they seem to have disappeared in the postwar years. Probably the AAR standards of 1939 along with increasingly gloomy prospects for passenger service foreclosed further experimentation. Whatever the cause, technology proceeded down the wrong track.[56]

The rediscovery of CEM did not originate with the railroads; it grew out of the work of Hugh De Haven in the 1940s and 1950s and its first practical application seems to have been in aircraft design, which is probably how researchers on railroad safety rediscovered the idea. However, while the FRA was studying crashworthiness as early as the mid-1970s, nothing came of the early work. Here again American companies and regulators were late to the party. European railroads seem to have focused on CEM at an earlier date, and automobile makers had been researching energy-absorbing designs since the 1960s. By 1996, *Automotive Engineering* was

The energy-absorbing features of new passenger cars should result in fewer injuries and fatalities in future accidents. (Courtesy David Tyrell, Volpe National Transportation Systems Center)

emphasizing that the Oldsmobile Cutlass came with "carefully engineered crumple zones."[57]

Modern American research dates from the 1990s, when FRA-funded engineers at the John A. Volpe National Transportation Systems Center began to design crush-resistant railroad cars using accident histories, computer modeling, and full-scale crash tests conducted at the Transportation Test Center (TTC). The outcome has been cars with roughly 3-foot crush zones at each end that will absorb about 2.5 million foot-pounds of energy each. These include an anti-climbing device and a pushback coupler that will keep cars in line. The coupler, attached with shear bolts, will begin to absorb some of the energy from the wreck. When the coupler reaches the end of its compression, there is a sliding sill that further absorbs energy. Finally, there are primary energy absorbers—metal tubes that crumple on impact like an accordion. At low to medium speeds these structures should prevent crushing of the passenger space. In addition, because they result in slower deceleration of the train—a gentler wreck—there will be a lower "secondary impact velocity"—that is, injuries from passengers being thrown about are less likely.

Crush-resistant cars can be intermingled with older equipment to improve the safety of the entire train, but if the train is composed of MU cars or has a locomotive pushing, the CEM car must lead, for if a conventional car leads it will absorb all the energy. Companies can also retrofit older equipment with the newer safety devices. Because it is the train rather than one or two cars that absorbs energy with CEM, collision severity is largely independent of train length (given weight). With conventional equipment, weight has mixed effects; it adds to kinetic energy but it might also make the cars stronger. With CEM equipment, weight simply increases crash

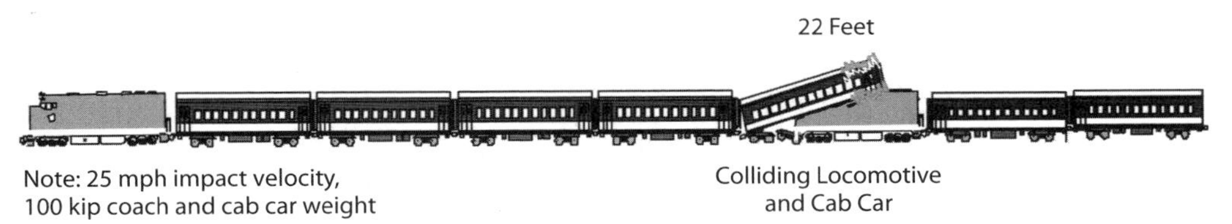

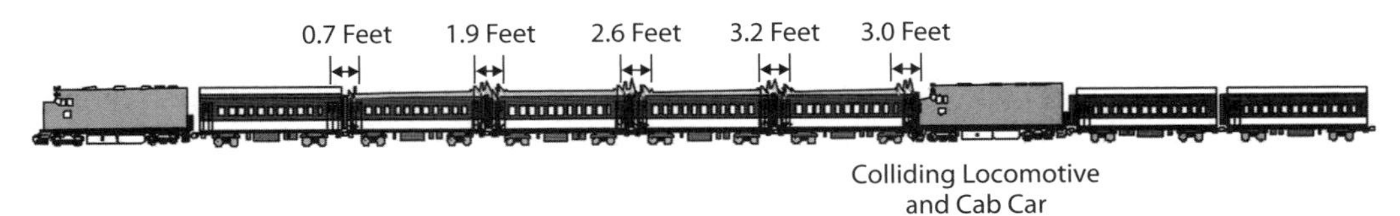

Figure 5.3. Crush Zones versus Conventional Passenger Cars in Collisions. (Elroy Martinez, David Tyrell, and Benjamin Perlman, "Development of Crash Energy Management Designs for Existing Passenger Rail Vehicles," ASME Rail Transportation Division *Proceedings* [November 13–19, 2004], Figure 15)

severity.[58] To see how CEM works consider Figure 5.3. In this figure, in a head-on collision between conventionally designed trains (top), the cab car absorbs virtually all the energy, telescoping and undoubtedly causing many casualties. In the bottom of the image, the crush zone on the cab car, which is weaker than the rest of the car, absorbs energy, as do the subsequent crush zones, leaving the remainder of the train intact.[59]

The FRA rules of 1999 reflect this new work. They allow an alternative approach for new Tier I equipment if it meets certain performance requirements.[60] APTA regulations issued about the same time essentially applied these standards to other carriers not regulated by the FRA. The FRA's Tier II requirements for trains that will travel over 125 mph (essentially those for Acela) are that each end of the train be capable of absorbing 13 mega joules (about 9.6 million foot-pounds) of energy without compromising passenger space. For this to do much good, the train must be either quite light or substantially slowed before any collision. In addition, a 30 mph collision with a standing train should not throw a medium-sized man forward in any car at a speed exceeding 25 mph and deceleration should not exceed 8g.[61]

One of the first purchases of cars incorporating CEM was by Metrolink. The spur had been a disastrous collision with an SUV near Glendale, California, on January 26, 2005, that killed 11 and cost the line $39 million to settle passengers' claims (Table A3.1). The company had been planning a purchase of new cars anyway, and in concert with the FRA and the APTA it specified MU train cars that would survive a head-on crash with a locomotive at 25 mph and yield secondary impact velocities for passengers of no more than 15 mph.[62]

The question arises: have these new standards for emergencies, interior design, and crashworthiness done any good? The answer appears to be: "probably." Table 5.1 suggests that fatalities per accident have certainly not declined in collisions during the past two decades, but it is likely there has not been time enough for the new equipment to make much of a difference; moreover, as in the past, their impact will be partially offset by rising train speeds. A head-on collision between a Metrolink commuter and a Union Pacific freight on September 12, 2008 (described above), illustrates both of these problems. The collision killed the commuter engine driver and 24 passengers because the locomotive telescoped about two-thirds of the first passenger car. The car dated from 1992 and had no CEM features. However, whether they would have made much difference is unclear, for the impact occurred at a combined speed of over 80 mph. A more hopeful example comes from the 2013 Metro North derailment/collision noted above that involved Kawasaki M8 cars with CEM features. Another example comes from a grade crossing wreck near Oxnard, California, on February 14, 2015. There, a Metrolink commuter composed of four cars, three of which had crush zones, collided with a Ford F-450 truck towing a trailer. The speed at impact was 56 mph; there were 32 injuries and the engine driver was killed. As of this writing, the NTSB has not released a report on the accident but given the speed of the collision, it seems plausible that but for the crush zones there might have been a higher death toll.[63]

Little Accidents

Train accidents have dominated concerns with passenger safety, yet what the FRA calls train incidents, and I term "little accidents," have been responsible for nearly as many deaths and more injuries. From 1958 through 2015, 377 passengers died in train accidents but 261, or about 69 percent as many, died in other ways.[64] Over the same period, injuries from little accidents are about 2.5 times those from train accidents. Yet the course of injury rates from such causes is confusing and suspect. They declined sharply down to about 1990 for reasons explained below; thereafter, they have risen sharply—especially since about 2005. Between 2006 and 2007, injuries to passengers from causes other than train accidents jumped about 60 percent in one year—from 779 to 1,253—and they have remained elevated ever since. Moreover, bruises, cuts, and sprains apparently comprised most of the increase. Thus, it seems likely that the sharp rise in injuries during these years reflects either an epidemic of slip-and-fall artists or a change in reporting criteria.[65]

The course of fatality rates from little accidents is clearer; they have declined continuously down to 2015 even as fatality rates from train accidents have stagnated for three decades. This decline is all the more impressive when one considers that the relative share of commuter passenger miles has been increasing. The average intercity rail journey is about nine times as long as the typical commute and so there are nine times the entraining and detraining per passenger mile associated with commuter traffic and thus nine times the risk of falling on your head.[66]

The central questions surrounding little accidents are why fatalities and injuries have been so common and why fatalities have been declining. The answers to these questions have a common source in how passengers enter and exit trains. A comparison with airplanes is instructive. At most airports, passengers enter a plane through a jet way that connects tightly to the plane, take a seat, and buckle up before the plane moves. The plane is then sealed for flight until it arrives at the jet way again and the passengers mostly remain seated. Contrast this with train travel—either on Amtrak or a commuter or rapid transit line. Passengers enter the train, sometimes by climbing up open steps, sometimes from a platform across a more or less considerable gap. The train starts whether they are sitting or not; they freely walk between cars and when the train slows for a station they get up and crowd to the door while the train is decelerating and then push out—again down steps or across the gap onto a platform.[67]

Thus, there are few chances to fall when getting on or off or riding in a plane and many opportunities to do so when riding a train. At most airports there is no chance one can get too close to a moving airplane on the runway, while at many train stations there are lots of chances to get too close to a moving train. Many examples breathe life into these generalities. Consider an Amtrak incident involving a 74-year-old woman on February 15, 2003, in Dade Country, Florida. She slipped while getting off the train, hit her head, and later died. On March 7 that year another passenger—this time a man age 62—died because he jumped from the end of a moving Amtrak train in Salt Lake, Utah. On February 23, 2005, a 54-year-old man tried the same stunt, leaping from a moving New Jersey Transit train—with the same result. On January 5, 2005, a 79-year-old man slipped while going down steps from an Amtrak train in Union Station, Washington, DC, landed on his head, and expired as a result. As these examples suggest, age sometimes plays a role in these tragedies—but not always. On July 4, 2004, a SEPTA train killed a 43-year-old man because he apparently stepped in front of it. On May 18, 2010, a 42-year-old man slipped on the platform and hit the side of a moving New Jersey Transit train, which killed him. Sometimes the tragedy reflects defective equipment. On the Long Island, on December 1, 1974, the car door caught a man and then dragged him off the platform, where he hit the third rail and was electrocuted. Investigation by the NTSB revealed that the company had experienced 22 such incidents involving train movement from 1969 to 1974. Doors are not supposed to do that and the NTSB recommended changes in company procedures along with better reporting and potential FRA regulation. At least some danger remains, however: on November 21, 2006, a New Jersey Transit train caught the clothing of a 49-year-old

Commuter train doors could be booby traps. The two images on the left show the door far enough open so the train cannot start; in the image on the far right, however, the passenger is still trapped but the train can move. Such an accident resulted in a man being dragged to his death on the Long Island on December 1, 1974. (National Transportation Safety Board)

man in the door as he was detraining and dragged him to his death. The FRA promptly issued a warning.[68]

Dozens of similar accidents involve getting on or off trains or being around them when they are moving. Yet if such tragedies remain all too common, they are far rarer than they once were, in part because passengers now mostly use stations that are safer than they were a generation ago. The sharpest decline in the fatality rate for little accidents occurred from about 1950 to the mid-1970s. While some of this "improvement" reflects the ICC's changes in reporting (see above) it also resulted because of the collapse of passenger traffic in rural areas and small towns where platforms were at grade, requiring passengers to use stairs to enter and exit the train. Moreover, in trains with low-level boarding, passengers typically operate the doors, which gives them the option of jumping on or off a moving train.

The massive transfusion of funds into Amtrak and commuter lines that began in the 1970s (see above) also rebuilt many suburban stations. About 1976, Amtrak's NEC spending budgeted $315 million for new and upgraded stations. In 1984, when Metro North began an $860 million upgrade, some of that money also went to raise station platforms. After

1990, the Americans with Disabilities Act (ADA) reinforced these trends with the requirement that train access be more handicapped accessible, which essentially required raised platforms with a small gap between train and station. In 1997, Boston's MBTA similarly built long raised platforms in stations. Raised platforms that were flush with car level reduced falls, and they made it harder for passengers to wander in front of trains or cross tracks.[69]

Yet raised platforms carried their own risks, and neither the FRA nor the commuter rail agencies seem to have been terribly concerned about passenger station safety until a raised platform on the Long Island Railroad resulted in the death of a young woman on August 5, 2006 (Table A3.1). She was 5′6″ tall and weighed 110 pounds, and while exiting a commuter train she fell into the 8-inch gap between the train and the (raised) Woodside station platform. She was also thoroughly drunk, with a blood alcohol level of 0.23, and although warned to stay under the platform, she crawled to the other side in front of an oncoming train that killed her. The NTSB investigation found that while this was the first such accident on the Long Island to result in a fatality, that railroad had averaged 50 to 60 accidents a year involving train-platform gaps; the Woodside station alone had averaged three to four a year. The board found that at some stations the gap could be 13 to 15 inches wide.[70]

The state public safety board recommended a series of improvements in station design and passenger warnings and the Long Island took a number of corrective measures. The FRA also responded with guidelines on how to address the safety hazard, and it held several workshops on right of way safety that addressed passenger risks. The APTA urged a Gap Safety Management Program on its members. Station upgrades have yielded many other modest safety improvements. There have also been a number of innovations to improve boarding safety on lines with both low- and high-level platforms. Virginia Rail Express has both centrally controlled doors and a centrally controlled plate (called a "trap") that extends across the gap at high platforms. Recently Amtrak has been testing mobile platforms to bridge the gap between train and station and while these are a response to ADA concerns, they should improve safety for other passengers as well. One study suggests large potential safety benefits from the spread of these and similar boarding improvements. All stations now have a warning zone near the edge of the platform for low-vision travelers; some stations boast improved lighting and public address systems; signs now warn of the dangers of closing car doors. Such improvements have yielded a gradual reduction in fatality rates from little accidents, and as they continue to spread it seems likely that the decline will continue.[71]

The trends in passenger safety in the years since World War II have been uneven and ultimately disappointing. Mortality rates from train accidents declined sharply, if unevenly, down to the 1980s but since that time have shown little trend. Most of this decline has resulted because train accidents have become rarer not because they have become less lethal. Injuries in train accidents have also fallen sharply since the early postwar years although, as noted, changing reporting norms may make such conclusions suspect. The decline in fatality rates from little accidents has recently exceeded that from train accidents, despite the shift to potentially higher-risk commuter traffic.

Throughout the period, the shift of traffic toward high-density well-maintained routes has diminished both train and little accidents while deregulation of the freight railroads also improved passenger safety. The large infusion of state and federal money that upgraded track, equipment, and station design has also been responsible for much of the decline in both train and little accidents. FRA regulations governing drug use reduced the likelihood of another Chase disaster. Yet the FRA has been slow to focus on passenger risks. The impact of its safety evaluations of Amtrak and a number of commuters appears to have been modest. The collective failure for nearly 30 years of the FRA, the APTA, Amtrak, and urban transit authorities to respond to NTSB warnings with a serious research program devoted to improving car safety represented a massive lost opportunity. A whole generation of passenger and transit cars was designed to obsolete safety standards and passengers died as a result. Increasingly after 1990, however, the FRA responded to NTSB prodding with research and regulations to improve the survivability of wrecks, and it has helped midwife the difficult birth of PTC. It seems likely the payoff to these policies and the recent focus on station safety will result in sharp improvements to passenger safety in the near future.

6

Look Out for the Train

Motorists and Trespassers, 1955–2015

Oh Almighty God, in whom
all men place their destiny . . .
deliver me from collisions
at the grade crossings and
ward off the subsequent
horrifying nightmares.

"A Railroad Engineer's Prayer," 1966

People outrun trains in movies. They also save the world, get the girl and own a talking dog.

CSX billboard, ca. 2001

By the 1960s, except for commuters, few Americans rode trains anymore. Their experience with railroads, therefore, came largely from waiting at grade crossings, and unfortunately, they did not always wait. On March 1, 1960, at about 1:00 p.m., the Atchison, Topeka and Santa Fe's *San Francisco Chief*, with 4 diesels pulling 11 cars with 76 passengers and 33 crew, departed San Francisco, bound for Chicago. As it neared Bakersfield, California, about 5:15 that afternoon, an oil tank truck and trailer carrying about 7,000 gallons of fuel oil lumbered onto a crossing guarded only by a sign, where the *Chief* struck it, traveling about 70 mph (see Table A3.1). One of the tanks wrapped around the first locomotive and exploded, immolating all three diesels and the first coach, and derailing all but the last two cars of the train. The holocaust killed the engine driver, the fireman, the truck driver, and 11 passengers, and injured 68 others. That disaster, of course, made headlines, while many more minor grade crossing tragedies at least made the news. By contrast, the 637 railroad trespassers killed that year merited scarcely any comment at all.[1]

Although Bakersfield was more deadly than most crossing accidents, it was by no means unique. The following year a train hit a school bus near Greeley, Colorado, killing 20 school children, and in 1967 another school bus-crossing accident in Waterloo, Nebraska, resulted in the death of 4 children. These accidents—where passenger trains or hazardous substances were involved, or which killed school children—were disproportionately important in driving the campaign to reduce crossing accidents. More typical in that it involved fewer casualties and all of them in the highway vehicle was the wreck on February 17, 1973, when an Illinois

A grade crossing accident at Bakersfield, California, on March 1, 1960, killed 17. The locomotive (*left center*) and another car are still on fire. Two telescoped passenger cars are on the right. The wreck precipitated a major campaign by the railroad unions to federalize crossing safety. (Courtesy Bancroft Library, University of California, Berkeley)

Central freight rammed a car that had crossed against the flashing light near Godfrey, Illinois, killing five teenage boys. Sometimes accidents killed whole families and more often than one might suppose, cars even ran into trains, as had occurred a decade earlier near Boron, California, at a crossing guarded only by a sign, taking the lives of two high school cheer leaders and an adult driver. Alcohol was often a factor. Mike Smith of the Indiana State Police recalled an accident where four teenagers—two girls and two boys—who had been out drinking, tried to beat a train to the crossing. "The engineer said he saw the front of the car dip down like they were going to stop and then it raised up and he saw them come across . . . That train was right on top of them." No one was killed but one of the boys was rendered a quadriplegic.[2]

Bakersfield was also a warning to the carriers and their train crews. To the carriers such wrecks were expensive. The law on liability for crossing accidents is complex, but juries often liked to "send a message" to the railroads. A survey of crossing accident liability done in the mid-1970s found that they cost the carriers about $35 million a year in awards and settlements. *Modern Railroads* estimated that adding in the other costs of such wrecks boosted the total to $75 million. Costs were rising, too; by the mid-1980s, awards and settlement costs amounted to $60 million and total costs were about $100 million a year. The Santa Fe's property damage alone from Bakersfield amounted to about $1 million and liability claims would have added enormously to that sum. The company promptly sued the trucking company, and Santa Fe President Ernest. S. Marsh sent an open letter to California's governor expressing concern over "such large volume transportation over the highways of dangerous explosives and inflammables."[3]

For train crews, Bakersfield symbolized the rising risks from crossing accidents, for while passengers in highway vehicles accounted for most of the casualties, typically 8 to 10 train crew died each year from such wrecks. Articles such as "Highway-Grade Crossing CARNAGE Continues Unabated" had been a regular fixture in *Locomotive Engineer* for years and they would continue, while engine drivers talked about the "tanker-truck demon." That journal reported the Bakersfield disaster to its readers as "Fiery Crash Kills 14." To the men and women who crewed the trains, especially those in the locomotive, grade crossing accidents were traumatic as well as deadly. To ride helplessly in a locomotive as it crushed a car, killing the occupants, was not something one forgot easily. Some experienced memory loss while others struggled to forget; one man recalled driving the locomotive that killed three young girls: "ten years later, I still wonder," he said.[4]

Bakersfield proved to be the final straw for the railroad brotherhoods. They requested support from the Association of American Railroads (AAR), and in May 1960 they petitioned the Interstate Commerce Commission (ICC) to investigate grade crossing accidents. This began the transformation of grade crossing accidents from a matter left primarily to states and railroads into a national highway safety matter, and it inaugurated a dramatic decline in such accidents. Accidents to trespassers, in contrast, have until recently received little public concern. This chapter begins with a review of crossing safety in the post–World War II years. As with other aspects of railroading, safety improvements have been the outcome of incentives along with regulations, organizational innovations, resources, and improving technology. We then turn to Federal policies from the 1970s on. The third section focuses on new approaches to crossing dangers that began in the 1990s, while the fourth section traces the efforts to prevent trespassing fatalities. The final section draws the threads of this and previous chapters together and provides an index of the social costs of railroading.

The Railroad and the Crossing

From the earliest days of railroading, even if you didn't work for a railroad or ride a train, the carriers' rights of way injected danger into your life. To travel anywhere, you had to cross the tracks. And so, as Chapter 1 depicted, casualties at crossings had always been a fixture of American railroading. The numbers of killed and injured fluctuated, reflecting the vicissitudes of working-class life, but crossing accidents rose down to about World War I as train miles and population increased. In the 1920s, crossing accidents exploded with the advent of auto-mobility, finally peaking about 1930 and then declining for the following three decades.

There was never a time when crossing accidents were purely a private matter. From the earliest days, states and localities had required railroads to warn and sometimes guard dangerous crossings. Otherwise crossing safety was up to individual citizens. While the injured or their families might sue for damages, most people wasted little sympathy on those who were hurt from trespassing, or failing to stop, look, and listen. Accordingly, with jury awards usually modest, railroads had weak incentives to guard crossings. By the twentieth century, however, states and cities began to play a more active role. Urban crossings were a traffic nuisance and a danger to children. Attitudes had also begun to change; although it has never entirely disappeared, the fault-based conception of crossing accidents retreated. As Barbara Welke has argued, freedom has increasingly come to mean freedom *from* rather than freedom *to*, and so drivers at crossings need protection from their own bad judgments. The new ideas resulted in part because engineers developed more complex views of causation that stressed the importance of visibility, weather, train speed, crossing alignment, and other conditions. By the 1930s, engineers were investigating crossing design and studying how to allocate crossing improvement dollars to achieve the highest safety payoff.[5]

It is useful to make an analogy between grade crossing guards and hazard warnings on consumer products. Both provide information but largely leave safety decisions up to individuals (although, of course, some crossing guards constrain motorists as well as inform them). The research on consumer products suggests that warning labels containing new information are more likely to be effective than exhortations. By analogy, passive crossing guards such as crossbucks and stop signs are likely to be largely worthless to local drivers for their only information is that there is a railroad crossing, which local people are likely to know anyway. Train warnings such as flashing lights, bells, and whistles, by contrast, contain new information—a train is imminent—while guards and gates can both warn of a train and try to prevent motorists from crossing. In 1970, a year typical of that decade, there were almost precisely 212,000 public grade crossings in the United States. Of these, almost 165,000 had at best crossbucks or similar signage and were therefore, virtually, unguarded. Of the nearly 47,000 crossings that warned of a train, only about one-fifth involved gates or watchmen. Thus, the dominant American approach to crossing guards was to provide information only—and that was valuable at only a small fraction of all crossings.

After the late 1950s, while crossings and train miles declined, automobile travel exploded. Rising automobile speed also contributed to dangers at crossings without active guards, which might provide inadequate visibility for drivers at higher speeds to be able to stop in time. By the mid-1950s, each year saw 3,500–4,000 accidents at public crossings (Figure 6.1). Such accidents were the dominant source of railroad fatalities during these years, typically taking 1,200–1,500 lives a year and injuring 3,000–4,000 others, and after 1959 both accidents and casualties, which had been declining since the 1930s, again began to rise (Figure 6.1).[6] By this time the carriers had economic incentives to guard crossings by

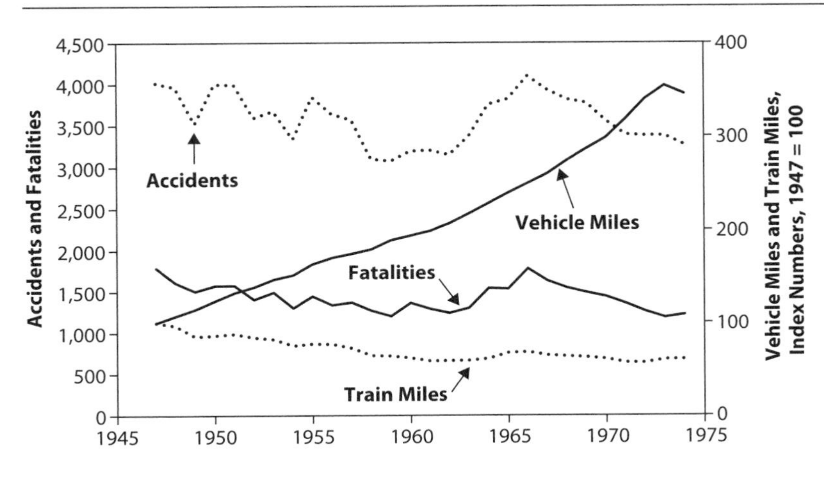

Figure 6.1. Grade Crossing Accidents, Fatalities, and Exposure, 1947–1974. (ICC/FRA, *Accident Bulletin*, and author's calculations)

installing flashing lights or automatic gates, for big companies made inviting targets in personal injury suits. However, the greatest payoff from such investments occurred where they replaced watchmen at already-guarded crossings. Thus, the Chicago & North Western claimed it saved approximately $100,000 a year in watchmen's earnings by investing about $364,000 in automatic gates at a set of Illinois crossings—earning a 28 percent return on investment, the company reported. Collectively, the carriers claimed to have invested their own money in about 6,500 installations of flashing lights during the 1950s, but about 2,500 of these simply replaced watchmen and their impact on safety must have been modest or perhaps even negative. At crossings with passive guards, upgrading might prevent expensive accidents, but flashing lights or automatic gates were also expensive to maintain. A better—that is, cheaper—solution, railroad managers believed, was to modify driver behavior through publicity campaigns.[7]

The likelihood of a grade crossing experiencing an accident depends on the number of train miles and vehicle miles, driver behavior, and other crossing attributes. Many crossings experience one or two trains a day and so the chance of meeting one is low—and the chance of an accident even lower. The psychological literature suggests that individuals often ignore low-probability risks and so they may pay no attention to the exhortation to "stop, look, and listen." Hence, the argument above that crossbucks provide essentially zero safety benefits to local drivers. Even at crossings with many trains and active guards, drivers may develop the habit of running the lights or gates, for the chances of being hit seem small enough to ignore. Careful crossing campaigns dated from the 1920s; their motive was not only to reduce accidents but also to spread the notion that the problem was motorist behavior, and that crossing accidents were a matter of highway, not railroad, safety. In the early 1950s, both the Atlantic Coast Line and the B&O instituted such programs, enlisting the newspapers and local civic groups, having watchmen observe crossings, and sending warnings to those who failed to obey the law. In 1961, the Illinois Central had a "near miss" program to encourage employees to report drivers who ran crossing lights; it published booklets and pamphlets, gave out bumper stickers, went to state fairs, supported a student essay contest, installed flashing lights on its locomotives, and improved crossing signage. Many other lines initiated similar programs and eventually they became the model for Operation Lifesaver (see below).[8]

Of course, if crossing accidents were a highway safety, rather than a railroad problem, then protection merited government funding. As noted, by the early twentieth century states were beginning to assume some of the financial responsibility for grade separation and crossing guards. By the 1960s, California, Illinois, Texas, Minnesota, and several other states had established separate accounts to fund guarding and elimination of crossings and some even helped the railroads with maintenance. In 1955, the Illinois Central and the Burlington ran through Centralia, Illinois, over 23 unguarded crossings. The following year, the carriers, along with the city and the state, jointly funded projects that moved track, built an overpass, closed some crossings, and guarded the rest. Such guarding, on sufficiently dangerous crossings, yielded a high safety payoff. When the Grand Trunk guarded 23 crossings in Illinois that had experienced 72 killed and 95 injured in 297 crossing-years, it cut the carnage to 5 killed and 2 injured in 255 crossing-years. Similarly, installation of gates on all crossings for a Southern Pacific commuter line between San Jose and San Francisco cut accidents 94 percent in 24 years even as highway traffic rose 6-fold.[9]

Federal money for grade crossing separation on Federal-aid highways had been available since 1916, and more came in the 1930s and 1940s. In the 1950s, federal spending for crossing elimination and guarding had averaged $60 million to $80 million a year. Along with state and railroad funding—and the decline in railroad track—these efforts slightly reduced the number of crossings and increased the proportion of those with active warnings from about 15 to 23 percent of the total between 1947 and 1975. Such actions stabilized accidents until the early 1960s (Figure 6.1). This must have been cold comfort for trainmen, however, for especially after 1959, as train miles declined crossing accidents per train mile rose (Figure 6.1). Moreover, a distressing number of these wrecks, like Bakersfield, involved hazardous substances—fuel oil, propane, gasoline—for transportation of these products by truck must have risen just as their shipment by rail was increasing. Indeed, on March 5, 1960, the dead from Bakersfield had not been buried when another train-tank truck explosion on the MKT near Chanute, Kansas, killed a father and a son who were engine driver and fireman. The rail unions had been fulminating about "Highway-Grade Crossing Carnage" for years. When the rail brother-

Risks to trainmen from crossing accidents rose during the postwar years. This dreadful accident occurred on November 27, 1956, when a fuel truck ran into a locomotive in Michigan City, Indiana. The engine driver and fireman were killed as were three maintenance of way workers when the wreck later exploded. (*Brotherhood of Maintenance of Way Employees Journal*, February 1956; courtesy Brotherhood of Maintenance of Way Employees)

hoods petitioned the ICC in 1960, they might reasonably have feared that their risks were increasing—and they would have been right.[10]

Moving at its usual stately pace, the ICC finally got around to hearings in 1961, taking more than 7,000 pages of testimony and exhibits. Some of its findings and recommendations displeased the trucking associations and the American Petroleum Institute, which led to further delays, and the commission's final report did not appear until January 1964. The report noted that the ICC had no power to regulate crossing safety, but it contained a number of useful suggestions including "the enactment of legislation to provide public funds" for grade crossing protection.[11]

While the ICC report had no immediate consequences, other developments kept public focus on crossing accidents. The High Speed Ground Transportation Act of 1965, which led to Metroliners in the Northeast Corridor, immediately directed attention to the many unguarded or poorly guarded crossings on that route. Spectacular crossing accidents had always been newsworthy, but when the National Transportation Safety Board (NTSB) opened for business in 1966 it began to probe such wrecks more deeply. Its study of the Waterloo, Nebraska, tragedy noted above raised a host of penetrating questions about school bus design, the effectiveness of train horns, and the need for active crossing guards. In 1967, at its urging—and sensitive to the headlines generated by the Everett crossing disaster in which a Buddliner hit an oil tank truck, immolating 11 people (Chapter 5)—the Department of Transportation (DOT) inaugurated a grade crossing task force. It surveyed state and railroad activities, estimated resource and information availability, and proposed an ambitious plan of action. The AAR and the rail unions also inaugurated a joint campaign to improve crossing safety signage and a "near miss" program to report to authorities vehicles that violated crossing laws.[12]

Crossing safety also benefited from the heightened concern over train accidents that led to the Federal Railroad Safety Act of 1970. In hearings over that act, witnesses noted the potential for crossing wrecks that involved hazardous materials. Indiana's Senator Vance Hartke (D), whose state included more than 10,000 grade crossings, observed that crossing accidents took more lives than aviation accidents. The railroads, of course, favored federal funding. Labor testimony also strongly supported federalizing crossing safety. "BLE Demands Passage of Rail Safety Bill," *Locomotive Engineer* informed its readers. The Volpe Committee, which had been formed to forge a consensus on legislation, recommended "an expanded concerted program of grade crossing safety be undertaken . . . [with] new sources of funding to finance an expanded grade crossing program." The act that passed required yet another study within a year.[13]

Federal Intervention

That study, conducted jointly by the Federal Railroad Administration (FRA) and the Federal Highway Administration (FHA), grew out of the efforts already under way at the DOT and provided a thorough assessment of the crossing accident problem as it appeared about 1970. It reviewed the trend in accidents and casualties, the number of crossings and protection, and the sources of funds. It also provided a number of economic analyses suggesting that—as noted above—prevention of crossing accidents could be extremely cost-effective. The study reported that federal spending on crossing protection had risen sharply from the $60 million to $80 million typical of the 1950s to an average of $139 million a year in 1967–1970—although it failed to note that about 64 percent of this increase simply made up for inflation. State spending averaged an additional $89 million a year over the same period, while the carriers spent about $7 million a year for guarding and removal and another $44 million a year for maintenance. Still, the report recommended a 10-year $759 million program to guard about 30,000 crossings. Rail labor enthusiastically supported such proposals. About the same time, in conjunction with the AAR, the FRA began to collect a detailed inventory of all public and private crossings in the United States, and in 1973, Congress responded to the FRA-FHA study with the Federal Highway Safety Act. That act specifically authorized funding from the Highway Trust Fund for the removal or guarding of crossings on the Federal-aid system of roads.

Moreover, in another section, for the first time it provided funds for guarding or removal of crossings on non-Federal-aid highways. The key to crossing safety, according to the experts, had always been the "Three E's"—engineering, education, and enforcement. They now had the resources for the first "E."[14]

Engineering Safer Crossings

Total authorized funding for fiscal years 1974–1976 was $425 million and additional money became available with the Highway Act of 1976. Federal funding largely severed the connection between local costs and local benefits for crossing improvements, thereby undermining the economic incentive to blame crossing accidents on drivers. When funding comes from localities (and to a lesser extent states) there is always the question: "why should we spend my tax dollars to protect fools from their folly?" Federal funding breaks this link. By fiscal year 1978, appropriations for crossing safety totaled $594 million. The new programs got off to a slow start, for many states were simply not prepared for the influx of new money. The programs had other problems, as the General Accounting Office (GAO) documented. Earmarking spending for crossings or other projects could sometimes prevent an efficient allocation of money, while the formula for apportioning funds to states (largely by area and population) ignored risks. Thus, the 1977–1978 appropriations provided Arkansas $520 per crossing for those not on Federal-aid highways, while Idaho got $117,000 per crossing. States also had different standards for treating identical risks. For example, the agency asked 6 states how they would safeguard an urban crossing that saw 18 trains and 14,000 automobiles a day—a high-density crossing. Of the six, three would have separated grades, two installed flashing lights, and one done nothing.[15]

In the 1978 Federal Aid to Highways Act Congress modified the formula for allocating funds among states, employing as criteria both population and number of crossings—which was a better reflection of risks. Yet segregating funds for crossing safety remained, along with the problem of how to allocate funds among crossings within states. The difficulty was that the large number of crossings and the rarity of accidents at most of them made it difficult to decide which ones merited protection. Researchers—mostly engineers—affiliated with the railroads and state highway departments had been studying the matter since the 1930s. They had developed a number of hazard indexes for ranking crossings, most of which reflected the combination of train miles and vehicle miles, with flashing lights, gates, and other devices assigned various "protection factors." While a number of states employed such formulas in the 1950s, the results were often contradictory due to other influences on accidents such as visibility, train speed, weather, and much else that might confound the analysis. In the 1960s, the development of computerized statistical packages improved the hazard prediction models, which the FRA-FHA (Federal Highway Administration) then integrated with resource allocation procedures to guide states in choosing which crossings to guard and how to guard them. While states were not required to employ these complex procedures, federal—and railroad—oversight probably improved the allocation process.

Return on Investment

Although as Figure 6.1 reveals, neither accidents nor fatalities declined markedly between 1947 and 1974, that does not mean that the efforts to improve safety during these years were ineffectual. Instead, they were simply too modest and were overwhelmed by the rise in traffic density. In fact, the number of public crossings fell about 17,000 over this period while an additional 15,000 were guarded. Statistical analysis for this period indicates that a 1 percentage point increase in the number of crossings guarded (about 488 crossings in 1974) yields a 1.36 percent decline in fatalities (17 in that year). Thus, guarding an additional 1,000 crossings would have avoided 35 fatalities. The cost of automatic protection at the time averaged about $11,400 per crossing, and so the cost of avoiding another fatality per year amounted to about $326,000 each. Reducing crossings had a lower safety payoff, apparently because those closed tended to be less dangerous. My research indicates a 1 percent elimination in total crossings in the period 1947–1974 reduced annual fatalities about 1.97 percent. Closing 1,000 crossings therefore would have saved about 12 lives a year. The high payoff to either guarding or elimination suggests that governments must have allocated resources toward the high-risk crossings.[16]

The federal crossing program that began in 1974 continues to this day and has appropriated large sums of money for crossing protection and removal. Ian Savage and Shannon Mok have calculated that the totals amounted to $8.5 billion by 2001 and additional appropriations have averaged $180 million to $220 million a year since. While the allocation of this money was never "optimal," states generally aimed it at the most dangerous crossings. Along with railroad and state funding, these investments generated sharp reductions in crossing accidents and casualties. Fatalities fell in half between 1975 and 1983 as did accidents. Thereafter, accidents and fatalities fell unevenly. Overall, a statistical analysis reveals that grade crossing fatalities per train mile fell about 3.9 percent a year from 1980 to 2015 even as automobile travel continued to rise.[17] Such accidents and casualties are almost entirely self-reported, with little checking by the FRA, and in 2004 *The New York Times* charged the carriers with numerous reporting failures. A report by the DOT inspector general did find significant underreporting of serious accidents to the Homeland Security Agency's National Response Center, which might have reflected confusing reporting requirements, but not to the FRA database. Thus, the improvements in safety are real, and research suggests that overall the program's benefits have been roughly twice its costs, at least up through 2001.[18]

The long-term campaign against crossing accidents is clearly an example of a reasonably efficient, highly effective program. However, even the best programs eventually hit diminishing returns. One researcher has estimated that for this most recent period, a 1 percent increase in guarded crossings would reduce fatalities about 0.31 percent. In 2010, guarding another 1,000 crossings would have increased that total by 1.5 percent. Since there were 222 crossing fatalities, the additional guarding would have reduced fatalities by $0.015 * 0.0031 * 222 =$ about .01 fatalities. The program to reduce crossing accidents by spending federal dollars for local safety improvements remains popular even as its marginal benefits diminish, for they seem free to local constituencies. Yet it seems clear that crossing safety should compete with other programs for safety dollars.[19]

The Great Whistle Ban Imbroglio

Crossing safety is but one of many good things, and when it has conflicted with peace and quiet, or otherwise proved costly to its recipients, the result has been a sharp conflict pitting the federal safety establishment against local citizens and state and local officials. While most major carriers were reducing their train miles during the 1970s and 1980s, the little Florida East Coast Railroad, which had become nonunion in the 1960s, was prospering. Its train miles actually rose 28 percent between 1975 and 1984, and for those living near its many grade crossings that was not good news. The increased sounding of the locomotive horn—two long blasts, a short blast, and then another long blast—precipitated a citizens' revolt. In 1984, the state legislature allowed localities to establish quiet zones that forbid the whistle's toot during nighttime hours on crossings that had active guards—gates or flashing lights. Train horns typically sound at between 104 and 114 decibels at 100 feet with a frequency range that interrupts normal conversation. Even a half mile away, horn noise remains louder than shouting. Noise, in short, was one of the hidden costs of crossing safety. Quiet zones were not a new idea, for some towns had had them for decades, but in 1991 a Florida legislator asked the FRA to investigate whether such zones compromised safety.[20]

The FRA duly studied the train horn ban and found that accident rates had jumped sharply at crossings where horns had been banned but not at other crossings. Similarly, on CSX, which had track near that of the Florida East Coast, but had no whistle bans (because, unlike the Florida East Coast, it was engaged in interstate commerce), there had been no surge in evening accidents. In July 1991, the FRA issued an emergency order overriding the Florida law and requiring the use of whistles. Accidents promptly returned to their old levels. The FRA also began a nationwide study of the topic. Shortly thereafter, it found that there were more than 2,100 such bans nationwide—in 27 states, with the greatest number being in Illinois and Wisconsin—and they were disproportionately at well-guarded crossings that featured gates or flashing lights. Where cities had instituted bans and then cancelled them, the FRA performed before and after studies, finding that cancelling the bans reduced accidents 38 to 59 percent. One wonders why. The horn provides no information about a train not already indicated by flashing lights unless drivers believe the lights are inaccurate—which is hard to imagine, given their low failure rate. Perhaps the blast of a train horn simply scares drivers in a way flashing lights do not.[21]

The railroads and their unions naturally hated quiet zones and sometimes simply ignored them, for they increased accidents and costs. At hearings on what became the Swift Rail Development Act of 1994 they urged federal preemption of the issue, and with no opposition, Congress required locomotives to sound horns at all crossings. It did allow the FRA to issue exemptions if requested by both the railroad and community (which gave the carriers a veto over the process) and if the community installed supplementary safety measures. The new law brought to the surface a host of issues: why did this need to be a federal matter? The issue of fault also reappeared: why should others sacrifice money or comfort to protect individuals from their own foolishness? That is, isn't there is a right to quiet as well as a right to safety? Such concerns had been implicit in grade crossing policies from the beginning, but since 1973 they had been submerged in a warm ocean of federal money. The Swift Act, however, was an unfunded mandate and the acceptable "supplementary safety measures" might well be expensive.[22]

The FRA, which had seen the problem coming, held hearings in early 1995. The result was an outpouring of public protests. As noted, many communities had had quiet zones for decades. Individuals—and property values—had adjusted to them. Sounding the horn would be a major change and might make houses near the track less desirable and less valuable. The growth in train miles, which had begun in the mid-1980s, added to the problem, while mergers sometimes rerouted and concentrated traffic, and commuter rail had been booming. Thus, many communities were discovering that train traffic was surging and train whistles were threatening an assault on peace and quiet. The result, as one headline announced, was that "Whistle Law Steams Suburbs."[23]

The federal law especially steamed up Illinois suburbs because to many it seemed like déjà vu all over again. Six years before the Swift Act, in June 1988, the state legislature had passed a similar law that had resulted in a similar explosion. Apparently, the state's lawmakers had intended to grandfather in existing quiet zones, but the wording was poor and so the law applied to all communities, some of which had had such zones for many years. "I'm extremely tired and extremely annoyed," one woman who lived about four blocks from the Burlington Northern tracks reported. "Who's going to pay for my nervous breakdown?" Another resident made a similar complaint: "You know I can't sleep. They blow the whistles all night long. They blow the whistles all day long. They blow the whistles on my way to work. I can't even read the newspaper. Who is going to pay my psychiatrist's bill?"

The bill's sponsor claimed his phone was ringing off the hook: "I've been called everything but my name," he said. The uproar resulted in an injunction and the legislature then dumped the issue in the lap of the Illinois Commerce Commission, which allowed exemptions for guarded crossings. Thus, for Illinois residents at least, Congress's actions must have seemed especially gratuitous.[24]

In 1996, Congress began to back-peddle; a new law required the FRA "to take into consideration the interests of communities with longstanding whistle bans," and mandated a one-year delay after the regulation became finalized. Finally, in 2000, the FRA delivered a notice of a proposed rulemaking. Although it suggested a number of possible "supplementary safety measures," the most likely candidate appeared to be four-quadrant gates and these were expensive. A series of 12 public hearings in 9 states failed to quell the rebellion. Next came congressional hearings in 1999 before the House Subcommittee on Ground Transportation at which the assembled representatives got an earful—from committee members themselves, from other representatives, from several senators, and many citizens and local officials.[25]

Ray LaHood (R), representative from Illinois and later to become secretary of transportation, termed the FRA rule "unworkable" and observed that he had to restrain himself from saying anything else. He seems to have overlooked that the rule largely stemmed from a congressional requirement. Bob Filner (D), representative from San Diego, wondered why trains going 5 mph had to blow horns in the middle of that city's downtown. Speaker of the House Dennis Hastert (R Illinois) suggested that the FRA might have learned from Illinois's 1988 experience. He testified that in his district one commuter line would have to sound its horn continuously for 40 miles. The Illinois Commerce Commission noted that many of the quiet zones had been in minority neighborhoods—thus raising the specter of environmental injustice. In addition, it argued that because public pressures would require states to install the additional safety measures to sustain the quiet zones, the result would misallocate safety resources toward these comparatively safe crossings, leaving unguarded others that were more dangerous.[26]

The mayor of Fond du Lac Wisconsin testified that his town saw 36 trains a day over 14 crossings, implying that horns would be sounded 504 times a day, or 21 times each hour. Rita Mullins, mayor of Palatine,

Illinois, raised the issue that others were loath to discuss: arguing that public officials needed to balance safety against quality of life and pointing out that some locations in Palatine were experiencing 150 trains a day. "What we are talking about here at its most fundamental level is an issue of federal preemption. A one-size-fits-all solution to railroad crossing safety is not the best public policy," she argued, to no avail. A woman from Antioch, Illinois, summed up the issues as many of that state's citizens saw them: "I'd like to have as much consideration given to me as a homeowner as the FRA wants to give me as a motorist," she observed. These complaints finally all boiled down to money. The mayor of Olmstead Falls, Ohio (pop. 9,000), claimed that FRA supplementary safety measures would cost the town more than $2 million, which amounted to 60 percent of its revenues. Had the federal government been willing to pick up the tab for the supplementary measures required for quiet zones, not a peep would have emanated from the suburbs.[27]

Little came of these hearings. The issue dragged on as the FRA took more public comments, published an interim final rule, and took comments on that. The final rule, which took up 76 pages of the Federal Register, appeared in 2005. It gave localities with existing quiet zones a longer period to comply (five to eight years) and allowed use of wayside horns as a substitute for locomotive horns in quiet zones. These are nearly as loud but concentrate the sound on the road immediately next to the crossing. Otherwise, it differed little from the proposed rule of 2000. As of September 2016, there were about 717 community quiet zones throughout the country. The episode is instructive for several reasons. First, although federal preemption may be desirable in crossing safety—after all, the safety of train passengers and crew in addition to that of local drivers is also at stake—neither Congress nor the FRA ever adequately made that case. Second, critics of public policies sometimes argue that the public demands unrealistic levels of safety. Yet in this case, when local communities faced the cost of safety—in either money or loss of amenities—they proved surprisingly flexible, demonstrating a willingness to trade safety for other amenities. It was Congress and the FRA, not the public, that refused to countenance any trade-offs.[28]

Education and Enforcement

Two of the "Three E's" of grade crossing safety, in addition to engineering, are education and enforcement. The careful crossing campaigns were a railroad organizational innovation that dated from the 1920s (see above) and anecdotal evidence suggested that they saved lives. In 1972, the Union Pacific in conjunction with its unions began such a campaign in Idaho and it was the origin of Operation Lifesaver, which ultimately became a nationwide program emphasizing education and enforcement to improve crossing safety. Because the program stressed driver behavior, it also raised again, if only by implication, the issue of fault, which the carriers could have hoped might influence juries. The UP activities seemed to cut fatalities in Idaho and the following year the company expanded the program to Nebraska with equally impressive results. Thereafter, it grew rapidly. In the 1980s, it became a national organization, enthusiastically supported by the railroad brotherhoods. By the early 1990s, there were branches in 49 states; the organization briefly affiliated with the National Safety Council, and eventually began to receive funding from both the FRA and the FHA. The basic approach included safety presentations to various audiences; advertising campaigns; displays at state fairs, booths, and malls; and messages on milk cartons. Individual railroads also continued their careful crossing campaigns under the umbrella of Operation Lifesaver. There was also an enforcement component. By the early 1990s, 35 state programs supported an "officer on a train" to demonstrate the seriousness of the problem to local police, and there were courses on accident investigation for local police departments. Ohio exemplifies a state that stressed both education and enforcement. In 1978, Operation Lifesaver began an active and persistent educational campaign in that state. It hired a full-time coordinator and 280 volunteers. It also put a trooper on a train—who reported violators to patrol cars—and it staged mock crashes in schools. The program helped to cut accidents at protected crossings from 377 that year to 93 in 1993.[29]

In the 1990s, the Norfolk Southern began to install a camera on its locomotives and at crossings, and the UP adopted similar procedures. By 2009, 95 percent of UP locomotives had such cameras. They yielded footage that was valuable in training, prosecution, and defense against liability claims. In 1994, the little Aberdeen, Carolina & Western, in collaboration with CSX and Operation Lifesaver, staged a crossing accident near Sanford, North Carolina, in which a locomotive hauling 5 cars plowed into the side of a 1989 Oldsmobile, shoving it 500 feet down the track and "killing" all 3 dummies in the car and—to the horror of the audience—decapitating one of them. About 2004, CSX also tried the blunt approach, erecting billboards splattered with eggs and labeled "cars hitting trains." By 2006, BNSF employees participated in teaching drivers' education classes; the company sponsored "officer on train" programs and offered courses in accident investigation. About the same time, the CN in conjunction with local police stepped up enforcement, monitoring crossings and ticketing miscreants.[30]

Sometimes individual accidents could be educational. On October 25, 1995, a crossing accident between a train and a school bus in Fox River Grove, Illinois, killed seven children. *Railway Age* thought it shocked the public—at least temporarily—into better behavior, for accidents in that state fell from 48 to 39 the following year. It also moved the state legislature to strengthen enforcement of crossing laws: getting caught driving around the gate now cost you $500 or 50 hours of community service. In Los Angeles, the MTA along with the sheriff's department installed cameras at crossings that provided a shot of the license plate and the face of a violator.[31]

However, individuals sometimes ignore familiar, low-probability risks, imagining "it can't happen to me" or "I am in control." Drivers may weigh an immediate saving in time against a small chance of disaster and choose to drive around a crossing gate or through a flashing light. Probably for this reason, grade crossing warning programs work best where there is a serious threat of fines. Like worker safety programs, they appear to be subject to a Red Queen effect—that is, without constant repetition, their effects wear off rapidly. In Idaho, for example, instituting Operation Lifesaver in 1972 cut fatalities from 19 that year to 11 in 1973, but when the program was discontinued deaths promptly rose to 20 in 1974 and then 22 in 1975. Similar results obtained in Nebraska: crossing deaths fell from 33 in 1974 to 19 with the institution of Operation Lifesaver in 1975 and then promptly shot up to 36 when the program ended in 1976. The day after the staged crash noted above, individuals in nearby towns were still driving around crossings. Eternal vigilance appears to be the price of crossing safety. A statistical study found that a 10 percent increase in the number of presentations in a state resulted in a 1.1 percent decrease in accidents, but did not test to see how long the effects persisted.[32]

New Approaches and New Technologies

In the early 1990s, the DOT initiated an ambitious Rail Highway Safety Action Plan, with the long-term goal of cutting fatalities in half within a decade. While the plan largely consisted of more of the same, one important innovation was a shift away from treating each grade crossing separately to a corridor approach that evaluated all the crossings—public and private—along a section of railroad. Such a program spread overhead costs and focused on a broader set of choices than the individual approach. Corridors also marked a shift in thinking away from guards and toward crossing elimination. For example, closing might make more sense than guarding if alternative crossings were easily available. As discussed above, the number of crossings had been shrinking since the 1950s as the carriers lost business and shed track. Public programs for grade separation or simply to close crossings abetted these trends: from 1975 to 2010, the number of public crossings declined about 40 percent.[33]

The dream of truly high-speed rail was the motivating force behind the corridor idea, for on such lines crossing accidents might yield catastrophic passenger casualties. This had been a continuing theme of public policy; in 1985, the NTSB warned of the need to focus on passenger accidents, and higher train speeds amplified the risks. The federal Intermodal Surface Transportation Efficiency Act (ISTEA, 1991) designated a number of high-speed rail corridors. Thus, in the mid-1990s, North Carolina in concert with various federal agencies implemented what it called a "sealed corridor" on about 173 miles of line that was part of a high-speed rail corridor and included 216 public and private crossings. The North Carolina goal was to allow train speeds of up to 110 mph, which required a massive upgrading or elimination of grade crossings.[34]

Corridors also spread to routes that were not designated high-speed rail, for while passenger casualties were not a risk on the freight railroads, liability awards were. In 1990, a jury awarded plaintiffs $15 million from a crossing accident on the Burlington Northern; a $4 million judgment hit the Norfolk Southern in 1996, while a 1997 case against Conrail totaled about $4 million. Other awards have been substantially higher. While further litigation sometimes reduced these sums, they must have been a great spur to crossing elimination. In California the Metrolink collision with an SUV on January 26, 2005, that killed 11 and cost $39 million (Chapter 4) spurred the development of a corridor on that line. In 2003, the state of Mississippi and the Kansas City Southern inaugurated a corridor project covering 147 miles of track with 85 crossings. Similar projects followed in Missouri and Louisiana. The UP, the BNSF, the KCS, RailAmerica, and the state of New York are all employing a corridor approach to try to close as many redundant crossings as possible and improve guarding on the remainder.[35]

The Problem of Private Crossings

The corridor approach for the first time focused public policy on private grade crossings. No one had ever counted them until the FRA survey described above, which found about 142,000 such crossings in 1975. Thereafter, the numbers fell slowly—to about 100,000 by 1990 and as of 2010 to 81,000. Surveys reveal that the largest number of these are farm crossings, while as of 2006 farm and industrial crossings combined made up about 84 percent of the total. Most have crossbucks or at most stop signs for guards. While there is little detail available about these crossings, they probably experience train density similar to that of equivalent public crossings (e.g., private farm crossings and rural public crossings), but much less highway traffic density. Accordingly, they have typically accounted for comparatively few accidents—in the 1970s, about 40 percent of all crossings were private but they accounted for only about 5 percent of fatalities. However, they are more likely to see heavy trucks and farm equipment than are public crossings, making any potential collision more dangerous. Moreover, as public crossing safety has improved, little has been done about private crossings even though their relative importance has increased, for as of 2015 they accounted for about 12 percent of crossing fatalities.[36]

The failure of public policy to address adequately the dangers of private crossings reflects both their on-average low risks and the legal difficulties surrounding the problem. There is little federal authority over private crossings; of the 50 states, only 28 appear to have regulatory authority over private crossing safety. The crossings themselves reflect a bewildering stew of ownership rights, easements, and undocumented grants. One carrier reported that it had 22,000 private crossings and in only one-fifth of them was there any written agreement between the carrier and the landowner. To the railroads, such crossings yield no benefits and potentially large liability costs; accordingly, they have sometimes voluntarily upgraded protection. They have also worked with landowners to close crossings. The Burlington Northern was able to close 5,750 private crossings between 2000 and 2014.[37]

Private crossings represent the threat of rare, but potentially high-consequence accidents. The FRA has developed regulations requiring sophisticated guards for public and private crossings where trains run up to 125 mph and it will not allow any such crossings where train speeds exceed that threshold (49 CFR § 213.347). Yet, with little federal money available for private crossings, it is not clear how this goal can be achieved.

Redesigning Crossings and Trains

The corridor approach was motivated by the distressing number of crossing collisions that occurred either because motorists didn't see the train coming at a passively guarded crossing—or somehow got stuck there—or drove through flashing lights or around gates into the path of a train at an actively guarded crossing. Traditionally these tragedies had been chalked up to driver error or, less kindly, to the "nut behind the wheel," but recently behavioral and engineering analysis has yielded a more complex understanding. For example, Herschel Leibowitz has argued that individuals have difficulty assessing the speed of large objects, which appear to move deceptively slowly. Thus, a cruise ship seems to move much more slowly than a small boat even if both are traveling at 20 knots. At the grade crossing, all of a sudden the train that seemed a comfortable distance away is on top of you. One response has been to make it easier to see and hear trains approaching the crossing and to judge their speed. A second tack involves redesign of crossings—to make them less dangerous or to make avoiding the guards more difficult.[38]

An early idea of the NTSB to make trains less deadly to automobiles in a collision by putting—essentially—a giant pillow on the front of the locomotive came to naught. A more fruitful approach made trains more visible by adding additional illumination to the locomotive and by requiring reflective materials on freight cars. The FRA had studied improved locomotive visibility as early as the 1970s, but nothing came of these early efforts. Beginning in the 1990s, the FRA and various railroads including the Alaska, Amtrak, the ATSF, the BN, CSX, and the SP experimented with strobe lights, ditch lights, crossing lights (that focused up to 15 degrees out

Locomotive cushions were an idea to improve grade crossing safety that did not pan out. (*Modern Railroads*, July 1976, courtesy *Railway Age)*

from the track), and other combinations. Nudged by Congress, the FRA issued a regulation requiring some such form of additional illumination by 1997. Enhancing visibility of freight cars was also an old idea that saw new life—as early as 1946, the Great Northern had used Scotchlite for its freight car logo. Initial experiences and studies found that the reflective material degraded too quickly to be useful, but investigations in the early 1990s revealed that manufacturers had found ways to improve visibility and durability. Visibility on the Alaska Railroad was a particular problem; nights were long, storms intense, and snowmobilers many. Service tests on that railroad and the Norfolk Southern investigating the most efficacious form of marking seemed to demonstrate that improved visibility might reduce accidents where motorists ran into trains. Although an AAR cost-benefit study indicated that a rule for freight car reflectorization was hopelessly cost-ineffective, with a nudge from Congress, the FRA mandated it in 2005.[39]

Some crossings were inherently dangerous. In 1967, the NTSB reported on a crossing near Sacramento, California, it termed a "booby trap." While flashing lights guarded the crossing, visibility was poor and warning times varied between 26 and 50 seconds. Worse, trains on nearby industrial tracks that were no danger to motorists would also trip the signal. Finally, when the railroad tested the signals weekly, lights and bells also flashed. Economic research suggests that some consumer safety devices such as childproof caps can have a "lulling effect," and clearly, some grade crossing warnings can too. In statistical terminology, this crossing generated large numbers of false positives—warnings when there was no danger—and so it "trained" people to run the light. On February 22, 1967, someone did so with a train coming, and the resulting accident killed nine

This thoroughly dangerous grade crossing is guarded only by a crossbuck and is obscured by vegetation. Federal, state, and railroad programs have gradually reduced the number of such dangerous locations. The warning sign that it will strand a lowboy is new and needed. (National Transportation Safety Board)

people. There were also crossings that had low clearance and could hang up a lowboy tractor-trailer on the tracks—as occurred on September 4, 1985, when an Amtrak train struck a lowboy stuck on the crossing near Schriever, Louisiana. A decade later a similar wreck occurred at a Maryland crossing where a MARC commuter train hit a stranded lowboy, injuring eight passengers. The disaster at Fox River Grove noted above resulted from another kind of booby trap: a traffic light stopped a school bus where it partially blocked a railroad crossing.[40]

At many passive crossings, dwellings or vegetation might obstruct drivers' vision or the crossing might occur at an oblique angle. Highway engineers have developed a visibility triangle (Figure 6.2) that relates train and driver speed to visibility requirements. Consider a driver going the speed limit at V2 in Figure 6.2; to stop in time (C) she needs visibility distance B along the track. A driver at V1 who has stopped at the crossing needs distance A to cross in time. Note that A increases with train speed and C with automobile speed. The NTSB reported an accident on April 5, 1996, near Calhoun, Louisiana, where a Kansas City Southern freight struck and killed a driver and her 8-year-old daughter; she could not see the train until she was 72 feet from the crossing—too late to stop. The NTSB also detailed an accident that resulted because the road met the railroad at a 35 degree angle and so the driver essentially had to look backward to see the train.[41]

The outcome of these and many other accidents and reports has been a gradual improvement in signage. Although as discussed above, the value of advance warning and stop signs over the traditional crossbucks seems likely to be low, companies have added them to both public and private passive crossings and posted approach warnings to humped crossings. Highway departments began to focus on proper linkage between traffic

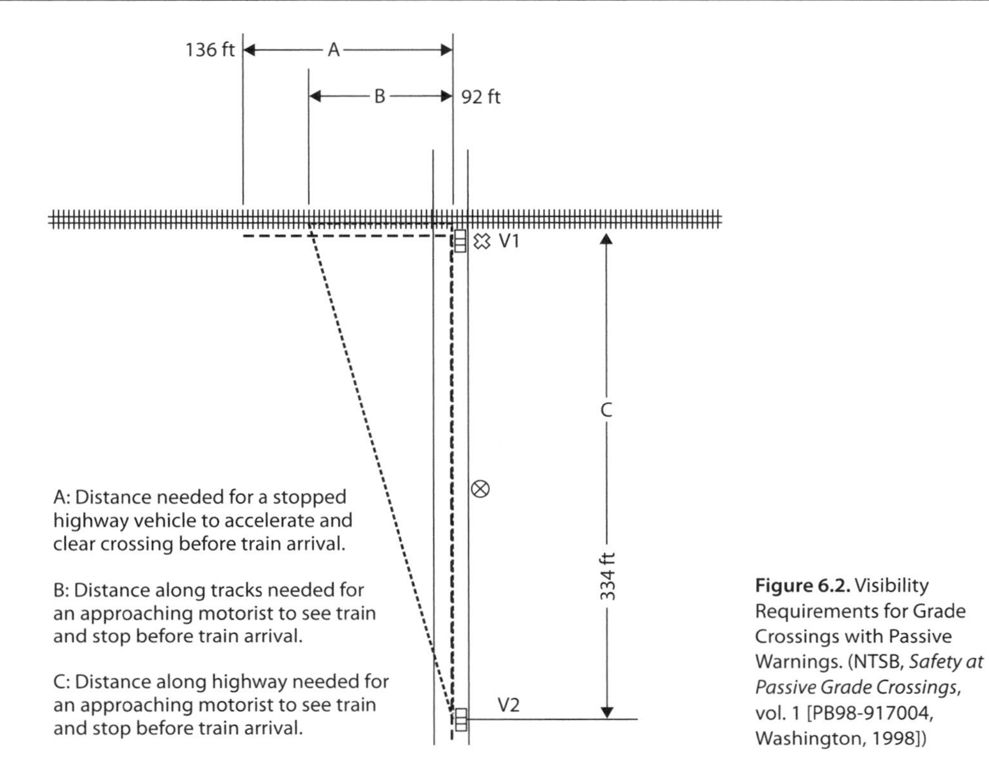

Figure 6.2. Visibility Requirements for Grade Crossings with Passive Warnings. (NTSB, *Safety at Passive Grade Crossings*, vol. 1 [PB98-917004, Washington, 1998])

and crossing signals. Recently, some carriers such as the KCS have installed blinking LED lights on traditional crossbucks while the BNSF is making similar solar-powered installations on private crossings. There have been other improvements as well.[42]

However, the most important developments have been improvements in active crossing guards. Surely one reason drivers go around crossing gates or run flashing lights is because they have experienced long waits for slow trains—which might have contributed to the Sacramento accident noted above—or because of a broken warning system. Constant warning time guards were developed in the 1960s as a joint venture of the Southern Pacific, Stanford Research Institute, and Marquand Industries (now Safetran), but they do not seem to have come into wide use until the 1990s. A defective active warning is a menace, for it might encourage the "lulling effect" noted above. While a signal that activates in the absence of a train encourages violations, one that occasionally—but rarely—fails to work when a train is coming is a death trap. As early as the 1980s, Texas had signs with telephone numbers to encourage motorists to report such dangers. By the 1990s, crossing guards were also incorporating sensors linking them to a system-wide network that allowed low-cost remote monitoring and would alert the carrier to a defective signal. Maintenance that is more prompt also reduces the likelihood that drivers might ignore the signal because it had malfunctioned in the past.[43]

Nevertheless, even if gates or lights always gave 20 seconds of warning and never malfunctioned, drivers would sometimes blow through them—indeed, on some occasions, literally—for in the early 1990s the BNSF was spending about $1 million a year to replace gates drivers had destroyed. On March 15, 1999, in Bourbonnais, Illinois, Amtrak Train 59, with 207 passengers and 21 crew and traveling about 70 mph hit a tractor trailer carrying steel and weighing about 37 tons that tried to beat it to the crossing (Table A3.1). The locomotive was equipped with ditch lights, there were lights and a gate at the crossing, and the weather was clear as the trucker drove around the guards. The wreck derailed the train, killing 11 and injuring 122 passengers and crew. Such an accident on a high-speed corridor might be orders of magnitude worse, and of course, the truck might be carrying gasoline. Such bad decisions sometimes flowed out of a bottle. On March 5, 1985, a motorist who survived after driving his truck around the gate and into the path of an Amtrak train near Fresno, California, registered 0.2 percent on a breathalyzer and was described by police as "staggering drunk." In another accident that took the lives of seven family members, the driver's blood alcohol level was an astounding 0.34 percent. As Gus Welty, editor of *Railway Track & Structures*, dourly observed, "No amount of education or enforcement or engineering is going to eliminate that segment of the population—careless, drunken, suicidal—that makes [crossing] accidents happen . . . In the long run . . . the idiots will always be out there."[44]

While Welty was right that engineering could not eliminate the "idiots," it could make their bad choices more difficult. As noted above, as late as 1970, improvements in crossing safety had largely consisted of providing better information to drivers, and active restraints such as gates were much less common. As long as the risks from grade crossing accidents centered on automobile occupants such an approach—which one might call "caveat driver"—was defensible. But with the growth in truck and hazmat traffic on road and rail, such accidents constituted a serious threat to train crews, passengers, nearly anyone near the track, and the dreams for high-speed rail. Use of gates had spread–they constituted about half of all active warning devices in 1996—but as noted, drivers sometimes drove around or crashed through them. Thus, while the "sealed corridor" in North Carolina and other potential high-speed routes closed as many crossings as possible, accidents such as Bourbonnais demonstrated that the remainder required far better seals. In response, suppliers developed long arm and four-quadrant gates along with median strips (so you cannot cross over to the other lane). Caltran and other commuter lines have also been installing four-quadrant pedestrian crossing gates. Four-quadrant gates brought their own problems, however, for they might trap a motorist on the crossing. One approach delays the exit gate a few seconds, but a vehicle stalled or otherwise unable to exit the crossing requires more sophisticated controls.

About this time crossing signaling also began to benefit from the same computer revolution that was allowing the carriers to link wayside safety devices to the locomotive cab. The result has been an increasing array of "intelligent" crossing guards. Some crossings now contain inductive coils linked to microchips that can raise the exit gate at an obstructed crossing. One initially employed on a high-speed Amtrak line near Mystic, Connecticut, could send a warning to the locomotive, which was equipped with cab signals. A stretch of high-speed rail between Chicago and St. Louis owned by Union Pacific also employs this technology. Inductive coils are high-maintenance, however, as they can break from frost heaves. Radar and other technologies may prove superior. Tentative assessments of these new guards suggest that four-quadrant gates and median barriers combined can reduce violations by up to 98 percent. However, the new technology may have other, offsetting effects, as one study found that it generated more aggressive driving (i.e., drivers speeded up when they saw the gates coming down).[45]

A host of other improvements is also coming through. LED flashing lights, which are brighter and cheaper, are replacing halogen, and companies are installing them in crossing surfaces too. Sophisticated control technology can warn of a second train coming or of delays. In Texas, the UP has been installing gates with photo-enforcement systems that have cut violations sharply. There are also spring-loaded gates that simply snap back when hit. Gates have also begun to come through with cables. Michigan has installed retractable barriers at some crossings, while the Wisconsin Railroad Commission has begun to install Stopgate, which the manufacturer advertised would stop a truck.[46]

Trespassers on the Right of Way

While grade crossing accidents have been a major concern of federal policy since the 1970s, the deaths of trespassers have engendered little concern from any level of government until quite recently. The death of a pedestrian involves no danger to third parties, whereas a crossing accident with a gas tanker risks train crew and sometimes passenger lives and a hazmat wreck might imperil the entire county. And trespassers, even more than individuals who are killed at crossings, seem to be the cause of their own demise. Finally, trespassers have traditionally been disproportionately working-class or underclass males—groups whose welfare has not always been of major concern to those in power. Yet as numbers of train and crossing accidents have fallen and worker safety improved, trespassing has begun to seem more important. In 1997, for the first time in years, fatalities from trespassing (other than at grade crossings) surpassed crossing deaths. In 2015, there were 511 people killed while trespassing on railroad property, and trespassers were the largest single category of railroad casualties, accounting for 62 percent of all railroad fatalities that year.[47]

Yet the number of trespassers killed has been declining for decades and is a faint reflection of the more than 5,000 killed a century ago. To grasp why trespass fatalities have fallen requires an understanding of why people trespass. Well into the twentieth century, trespassing on railroad property had been an integral part of working-class life. Transient workers—"hobos"—were mostly men and boys who rode trains looking for work because passenger fares put such travel beyond the reach of those with working-class incomes. In most cities, trains went right through the streets and pedestrians crossed anywhere. Rural stations were at grade and open, so people could easily cross tracks in front of incoming trains. And before the automobile, individuals of modest means walked—sometimes long distances—because the tracks were a natural highway that was straighter and better maintained than winding dirt roads. The railroads disliked trespassers because they sometimes vandalized trains and if injured might yield a lawsuit. But trespassing was an informal working-class property right. Despite the carriers' efforts, trespassing remained ubiquitous for decades, and every year it yielded a bounty of death and injury. In the 1930s, for example, the carriers typically killed more than 2,000 trespassers a year.[48]

Economic development has steadily reduced this carnage. In the 1930s, about one-fifth of all trespassing fatalities were individuals riding trains while modern data reveal that only about 5 percent of casualties result from that cause. The day of the hobo and the casualties attendant on that lifestyle have largely receded into history. This evolution reflects deep-seated changes in the nature of work. The sensational rise in agricultural labor productivity after World War II, especially because of harvest mechanization, has reduced the need for farm workers and their numbers have slowly declined. Migrants who follow the crops have become rarer still, and most of those remaining now drive cars.

Thus, most trespass casualties result not from individuals illegally riding trains but from walking the tracks, and here too, motives have changed. Cities have largely removed railroads from the streets, and with rising incomes working-class men and women now drive. The investment in raised station platforms and fencing during the 1970s and 1980s discussed earlier (Chapter 5) has made shortcuts across the tracks more difficult, while passenger traffic has been shifting to these newer urban and suburban stations that are more likely to have such safety features. By 1950, trespassing deaths had fallen to about 1,200 a year and they continued to decline down to about 1975, after which they have largely stabilized (Figure 6.3). This decline has resulted mostly from the long-term reduc-

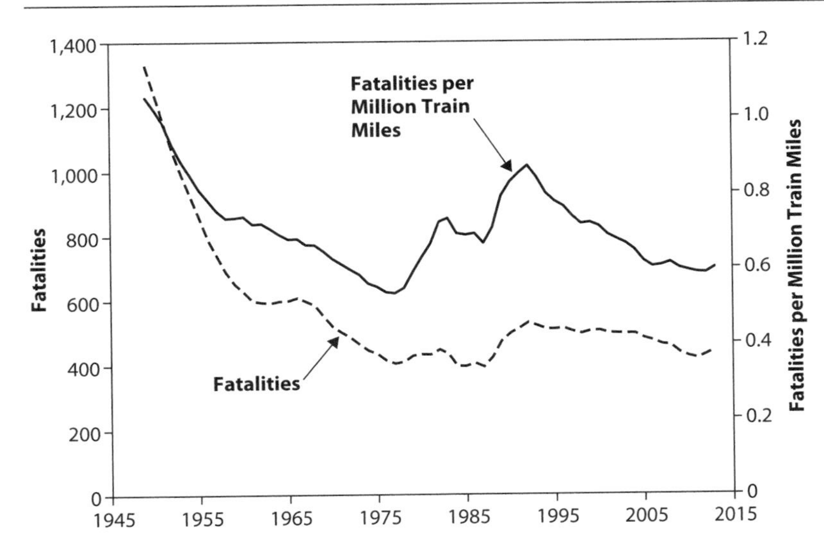

Figure 6.3. Trespassing Fatalities and Fatalities per Million Train Miles, Five-Year Moving Averages, 1947–2015. (ICC/FRA, *Accident Bulletin/Safety Statistics*, and *Rail Highway Crossing Accident/Incident and Inventory Bulletin*) Figures exclude trespassers killed at grade crossings.

tion in train miles and changes in the structure of the economy and level of well-being, although the upgrading of urban and suburban stations has also played a role. Fatalities per train mile also demonstrate a long-term decline broken by a puzzling rise between 1977 and 1991. Train miles were declining and shifting away from low-density lines during these years. Perhaps such lines also yielded little exposure to trespassers. In any event, a statistical analysis reveals that between 1990 and 2015 fatalities to trespassers per train mile have fallen about 1.5 percent a year.[49]

Studies of the demographics of trespasser casualties reveal that the great majority are males; they are younger than average but—contrary to headlines—few are children under 14. Most are also comparatively low income. Moreover, many trespassing casualties are the outcome of significant behavioral or personality problems. There is little doubt that suicide and alcohol have always proved important sources of trespasser fatalities and they remain so. Thus, in a study of about 1,000 trespassing fatalities between 2005 and 2010, a retrospective coroner's report yielded not quite 28 percent as suicides. Because the railroads did not report suicides to the FRA, this means that the official data on trespassing understate the true number of fatalities.[50] In addition, more than half of all the deaths in the sample involved alcohol or drugs in some way (these categories overlap). The two most common activities of the deceased were walking the track or "sleeping, laying, reclining, lounging [or] sitting on [the] track." Another study of metropolitan Chicago from 2004–2012 that covered all railroad fatalities other than employees and in automobiles found that 47 percent were suicides. There also appeared to be clusters in these suicides, suggesting that some were copycat and by implication that preventing one might prevent others.[51]

An ancient motivation for trespassing is that the tracks can be a barrier to travel. In Elko, Nevada, until the Western Pacific relocated its tracks they separated the brothels from the rest of town, resulting in the deaths of many eager young men. Sometimes the railroad runs near a school or through a park—which seems to have been the cause of the death of James Boyle, who was killed by a commuter train near a park in Santa Ana, California, in 1988—or the tracks may lie between residences and a pool or a shopping center. In 1994, a police officer explained why there were so many trespassing fatalities in one of the communities outside Boston: "parents park on Route 135 and take their kids across the railroad tracks to get to the lake." Another area where an 88-year-old man was killed was "a popular crossing place to a shopping area." Stations too remain popular spots for trespass. The author rides the MARC between Baltimore and Washington from time to time and most of the suburban stations, while they have "approved" places to cross the tracks, have no means to prevent trespassing at "unapproved" places. An example comes from a small station that serves both passenger and freight rail in Del Mar, California. On December 6, 1990, a woman passenger, hearing a freight whistle and mistaking it for her train, took a shortcut from the parking lot across the tracks where the train struck and killed her and a Good Samaritan trying to rescue her. This tragedy might well have happened even at an approved crossing. Had that been the case, the fatalities would not have been trespassers and would have been buried in the FRA's "other non-trespassers" category. (Before 1957, one of the deaths would have been included in "travelers not on trains.") Thus, pedestrian casualties exceed "trespassers" by some unknown number.[52]

Trespassing casualties also sometimes result because the track is an aid to recreation—an "attractive nuisance" as the law sees it. Such motives are not new—walking the track on a beautiful day was a popular pastime 100 years ago and for some it remains so today. The author recalls walking the tracks of the Boston & Maine as a boy while hunting pheasant—and jumping off just before the train arrived. Apparently, recreation was the motive that lured two Boston University students to the tracks early on the morning of February 8, 2005, where an MTA commuter hit and killed them. New forms of recreation have also emerged. If the old-time hobo is largely gone there are now "yuppie hobos"—college students or middle-class individuals who hop trains for fun—and sometimes tragedy, as occurred in 2011 when a Colorado College student lost both her legs trying to board a moving freight. Riding the top of a train—"train surfing"—can also lead to an unpleasant end: recently one man riding the top of a Metro-North commuter line touched the pantograph and the resulting 12,000 volts burned him to death. Small children find adventure on the rails, or teenagers congregate there to avoid adults.[53]

Sometimes individuals explore attractive but dangerous nuisances such as rail yards. Thus, in 1978, when two boys climbed a freight car in a Conrail yard north of Washington, they got too close to the 12,000-volt line; the current badly burned one of them as it took a shortcut to ground. Almost the same accident occurred again in 2002, when two 17-year-old boys made the same error with the same consequences in Amtrak's Stroudsburg, Pennsylvania yards; a jury awarded the victims $24 million. Trestles are a popular spot for fishing or swimming, and a death-trap if a train comes. In the 1990s, bungee jumping companies began organizing—and charging for—jumps from Union Pacific bridges, precipitating a "war" with the railroad. Railroad tracks also make good trails for ATV or horse back or snowmobile riders, or for cross country skiers, recreational railbike riders, and hunters. Track motorcar clubs also enjoy scenic outings on the rails. Sometimes recreationalists get permission to use the tracks, but not always: bridges and ATV riding accounted for about 3 percent of fatalities in one of the demographic studies noted above. Recently, taking graduation and wedding photographs on the railroad track has become fashionable, with occasionally fatal results. Being struck by a train seems like a bad start to a marriage.[54]

These examples help explain why trespassing safety has failed to improve in recent years. The gains that resulted from the decline in hobos and use of tracks for travel to work have largely ended, even as recreational use and commuter station traffic has expanded with rising population and affluence. In addition, trespassing has become more dangerous. Traffic density is rising and trains are faster—sometimes much faster—than in the past. Moreover, they are frighteningly quiet. Diesels make far less noise than did steam locomotives and continuously welded rail (CWR) has done away with the familiar click-clack. Rail lubricants (Chapter 3) reduce friction and noise, while several companies now advertise methods of reducing track noise even more, which will surely lengthen the casualty list. As with locomotives that whistle for crossings, noisier is safer. Commuter trains in push mode with the locomotive at the rear are especially deadly. In October 1990, an Amtrak commuter with the engine pushing and traveling at 85 mph came out of the darkness near Encinitas, California, and killed two out of a group of five trespassers who never heard it. Silent trains and trespassers with headphones are a particularly lethal combination: an MBTA commuter killed 10-year-old Natasha Shillingford who was walking the tracks on May 25, 1994. She and her friends were wearing headsets and never heard a thing.[55]

While it is probably best to treat suicides by train, and alcohol- or drug-related deaths as public health rather than railroad problems, there are cost-effective policies that might diminish trespassing casualties. Operation Lifesaver already campaigns against trespassing. Yet like running a grade crossing, trespassing is a low-risk activity and here too education probably works best combined with serious threats of fines. After the Del Mar accident described above, owners of the parking lot fenced off the shortcut, but within a year trespassers had found another one. As this incident suggests, and as one FRA study demonstrates, such deaths are not sprinkled randomly across the rail network but rather cluster in high-risk spots. These include not only stations, but also bridges or track near parks. Thus, focusing policies on such areas can increase their payoff. Raised platforms or passenger bridges and tunnels at stations help protect passengers from entering the fatality rolls as trespassers or "other nontrespassers" because they raise the cost or reduce the payoff to risk-taking. Another imaginative response by one commuter line to the chronic fence hopping by well-dressed morning commuters was to grease the fence, thereby raising their cleaning bills. Moreover, some of the "intelligent technology" being designed to make crossings safer may also provide a low-cost way to improve enforcement and so to deter trespassing. Intelligent videos at stations, at tunnel entrances, across from parks, and on bridges and trestles can warn of pedestrians on tracks—and warn trespassers that they are being recorded.[56]

The Social Costs of Railroading

In this and previous chapters, railroad accident injuries and fatalities have usually been expressed relative to some measure of exposure. Thus, trespassing fatalities can be expressed per train mile while worker and passenger deaths are presented relative to employee hours and passenger miles, respectively. An alternative approach is to see railroads as producing both "goods" and "bads" and to inquire into how that mix has changed over the years. The goods, of course are transportation outputs, while the bads are mostly casualties and environmental damages. This book has focused on one of the group of the bads—injuries and especially fatalities—and a relevant question is: how much in the way of bads has society had to put up with to get a unit of goods? For the years 1975–2014, I have constructed an index that expresses fatalities relative to railroad output—which is simply ton-miles and passenger miles weighted by their respective values in 2010. The index, arbitrarily set to 100 in 1975, registers 24 in 2014. By this measure, the social costs of railroading in 2010 were about one-quarter as great as they were in 1975. Of course, this index ignores injuries and other social costs such as environmental damages, but as this book has demonstrated, injuries too have fallen sharply. So have at least some forms of environmental damages. To take but one example, as a result of increasing locomotive efficiency, larger, more efficient cars, and the spread of rail lubrication, diesel fuel consumed by the freight railroads—and thus their contribution to air pollution—has been stable between 1975 and 2014 even as output has more than doubled.

The spectacular decline in grade crossing accidents and casualties in the six decades since World War II resulted from the interplay of incentives, organizations, resources, regulations, and technology. It also reflected major shifts in attitudes toward accident responsibility that had been in process for decades. By the 1950s, older approaches that had focused largely on drivers were being modified as engineers began to investigate how crossing guards affected accident probabilities. As their liability costs rose, the railroads' incentives to reduce accidents increased, and they became less interested in depicting accidents as matters of individual responsibility than in portraying them as highway rather than railroad problems. Down to the mid-1960s, the railroads, aided substantially by state and local funding and modestly by federal money, cut the number of crossings sharply and guarded others, but after 1959 these efforts fell short as highway traffic exploded and crossing accidents ballooned.

Federalization came to crossing accidents, as it did to many other areas of health and safety, about 1970. The large infusion of federal resources greatly expanded the program, which, along with better guarding technology and organizational innovations such as Operation Lifesaver, has dramatically reduced crossing fatalities and injuries. The rail unions—especially the Locomotive Engineers, whose safety was directly involved—supported federalization, as did advocates of high-speed rail. By making crossing safety largely free to localities, federal funding undercut opposition, yet federalization was not costless as the opponents of whistle bans discovered. The extent of opposition when localities had to pay for safety in either decibels or dollars suggests that a generous lubrication of federal money was necessary to achieve the safety gains. Yet diminishing returns are a fact, and since the 1990s, the safety benefits from crossing improvement have been much less.

In contrast to crossing safety, the reduction in trespasser fatalities owes little to public policy, for—except where they are children—trespasser deaths seem to elicit little sympathy. Rising incomes and the decline in migrant labor reduced railroad trespassing deaths down to the 1980s, but in recent years recreational trespassing seems to be on the rise. While many deaths are suicides, many others involve drugs and alcohol, sometimes in combination with singularly foolish behavior, and they are difficult to prevent. Yet there are glimmerings of improvement. Better-designed stations and improved technology such as video recorders, along with a policy focus on comparatively high-risk areas of track, all show promise.

Crossing and trespassing fatalities are part of the social costs of delivering rail transportation. Collectively, these and other fatalities have fallen sharply relative to transportation output since 1970, as have injuries and the carriers' contributions to air pollution. In the past 35 years, railroads have helped make us safer and cleaner as well as richer.

Conclusion

In 1978, the director of Accident Prevention and Safety of the Canadian National Railways spoke to the railway bridge and building engineers. "Accidents are bad for business," he told them. That, as the previous chapters have described, has been generally true on the railroads. And the reverse has been true as well; what has been bad for business has been bad for safety. That is part of the reality behind the claim that richer is safer—an observation that is consistent with both the available data and common sense. When I was born—in rural New England before the age of antibiotics—stepping on a rusty nail in the barnyard risked death from blood poisoning. Now you get a shot. Americans' life expectancy has risen steadily since World War II. While much of the gain has resulted from the retreat of infectious diseases, accidental death rates have fallen as well. The decline in railroad deaths has made a modest contribution to that improvement. We tend to take these gains for granted; they are part of Americans' birthright. In fact, however, improvements in public health and safety have always been contingent on the effective workings of private markets and public policy. Some of the value of studying the history of railroad safety is that it provides a glimpse into when these institutions have worked well and when they have not.

This long view of railroad safety yields a number of insights. One is—to repeat an earlier point—the contingency of these events. There has surely been progress in railroad safety, for what other term can one use to describe the decline in worker fatality rates from around 1 per million employee hours in 1890 to about 0.02 in 2015? But it has by no means been linear. The steady improvements in railroad safety that characterized the first half of the twentieth century proved far less inevitable than contemporaries must have imagined. Contingency implies that things might have been different. A plausible counterfactual, as I have suggested, is that if economic deregulation had occurred a generation earlier—in 1950, say, instead of 1980—there would surely have been far fewer train accidents in the intervening years while many more railroad workers would have lived to collect their pensions.

From the earliest beginnings, railroad safety has gone through three more or less distinct phases. Down to the mid-1950s, most aspects of safety improved: train accidents declined, workers' risks fell, and grade crossing and trespassing dangers eventually receded. Then the good times largely ended as train wrecks skyrocketed, worker risks rose, and grade crossing dangers increased. Finally, the dangers began to recede again in the 1970s or 1980s, depending on which safety aspect one chooses to use. Because the safety of railroading reflects the railroads' business fortunes, I have embedded this story within a Schumpeterian framework. In Joseph Schumpeter's vision capitalism evolved through creative destruction—the result of entrepreneurial striving that results in new firms, new goods, and new production processes that undermine the market position of older firms and technologies. In the early days, of course, the railroads were a new and mightily disruptive technology that was both enormously creative and at the same time destructive of older forms of transportation, eastern agriculture, and life and limb as well.

Yet accidents provided their own antidote, for they were expensive—especially if they harmed passengers—and so they created strong incentives for better safety. As the railroads evolved, safety gains reflected the interplay of innovations in technology and institutions, and of a massive commitment of resources. The carriers developed increasingly complex managerial hierarchies and rules of operation, as well as industry-wide associations to help facilitate interchange of equipment and standardization of technology. And they formed what I have called a "technological community" as well—a loose association of technologists and publicists that digested the lessons from accidents and developed and diffused new methods. As a result, train accidents diminished and passenger and worker safety improved from at least the late nineteenth century on.

Yet by the 1890s, the railroads had stirred up a hornet's nest of opposition. What I term the "robber baron narrative" not only denounced the carriers as monopolies and political fixers, it also depicted railroad accidents as the result of callous indifference to public welfare. These views led to increasingly strict economic regulation by the states and the Interstate Commerce Commission (ICC) after 1906. State legislators, competitors, unions, and others all began to shape decisions on rates, employment, abandonment, and new equipment. A sprinkling of federal safety legislation originated with the Safety Appliance Act of 1893, which mandated semiautomatic couplers and air brakes. But while other laws and regulations followed in the twentieth century, safety remained largely unregulated under a regime I have termed "voluntarism."

By the 1920s, the carriers no longer monopolized transportation; they had become an old technology and were themselves beset by a wave of new competition in the form of automobiles, trucks, pipelines, and airlines. Yet the railroads were like Gulliver among the Lilliputians—bound by many small decisions that collectively hindered their ability to compete. Voluntarism presumed the companies' ability to deliver acceptable safety results and it was undermined by the combination of economic regulation and new competition that eroded their profitability.

The 1960s and 1970s were transitional decades for railroads and accordingly for their safety. Critics had always claimed that the carriers traded safety for profits, but by 1960 they could deliver neither. From World War II on, nominal profit rates had hovered between 2 and 4 percent, and adjusted for inflation they turned sharply negative after 1965. Financial deterioration reduced resources and undermined morale and incentives for safety. Thus, the safety collapse that began in the 1950s reflected economic regulation at all levels of government rather than inadequate federal safety legislation. In the years down to 1980, it is at least an open question whether public policies on balance prevented more injuries and fatalities than they caused.

Yet managers at a number of carriers thought they saw a way through this slough of despond. Led by Downing Jenks at MoPac, Bill Brosnan at the Southern, L. Stanley Crane at Conrail, and others, and urged on by *Modern Railroads*, they began to fight the new competition with better, cheaper service. Ironically, one prong of the attack involved an uncoordinated charge into very large freight cars that worsened problems for a time. Yet it was in this quest for better railroading that lay the origins of safer railroading. To respond to the new competition required attention to costs and better service, and both required far more attention to agency problems and safety. The carriers created new organizational structures; internal restructuring allowed a broader view of accident and injury costs, while unit trains and inter-company run-through arrangements helped speed traffic and avoid dangerous switching. These developments not only reduced train accidents, they improved worker safety as well.

I have argued that a Red Queen effect often bedevils safety institutions—that is, they need a shakeup from time to time just to maintain the status quo. Work accident rates had risen throughout much of the 1950s and 1960s as the carriers' infrastructure deteriorated, morale declined, and safety departments fell into decay. Companies abandoned worker committees and overemphasized exhortation and discipline but failed to engage new risks. The rejuvenation of safety work owed little to public policy and much to the carriers' increasing determination to compete, along with the dawning awareness that improving service and worker safety was part of a package. The rising costs of work accidents reinforced this commitment, and carriers began to see employee injuries as part of a broader loss and damage problem.

Still, the great safety turnaround did not begin until partial economic deregulation arrived from 1976 to 1980. The policy changes inaugurated by the "Four-R Act" of 1976 facilitated these struggles to compete in several ways. Because they provided the carriers the prospect of a profitable future, they highlighted the importance of incentives in improving safety.

Along with tax changes, economic deregulation also provided resources for reinvestment as the carriers shed marginal track and passenger service and reallocated resources to profitable routes. They responded to intermodal competitors with their own gale of innovation. Consolidation through merger and abandonment encouraged more run-through trains and concentrated maintenance on high-traffic lines. Competitive pressures resulted in vertical disintegration as the railroads increasingly contracted out a range of activities from maintenance to office work.

The carriers also invested in what amounts to a technological revolution in the years after 1980. Modern Americans, when they use the word "technology" often seem to equate it with the computer-Internet revolution. Indeed, railroading has always been about information, communication, and control, and the computer-Internet revolution has dramatically lowered the carriers' costs of acquiring information. However, there has been much more to the carriers' improved technology than just computerization. Premium steel and curved plate wheels were two among many innovations that reduced track and equipment derailments. Along with their suppliers, the carriers developed, tested, and adopted a host of sophisticated remote sensors to monitor bearings, wheels, load shifting, and much else. Collectively these generate a flood of low-cost, real-time information on track and equipment. All of these and the many other improvements described in the previous chapters have been undramatic and incremental: no great improvement such as the air brake of 1890 or the diesel of 1950 has revolutionized railroading since 1965, although perhaps electrically controlled brakes and positive train control (PTC) may do so in the future. In total, however, these incremental technological improvements have revolutionized safety.

The needs of competition pushed railroads to become less insular as well and this too reflects a kind of organizational innovation. Men and women with PhDs and expertise in industrial engineering arrived. They employed the newly available data in computerized engineering and economic models that have upgraded maintenance efficiency and safety. Companies hired managers from outside the industry and upgraded education for wage earners and executives alike. As equipment and track failures receded managers began to focus on human failures. They contracted with DuPont and other safety consultants and imported behavior-based safety. The result has been a new emphasis on training, work planning, and fatigue management. Total Quality Management (TQM) also debuted in the 1980s with its emphasis on the costs of poor quality, zero defects, continuous improvement, and employee involvement.

The results of these investments and innovations have included not only sharp reductions in train accident and worker casualty rates; they have reduced social costs as well. Railroad productivity has skyrocketed, and so transportation of people and goods generates fewer fatalities and injuries and less pollution per person and per ton than it did a half century ago.

The above story, painted in broad brush strokes, makes railroad safety even to this day largely an achievement of market forces and private enterprise. The message is, quite simply, that with proper incentives private markets generate powerful motives for better safety. Moreover, as the quarter century after about 1955 demonstrated, another message is how difficult it is to improve safety when the incentives and resources are lacking.

Painting with a 4-inch brush as has been done above leaves out a lot of important detail. In particular, in 1966, a new safety regime arrived and from then on public policy has increasingly shaped the carriers' actions. After that date, the Federal Railroad Administration (FRA) took over safety regulation from a geriatric ICC while the National Transportation Safety Board (NTSB) became the major source for accident investigations. A century earlier, Charles Francis Adams had thought that sunshine was the best medicine for railroad accidents and the NTSB is his descendant. Failure can be instructive but that requires an instructor, and the NTSB's accident investigations provided invaluable information, sometimes even amounting to a research agenda. The agency also focused debate on particular problems and solutions. Its establishment has been one of the most consequential federal policies for railroad safety in the past half century. After 1970, with the Federal Railroad Safety Act of that year, there is little that falls outside the jurisdiction of the FRA. Finally, congressional oversight of railroad safety, which was virtually nonexistent before 1965, has exploded since that time, with hearings nearly every year, often on topics suggested by the NTSB.

This new regime reflected a head-on collision between the spectacular increase in hazmat wrecks that began in the late 1960s and deep-seated changes in attitudes on the proper role of government that probably originated in the New Deal and came to full fruition in the 1960s. The disasters of 1968 and 1969 at Dunreith, Laurel, Crete, and Glendora frightened Congress and the public and ensured passage of the Federal Railroad Safety Act of 1970. Here again, we get a glimpse of the Red Queen, for the eruption of hazmat problems in the late 1960s apparently caught the Association of American Railroads' (AAR) tank car committee unawares.

Timing matters: had the upsurge in hazmat wrecks occurred a decade earlier, the public response might well have been less dramatic, for the burden of proof was then on those who would intervene in private markets. However, by the 1960s, the default position had become that the market could not be trusted to make the appropriate decisions. In 1974, when Representative John Dingell (D Michigan) suggested that one way to improve the Penn Central's safety was to jail its managers, he spoke from that perspective. So did Congressman Fred Rooney (D Pennsylvania), when he told Congress in 1976 of the need to persuade railroads that safety was cost-effective.

Yet the assumption that eroding safety was the result of indifferent or ignorant managers reflected an almost complete misunderstanding of the problem. This proved a difficult lesson for Congress to master. Thus, the first response to the rise in accidents was to expand safety regulations. These proved to be insufficient, and it took a decade until the importance of economic deregulation finally became clear to decision makers. Hazmat worries contributed to the formation of the Transportation Technology Center (TTC) and more recently to the federal mandate for PTC. Indeed, as I have argued, public concerns may demand a reduction in not only the rate of hazmat accidents but their absolute level as well. The carriers may have to run a little faster just to stay in the same regulatory spot.

Sometimes safety regulations have benefited the carriers. The 1908 law governing transport of hazardous materials fell into that category and, indeed, the railroads pushed for its enactment. In modern times, companies supported strict rules on random tests for drugs and alcohol, and these overrode union intransience in a way the carriers probably could not have. More recent rules on the use of cell phones probably fall into this category as well. Overall, however, while the post-1970 safety regime has generated a blizzard of regulations and laws, it has proved difficult to find their impact in the accident and casualty statistics. That is a reflection of the law of diminishing returns: with most aspects of railroading already quite safe, there is no Big Thing that will yield dramatic improvements. That no Big Thing exists, I have argued, is because the private market puts a high price on accidents and injuries. For example, it is hard to find the imprint of recent regulations on maintenance of way work in the fatality statistics because the carriers and equipment suppliers have long been concerned with the topic.

Of course, this does not mean that the carriers never make mistakes, for they are large bureaucracies beset by agency problems and run by fallible human beings without perfect foresight. It does mean, however, that safety regulations largely police a safe beat; they are like the cop charged with preventing people from destroying their own property: self-interest will ensure that she has little to do. Thus, while the FRA is far more engaged than the ICC ever was, and federal regulations abound, the agency has generally ruled with a light hand, and so modern times are not as dramatically different from the era of voluntarism as it might seem at first blush. In fact, most federal safety regulation after 1965 has functioned as a backup. That does not mean it has been irrelevant; rather, it has been a second line of defense.

The FRA and the NTSB have also performed a number of what, following Mancur Olson, I have called "market-augmenting activities." While observers sometimes overestimate the effect of safety regulation, it is easy to overlook the value of these more subtle policies. We saw the importance of NTSB accident investigations. The FRA midwifed the TTC, which for the first time allowed real-world analysis of the potential interactions of new technologies. It was surely the most significant development in railroad research since the inauguration of the first testing laboratory by the Pennsylvania Railroad in 1875. The initial rise in federal research after 1970 may also have spurred the AAR to step its game up while federal funding continues to support a host of valuable projects such as the long-running investigation of crash energy management (CEM). The FRA's research and company safety audits have encouraged the spread of innovative programs such as Confidential Close Call Reporting System (C3RS) and behavior-based safety. The Rail Safety Advisory Committee has also helped build trust and encouraged cooperation among labor, management, and other interested parties.

The FRA also played a constructive role in waiving antiquated regulations that discouraged electro-pneumatic brakes and there is something in that story for both conservatives and liberals. Men and women of a conservative persuasion worry that regulation can stifle innovation, for antiquated rules such as those requiring inspection of self-diagnosing brakes simply raise costs. Liberals, however, may note the FRA's flexibility. The agency has also issued regulations ensuring interoperability for PTC when it finally arrives. Laws and regulations also come at a price of course. Because most FRA regulations mimic self-interest, their costs have not been large, although the carriers have recently begun to complain about their paperwork burden.

The decades after about 1975 have also seen the public takeover of passenger travel as Amtrak has largely replaced the private common carriers for long-distance traffic, while states and municipalities provide commuter service. By removing the albatross of passenger traffic from the freight railroads, Amtrak helped make them viable. In a backhanded way, therefore, because it freed up their resources, it also contributed to their safety. In the years after the Penn Central bankruptcy, federal and some state funds not only rebuilt the Northeast Corridor, they rejuvenated the commuter lines as well. Yet government control has not been a panacea, for while passenger travel is safe, most of the gains occurred before the mid-1980s and improvements since then have been quite modest. Foot-dragging in the design of safer passenger coaches, despite NTSB prodding and the development of crashworthy automobiles is disconcerting. The public sector has been similarly slow to focus on non-train accidents.

The area where public policy has had the greatest impact has been in providing resources for grade crossing improvements. Although the spectacular decline in crossing fatalities from roughly 1,600 a year in 1965 to 240 in 2015 owes something to the development of freeways, Mothers Against Drunk Driving, Operation Lifesaver, and better automobiles, it is hard to imagine such gains without the federal funding for crossing removal and improved guarding. Certainly, states and localities have proved reluctant to bankroll many projects. As other researchers have shown, federal investments have also had a high payoff. The benefits of crossing safety accrue to train passengers and crew as well as local automobile and pedestrian travelers, which supports the case for federal funding. Still, crossing safety may well have hit diminishing returns and undoubtedly should compete for funds with other potential safety projects.

What of the future? Previous chapters have demonstrated that since 1980 train accidents, worker fatalities, and deaths at grade crossings, all expressed relative to exposure, have declined steadily at 2 to 4 percent a year, and even trespassing fatalities have fallen since 1990. Only passenger fatality rates have stagnated. While it is worth remembering that in 1955 no one imagined that a half century of improving railroad safety would abruptly end, there is reason for continued optimism. Passenger safety, which has shown the least improvement, should benefit from the recent focus on little accidents, and the advent of better crossing guards, PTC, and more crashworthy equipment. The use of remote sensors and controls—including drones—will continue to reduce risks. With the railroads' commitment to closing private crossings, along with continuing federal financial support and more sophisticated guarding, grade crossing safety seems likely to improve. There has been a focus on trespassing as well. More fundamentally, the broad erosion of safety that began in the 1950s reflected the collapse of the railroads at a time when there was little grasp of just how debilitating economic regulation had been and when railroad safety had far less institutional support. Despite worries by the railroad press that Congress will reinstitute economic regulation, that appears unlikely; nor are the FRA and the NTSB going away. Thus, it seems inconceivable that such a turnaround could again occur and, far more likely, that railroad safety will steadily improve.

There are, however, some clouds on the horizon. As this book has argued, since about 1980, with adequate financing, the freight railroads have had both the motivation and expertise to make good safety decisions while until recently, regulatory costs have been modest, for Congress and the FRA have largely reinforced private decision-making. While this package has been responsible for steady safety gains, the recent laws and regulations governing PTC and hazmat shipping move in a different direction, for both mandate large and apparently cost-ineffective expenditures. Specifying minimum crew sizes may come with a stiff price tag as well. Inevitably, these occur at the expense of other, perhaps better, safety investments elsewhere, and it is questionable whether Congress can allocate the carriers' money more wisely than they do. Although a safety collapse seems incredible, large-scale misallocation of resources might reduce future safety gains, and it is worth remembering that richer is only safer when decision makers use the resources wisely.

Appendix 1
Train Accidents That Shaped Railroad Safety, 1831–1955

Table A1.1 lists 15 train accidents that significantly shaped railroad safety. That does not mean that there is any direct connection between the particular wreck and a particular law or regulation. More often the accident is one of several that highlights a problem or suggests a solution.

Table A1.1. Fifteen Important Train Accidents, 1831–1955

Date	Type of Accident	Railroad and Location	Casualties*	Significance
1831, June 17	Boiler explosion	South Carolina; Charleston, SC	1 killed	First recorded train accident
1853, March 4	Rear collision	PRR; Mount Union, PA	7 killed	Failure to protect train typifies problems of worker misfeasance
1853, May 6	Bridge accident	New Haven; Norwalk, CT	46 killed	Open drawbridge; one of the first large-scale disasters. A state law eventually required stopping at such bridges
1856, July 17	Head-on collision	North Pennsylvania; Camp Hill, PA	66 killed	Exemplifies weaknesses of Timetable and Train Order system
1871, August 16	Rear collision	Eastern; Revere, MA	29 killed	Massachusetts commissioners force improvements in operating practices
1877, December 27	Bridge collapse and fire	Lake Shore; Ashtabula, OH	81 killed	Worst accident of the nineteenth century; begins movement to improve bridge safety
1887, January 4	Head-on collision and fire	Baltimore & Ohio; Republic, OH	13 killed	One of several leading states to ban stoves
1888, October 10	Rear collision	Lehigh Valley; Mud Run, PA	65 killed	Exemplified continued sloppy operating practices
1904, September 24	Head-on collision	Southern; Hodges, TN	63 killed	One of many wrecks leading to public outcry, block signals, and Safety First
1911, August 25	Derailment from transverse fissure	Lehigh Valley; Manchester, NY	20 killed	Results in major rail research
1913, July 30	Rear collision	PRR; Tyrone, PA	1 killed	One of several wrecks demonstrating the superiority of steel cars
1918, July 9	Head-on collision	Nashville, Chattanooga & St. Louis; Nashville, TN	101 killed	One of several collisions about this time leading the ICC to order automatic train control in 1922
1922, August 12	Grade crossing accident	Minneapolis, St. Paul & Sault St. Marie (the Soo); Annandale, MN	10 killed	Passenger train with oil truck. Symptomatic of the emerging crossing problem

(continued)

Table A1.1. (continued)

Date	Type of Accident	Railroad and Location	Casualties*	Significance
1945, August 9	Rear collision	Great Northern; Michigan, ND	34 killed	In train order territory; one of several leading the ICC to order expansion of block signals and train control in 1947
1950, November 22	Rear commuter collision	Long Island; Richmond Hill, NY	79 killed	At low speed; it revealed gaps in safety procedures and weaknesses in cars

Sources: Robert Shaw, *Down Brakes* (Chicago: Macmillan, 1961); ICC Accident Reports, http://dotlibrary.specialcollection.net.

*Before the twentieth century, injury data are unavailable or unreliable.

Appendix 2
Adjusting Train Accidents for Inflation and Reporting Changes, 1947–1978

This appendix explains the use of regression analysis to estimate the impact of changing prices and reporting thresholds on the reporting of collisions and derailments. It also contains both the reported and corrected data expressed relative to train or freight car miles. Because the accident data are count variables I employed the negative binomial regression routine in Stata. I first created the variable "CFactor," which is an index of the railroad prices divided by the reporting threshold, with the 1947 value set to 1. I regressed collisions and derailments separately on "correct" along with various controls, employing the logs of all variables. The results are in Table A2.1.

Table A2.1. The Impact of Inflation and Reporting Changes on Derailments and Collisions, 1947–1978

	Derailments*		Collisions*	
Variable	Coefficient	Z-Score	Coefficient	Z-Score
Log(Freight Car Miles)	3.42	7.45	—	—
Log(Train Miles)	—	—	2.30	5.42
Log(CFactor)	0.509547	3.23	0.6470062	4.22
Trend	0.053	7.90	0.035	3.23
Constant	23.09	–5.92	23.61	–4.01

Source: Author's calculations based on data from ICC/FRA, *Accident Bulletin*.
*Based on 32 observations.

I estimated the corrected figures as follows: corrected derailments = exp((log(derailments) – 0.509547 * log(CFactor)). The corrections for collisions were estimated the same way and both results are in Table A2.2. I employed these same procedures for the period 1979–2010, but the corrected figures track the reported data closely and so results are not shown.

Table A2.2. Derailments and Collisions Reported and Adjusted, 1947–1978

	Derailments			Collisions	
Year	Reported	Corrected	CFactor	Reported	Corrected
1947	9,404	9,404	1	4,451	4,451
1948	6,791	8,328	0.670	3,112	4,032
1949	4,867	6,033	0.656	2,077	2,728
1950	5,980	7,181	0.698	2431	3,067
1951	6,611	7,867	0.711	2,656	3,312
1952	5,783	7,016	0.684	2,514	3,213
1953	5,011	6,237	0.651	2,319	3,062

(continued)

* The railroad price index is from AAR, *Railroad Facts*, various years.

Table A2.2. (continued)

	Derailments		Collisions		
Year	Reported	Corrected	CFactor	Reported	Corrected
1954	4,109	5,008	0.678	1,796	2,309
1955	4,857	6,072	0.645	2,244	2,980
1956	5,369	6,426	0.703	2,393	3,006
1957	2,684	4,396	0.380	1,273	2,382
1958	2,579	4,401	0.402	976	1,759
1959	2,850	4,443	0.418	1,082	1,901
1960	2,918	4,485	0.430	989	1,707
1961	2,671	4,043	0.443	982	1,662
1962	2,830	4,232	0.454	999	1,665
1963	3,170	4,720	0.458	1,092	1,810
1964	3,399	5,000	0.469	1,229	2,006
1965	3,869	5,518	0.498	1,380	2,166
1966	4,447	6,216	0.518	1,552	2,375
1967	4,960	6,700	0.554	1,522	2,230
1968	5,487	7,209	0.585	1,727	2,442
1969	5,960	7591	0.622	1,810	2,461
1970	5,602	6,811	0.681	1,756	2,251
1971	5,131	5,984	0.739	1,529	1,859
1972	5,509	6,149	0.806	1,348	1,550
1973	7,389	7,775	0.905	1,657	1,768
1974	8,513	8,371	1.034	1,551	1,518
1975	6,331	8,975	0.504	1,002	1,561
1976	7,970	10,722	0.559	1,372	2,000
1977	8,073	11,978	0.461	1,363	2,249
1978	8,763	12,477	0.500	1,476	2,311

As discussed in the text, both collisions and derailments are usually expressed relative to some measure of exposure and so I have taken the reported derailment data from Table A.2.2 as well as the figures corrected for inflation and reporting requirements and in turn expressed them relative to freight car miles. Similarly, I have expressed the reported and corrected data for collisions relative to train miles. Both sets of figures are contained in Table A2.3.

Table A2.3. Adjusted Derailments and Collisions per Freight Car Mile and per Train Mile, 1947–1978

	Derailments per 10 Million Freight Car Miles		Collisions per Million Train Miles	
Year	Reported	Corrected	Reported	Corrected
1947	2.86	2.86	3.22	3.22
1948	2.12	2.60	2.33	3.02
1949	1.71	2.11	1.77	2.32
1950	2.01	2.41	2.04	2.57
1951	2.12	2.52	2.19	2.73
1952	1.89	2.29	2.17	2.77
1953	1.63	2.02	2.05	2.70
1954	1.43	1.74	1.73	2.23
1955	1.56	1.95	2.10	2.79
1956	1.70	2.03	2.26	2.84
1957	0.87	1.43	1.27	2.38
1958	0.92	1.57	1.10	1.98
1959	1.00	1.55	1.22	2.15
1960	1.03	1.58	1.16	2.00

Table A2.3. (continued)

	Derailments per 10 Million Freight Car Miles		Collisions per Million Train Miles	
Year	Reported	Corrected	Reported	Corrected
1961	0.98	1.48	1.21	2.06
1962	1.01	1.52	1.23	2.05
1963	1.12	1.67	1.34	2.22
1964	1.17	1.72	1.46	2.38
1965	1.31	1.86	1.48	2.33
1966	1.49	2.08	1.65	2.52
1967	1.64	2.21	1.70	2.49
1968	1.82	2.40	1.97	2.79
1969	1.96	2.50	2.09	2.85
1970	1.87	2.28	2.09	2.68
1971	1.76	2.05	1.95	2.37
1972	1.82	2.03	1.73	1.98
1973	2.36	2.49	1.99	2.13
1974	2.77	2.73	1.86	1.82
1975	2.29	3.25	1.33	2.07
1976	2.79	3.76	1.77	2.58
1977	2.81	4.17	1.82	3.00
1978	3.02	4.30	1.96	3.07

Appendix 3
Accidents That Shaped Railroad Safety, 1960–2010

The following 16 train accidents and 1 train incident mark important milestones in modern American railroad safety. Some of them directly influenced thinking, laws, and regulations about safety matters. Others are representative of classes of accidents that had similarly important results. All of these are discussed, at least briefly, in the text.

Note that casualty figures are approximate as different sources report different numbers—especially for injuries. Each of these accidents is far more complex than can be indicated here. The interested reader can find more complete descriptions at the National Transportation Safety Board (NTSB) website (www.ntsb.gov); the Federal Railroad Administration (FRA) website (www.fra.dot.gov), and the Department of Transportation's (DOT) Historical ICC Accident Reports (http://dotlibrary.specialcollection.net). The latter source also contains some early NTSB reports.

Table A3.1. Seventeen Important Train Accidents and Incidents, 1960–2015

Date	Type of Accident	Railroad and Location	Casualties	Significance
1960, March 1	Grade crossing with oil truck	ATSF; Bakersfield, CA	14 killed, 68 injured	Set in motion the modern federal program to reduce crossing accidents
1968, January 1	Derailment; hazmat	PRR; Dunreith, IND	Zero killed; 5 injured	First in a series of hazmat derailments that supported the Federal Railroad Safety Act of 1970
1969, February 19	Derailment; hazmat	CB&Q; Crete, NE	3 killed; 53 injured	
1969, September 11	Derailment; hazmat	Illinois Central; Glendora, MS	Zero killed; 1 injured	
1969, October 6	Derailment; hazmat	Southern; Laurel, MS	2 killed; 33 injured	
1972, October 30	Rear collision	Illinois Central Commuter; Chicago, IL	45 killed, 332 injured	Early disaster leading the NTSB to urge automatic train control and safer passenger cars
1978, February 22	Derailment; hazmat	L&N; Waverly, TN	16 killed; 25 injured	Speeded up tank car regulation and research
1978, February 26	Derailment; hazmat	Atlantic & St. Andrews Bay; Youngstown, FLA	13 killed; 147 injured	Speeded up tank car regulation and research
1987, January 3	Rear collision	Amtrak/Conrail; Chase MD	16 killed; 74 injured	Contributed to federal regulations for employee drug and alcohol testing

Table A3.1. (continued)

Date	Type of Accident	Railroad and Location	Casualties	Significance
1988, January 14	Head-on collision	Conrail; Thompson-town, PA	4 killed; 2 injured	The NTSB raised the problem of fatigue
1993, January 18	Head-on collision	Northern Indiana Commuter; Gary, IND	7 killed; 95 injured	Touched off modern work on crashworthy equipment
1996, February 1	Runaway and derailment	ATSF; Cajon CA	2 killed; 32 injured	The FRA mandated end-of-train devices that could apply brakes
1999, March 15	Grade crossing with semi-truck	Amtrak; Bourbonnais, IL	11 killed; 122 injured	Contributed to sealed corridor approach; need for intelligent guards
2005, January 6	Collision on side track; hazmat	NS; Graniteville, SC	9 killed; 75 injured	Contributed to positive train control (PTC) mandate in Railroad Safety Act of 2008
2006, August 5	Train incident	Long Island; Queens, NY	1 killed	Passenger fell into train-platform gap. Symbolized a shift in regulatory focus toward non-train dangers
2005, January 26	Grade crossing with SUV	Metrolink; Glendale, CA	11 killed; 42 injured	Metrolink responded, buying crash energy management (CEM) design cars
2008, September 12	Passenger/freight head-on collision	Metrolink; Chatsworth, CA	25 killed; 102 injured	Engine driver was texting; contributed to PTC mandate and cell phone regulations

Abbreviations

AAR — Association of American Railroads
ACSES — Advanced Civil Speed Enforcement System
ADA — Americans with Disabilities Act
APTA — American Public Transit Association
ARA — American Railway Association
ARBBA — American Railway Bridge and Building Association
AREA — American Railway Engineering Association
AREMA — American Railway Engineering and Maintenance of Way Association
ASME — American Society of Mechanical Engineers
ATSF — Atchison, Topeka & Santa Fe Railroad
B&M — Boston & Maine Railroad
B&O — Baltimore & Ohio Railroad
BART — Bay Area Rapid Transit
BLE — Brotherhood of Locomotive Engineers
BLET — Brotherhood of Locomotive Engineers and Trainmen
BLEVE — boiling liquid expanding vapor explosion
BN — Burlington Northern Railroad
BNSF — Burlington Northern, Santa Fe Railroad
C&O — Chesapeake & Ohio Railroad
CAB — Civil Aeronautics Board
CBO — Congressional Budget Office
CEM — crash energy management
CFR — Code of Federal Regulations
CN — Canadian National Railways
COFC — container on flatcar
CP — Canadian Pacific
CT — *Chicago Tribune*
CTC — Centralized Traffic Control
CWR — continuously welded rail
D&H — Delaware & Hudson
DART — Dallas Area Rapid Transit
DOT — Department of Transportation
ECP — electronically controlled pneumatic brakes
EOT — end-of-train
EPA — Environmental Protection Agency
FAA — Federal Aviation Administration
FAMES — Fatality Analysis of Maintenance of Way Employees and Signalmen
FAST — Facility for Accelerated Service Testing
FCC — Federal Communications Commission
FELA — Federal Employers Liability Act
FEMA — Federal Emergency Management Agency
FHA — Federal Highway Administration
FRA — Federal Railroad Administration
Frisco — St Louis, San Francisco Railroad
FTA — Federal Transit Administration
GAO — Government Accountability Office
GRMS — Gage Restraint Measuring System
HAL — heavy axle load
HML — Hagley Museum and Library
ICC — Interstate Commerce Commission
ICG — Illinois Central Gulf Railroad
IEEE — Institute of Electrical and Electronics Engineers
ISTEA — Intermodal Surface Transportation Efficiency Act
KCCU — Kheel Center, Cornell University
KCS — Kansas City Southern Railway Company
L&N — Louisville & Nashville Railroad
LAT — *Los Angeles Times*
LE — *Locomotive Engineer*
LPG — liquefied petroleum gas
L/V — lateral to vertical

Metro Metropolitan Area Transit Authority
MKT Missouri-Kansas-Texas
MoPac Missouri Pacific Railroad
MARC Maryland Area Regional Commuter
MARTA Metropolitan Atlanta Rapid Transit Authority
MBTA Massachusetts Bay Transportation Authority
MR *Modern Railroads*
MTA Metropolitan Transport Authority
MTB Materials Transportation Bureau
MU multiple unit
N&W Norfolk & Western
NA National Archives
NAS National Academy of Sciences
NEC Northeast Corridor (railway line)
NS Norfolk Southern Railroad
NSC National Safety Council
NTSB National Transportation Safety Board
NYT *The New York Times*
OSHA Occupational Safety and Health Administration
OTA Office of Technology Assessment
P&LE Pittsburgh & Lake Erie
PCF possible contributing factor
Pennsy Pennsylvania Railroad
PERT Program Evaluation and Review Technique
PIH poisonous inhalation hazard
PR *Progressive Railroading*
PRESS Passenger Rail Equipment Safety Standards
PRR Pennsylvania Railroad
PRRC Pennsylvania Railroad Collection
PTC positive train control
RA *Railway Age*
RG Record Group
REM *Railway Engineering and Maintenance*
RLC *Railway Locomotives and Cars*
RPI Railway Progress Institute
RSC *Railway Signaling and Communication*
RTS *Railway Track & Structures*
SACP Safety Assurance and Compliance Program
SEPTA Southeast Pennsylvania Transit Authority
SOFA Switching Operations Fatality Analysis
Soo Minneapolis, St. Paul & Sault St. Marie Railroad
SP Southern Pacific Railroad
STOP Safety Training Observation Program
TGC track geometry car
TOFC trailer on flatcar
TOPS Total Operations Processing System
TQM Total Quality Management
TRR *Transportation Research Record*
TTC Transportation Technology Center
TTD Track-Train Dynamics
UP Union Pacific Railroad
UTU United Transportation Union
WP *The Washington Post*
WSJ *The Wall Street Journal*

Notes

Introduction

1. NTSB, *Collision of Norfolk Southern Freight Train 192 with Standing Norfolk Southern Local Train P22 with Subsequent Hazardous Materials Release at Graniteville, South Carolina, January 6, 2005* (RAR 05/04, Washington, 2005); NTSB, *Collision of Metrolink Train 111 with Union Pacific Train LOF65-12, Chatsworth, California, September 12, 2008* (RAR 10/01, Washington, 2010). These and the other accidents described are available at www.ntsb.gov.

2. Aaron Wildavsky, "Richer Is Safer," *Public Interest* 60 (Summer 1980): 23–39, and "Why Health and Safety Are Products of Competitive Institutions," in *Market Liberalism: A Paradigm for the 21st Century*, ed. David Boaz (Washington: Cato Institute, 1993), 379–387.

3. Symes is from US 85th Cong., 2d Sess., Senate Committee on Interstate and Foreign Commerce, Hearings, *Problems of the Railroads Part I, Testimony of the Railroads* (Washington, 1958), 125.

4. For an introduction to Schumpeter's ideas, see http://www.econlib.org/library/Enc/CreativeDestruction.html. Good discussions of the railroad in American life are H. Roger Grant, *Railroads and the American People* (Bloomington: Indiana University Press, 2012), and Maury Klein, *Unfinished Business: The Railroad in American Life* (Hanover, NH: University Press of New England, 1994). See, too, George Douglas, *All Aboard* (New York: Paragon, 1994), which contains the quotation on xv. See also my "Railroad Cartoons," http://sophia.smith.edu/~maldrich.

5. "Railroad Epitaphs," *Saturday Evening Post* 35 (December 27, 1856): 8.

6. Constance Nathanson, *Disease Prevention as Social Change* (New York: Russell Sage Foundation, 2007), argues that public health advocates require heroes and villains; I have applied her idea to safety as well.

7. A top hit of 2014 was St. Vincent's "Digital Witness." Readers whose musical tastes have not been permanently stuck in the 1950s may find it helpful to know that "Maybelline" tells the story of a car chase between a Cadillac and a V-8 Ford, while "Tell Laura I Love Her" recounts an automobile accident from speeding.

8. "Positive Train Control: Oh What a Mess," *Trains* 74 (March 2014): 16, General OneFile web.

9. Robert Higgs, *Crisis and Leviathan* (New York: Oxford University Press, 1987).

10. The new system of the 1960s is described in James Q. Wilson and John J. Dilulio Jr., *American Government: The Essentials: Institutions and Policies*, 12th ed. (Boston: Wadsworth, 2011), Chapter 17. The quotation is from "UTU's Thompson Decries Return of Robber Barons," UTU *News* 38 (May 2006): 1.

11. Stephen Kelman, *Regulating America, Regulating Sweden: A Comparative Study of Occupational Health Policy* (Cambridge, MA: MIT, 1981); David Vogel, *National Styles of Regulation: Environmental Policy in Great Britain and the United States* (Ithaca, NY: Cornell University Press, 1986). For an example of labor criticism, see "FRA Issues Questionable Remote Control Safety Audit," Locomotive Engineers and Trainmen *News* 18 (June 2004): 1.

12. Mancur Olson, *Power and Prosperity* (New York: Basic, 2000). See also Omar Azfar, *Market Augmenting Government: The Institutional Foundations for Prosperity* (Ann Arbor: University of Michigan Press, 2003). My use of the term is somewhat broader than in these works.

13. Richard Saunders, *Main Lines: Rebirth of North American Railroads, 1970–2002* (DeKalb: Northern Illinois University Press, 2003), 381.

Chapter 1. The Long View

1. The classic discussion of Americans' ambivalence toward the new technology that railroads

represented is Leo Marx, *The Machine in the Garden* (New York: Oxford University Press, 1964). For more detail on the materials in this chapter, see my *Death Rode the Rails: American Railroad Accidents and Safety, 1828–1965* (Baltimore: Johns Hopkins University Press, 2006).

2. Because fatalities were measured with greater accuracy than injuries, they are used exclusively in this chapter. Social costs, of course, include such things as environmental damage as well as fatalities (and injuries).

3. Thomas Grattan, *Civilized America*, 2d ed., vol. 1 (London: Bradbury and Evans, 1859), 161; Charles Weld, *A Vacation Tour in the United States and Canada* (London: Longman, Brown, Green and Longmans, 1855), 161.

4. Frederick Marryat, *A Diary in America*, part 2, vol. 1 (London: Longman, 1839), quotations on 26–27; Michel Chevalier, *Society, Manners and Politics in the United States* (Boston: Weeks, Jordan, 1839), 271. For a superb discussion of the ways in which geography and resources shaped American railroads, see James Vance, *The North American Railroad: Its Origin, Evolution and Geography* (Baltimore: Johns Hopkins University Press, 1995).

5. For a historical discussion of internal principal-agent problems, see Naomi Lamoreaux et al., "New Economic Approaches to the Study of Business History," *Business and Economic History* 26 (Fall 1997): 57–79.

6. Nathan Rosenberg, "The Direction of Technological Change: Inducement Mechanisms and Focusing Devices," *Economic Development and Cultural Change* 18 (October 1969): 1–24.

7. The best discussion of early attitudes is Craig Miner, *A Most Magnificent Machine: America Adopts the Railroad, 1825–1852* (Lawrence: University of Kansas Press, 2010), Chapter 7.

8. "The Railroad Slaughter at Norwalk," New Orleans *Times-Picayune* (May 18, 1853); "The Railroad Murder," *Albany Evening Journal* (August 18, 1853); "The W. & P. Railroad Disaster," *New York Weekly Herald* (August 20, 1853). "The ground was strewn . . ." comes from Miner, *A Most Magnificent Machine* (n. 7), 132.

9. *New York Observer and Chronicle* (February 8, 1955): 48. The reader interested in the portrayal of accidents in cartoons should consult my website: "Railroad Cartoons," http://Sophia.Smith .edu~maldrich.

10. "The Norwalk Calamity," *New York Weekly Herald* (May 28, 1853). On liability, see James Ely, *Railroads and American Law* (Lawrence: University of Kansas Press, 2001).

11. For a detailed discussion of the introduction of the telegraph to early railroading, see Benjamin Schwantes, "Fallible Guardian: The Social Construction of Railroad Telegraphy in 19th Century America" (PhD diss., University of Delaware, 2008). That failure is instructive is from John Dewey but modern engineers agree. See, for example, Henry Petroski, *Success through Failure: The Paradox of Design* (Princeton, NJ: Princeton University Press, 2008).

12. Steven Usselman, *Regulating Railroad Innovation: Business, Technology and Politics in America, 1840–1920* (New York: Cambridge University Press, 2002), is the best overall study of the institutions shaping railroad technology.

13. No history of the railroad supply industry exists. Two excellent modern discussions of locomotive production are John K. Brown, *The Baldwin Locomotive Works, 1831–1915: A Study in American Industrial Practice* (Baltimore: Johns Hopkins University Press, 1995), and Albert Churella, *From Steam to Diesel: Managerial Customs and Organizational Capabilities in the Twentieth Century Locomotive Industry* (Princeton, NJ: Princeton University Press, 1998). The equipment industry is discussed in John White, *The American Railroad Freight Car: From the Wood-Car Era to the Coming of Steel* (Baltimore: Johns Hopkins University Press, 1993), and his *The American Railroad Passenger Car* (Baltimore: Johns Hopkins University Press, 1978).

14. A modern review of the literature on offsetting behavior is James Hedlund, "Risky Business: Safety Regulation, Risk Compensation and Individual Behavior," *Injury Prevention* 6 (June 2000): 82–90.

15. The classic source on nineteenth- and early twentieth-century railroad technology is Steven Usselman, *Regulating Railroad Innovation* (Cambridge: Cambridge University Press, 2002).

16. Thomas McCraw, *Prophets of Regulation* (Cambridge, MA: Harvard University Press, 1984), is the best single source on the Massachusetts Railroad Commission, which he terms a "sunshine commission." "Voluntarism" is on 27 and "gun behind the door" on 35.

17. "No Excuse for Dreadful Railway Holocausts," *The Boston Globe* (February 17, 1887).

18. The literature on the origins of federal regulation is vast. For a sampling, see Gabriel Kolko, *Railroads and Regulation, 1877–1916*

(New York: W. W. Norton, 1965); Stephen Skowronek, *Building a New American State: The Expansion of National Administrative Capacities* (New York: Cambridge University Press, 1982); and Ari and Olive Hoogenboom, *A History of the ICC: From Palliative to Panacea* (New York: W. W. Norton, 1976).

19. The carriers opposed legislation that would have mandated a specific coupler technology. There was little opposition to a performance law that simply required cars to couple automatically. While the new equipment sharply reduced worker fatalities and injuries, it did not do so immediately. This was partly because it was phased in but also because it required a good deal of learning by using. It also led to some offsetting behavior—for example, coupling cars in yards at faster speeds.

20. The classic discussion of this case and its consequences is Albro Martin, *Enterprise Denied: Origins of the Decline of American Railroads, 1897–1917* (New York: Columbia University Press, 1971). See also K. Austin Kerr, *American Railroad Politics, 1914–1920* (Pittsburgh: University of Pittsburgh Press, 1968).

21. None of these acts made major contributions to safety because none dealt effectively with important risks. The Hours of Service Act was poorly designed to deal with fatigue (Chapter 4), while the risks to trainmen from cleaning ash pans and other dangers from locomotives were comparatively minor. "A Third Man on the Engine," RA 35 (February 13, 1903): 7.

22. The literature on risk perception has become large. An early statement is Amos Tversky and Daniel Kahneman, "Judgment under Uncertainty: Heuristics and Biases," *Science* 185 (September 27, 1974): 1124–1131. For a good introduction, see James Flynn et al., *Risk Media and Stigma* (London: Earthscan: 2001), and Paul Slovic, *The Perception of Risk* (London: Earthscan, 2000). See also Timur Kuran and Cass Sunstein, "Availability Cascades and Risk Regulation," *Stanford Law Review* 51 (April 1999): 683–768. See also Cass Sunstein and Richard Zeckhauser, "Overreaction to Fearsome Risks," *Environmental and Resource Economics* 48 (March 2011): 435–449. On risk amplification and attenuation, see Nick Pidgeon et al., eds., *The Social Amplification of Risk* (New York: Cambridge University Press, 2003). Lynde is quoted in "Responsibility for Railway Wrecks," *Lexington Herald* (September 14, 1904).

23. ICC, *In Re Investigation of Accident on the New York, New Haven & Hartford Railroad at Westport, Conn., on October 3, 1912* (Washington, 1912); "A Morgan Rockefeller Road," *Boston Journal* (November 20, 1912); "Commuters Call Road a 'Cancer,'" *Springfield Union* (November 28, 1912). For details, see my "Another Wreck on the New Haven: Accidents, Risk Perception and the Stigmatization of the New York, New Haven & Hartford Railroad, 1911–1914," *Social Science History* 39 (Winter 2015): 613–646.

24. The ICC did not get the power to regulate hazards other than explosives until 1921.

25. "The New Haven and New England," RA 53 (December 20, 1912): 1171. For railroads' increasing focus on public relations, see Ray Stannard Baker, "Railroads on Trial," *McClure's* 26 (March 1906): 535–549; Scott Cutlip, "The Nation's First Public Relations Firm," *Journalism Quarterly* 43 (March 1966): 269–280; and Mark Aldrich, "Public Relations and Technology: The Standard Railroad of the World and the Crisis in Railroad Safety," *Pennsylvania History* 74 (Winter 2007): 74–104. For studies of the emergence of corporate public relations, see Alan Raucher, *Public Relations and Business, 1900–1929* (Baltimore: Johns Hopkins University Press, 1968); Richard Tedlow, *Keeping the Corporate Image: Public Relations and Business, 1900–1950* (Greenwich, CT: JAI, 1969); and Roland Marchand, *Creating the Corporate Soul* (Berkeley: University of California Press, 1998).

26. That labor contracts may involve a gift exchange relationship is an old idea in both sociology and economics. See, for example, F. J. Roethlisberger and W. J. Dickson, *Management and the Worker* (Cambridge, MA: Harvard University Press, 1947). For modern examples, consult Avner Offer, "Between the Gift and the Market: The Economy of Regard," *Economic History Review* 50 (August 1997): 450–476.

27. "Trespassers Killed and Injured on the New York Central," RA 55 (August 15, 1913): 267–269. For a brief history of the Good Roads Movement, see Bruce Seely, *Building the American Highway System: Engineers as Policy Makers* (Philadelphia: Temple University Press, 1987).

28. In the interest of brevity and readability I have omitted a number of minor safety rules that the ICC promulgated. For example, in 1921 it instituted a performance requirement for locomotive headlights and mandated a number of other modifications in locomotive design that improved safety. See my "Safe and Suitable Boilers: The Railroads, the Interstate Commerce Commission and Locomotive Safety, 1900–1945," *Railroad*

History 171 (Winter 1994): 23–44. In addition, the Signal Inspection Act came in 1937, but it had more to do with employment maintenance than safety.

29. "Evolution in Chilled Wheel Rims," RA 124 (June 26, 1948): 86; "New York Railroad Club; Signal Glass," RA 56 (February 27, 1914): 432.

30. For the economic gains associated with many of these changes, see, for example, "Building Permanent Bridge Reduces Maintenance Forces," REM 21 (March 1925): 92–93; "What is the Economic Weight of Rails?" RA 88 (March 24, 1930): 1231–1237.

31. For studies of track-train dynamics, see "A Scientific Study of Railway Track Under Load," RA 68 (March 5, 1920): 670–673; "Causes and Prevention of Freight Car Derailments," RA 80 (January 16, 1926): 234–236; "Causes and Prevention of Freight Car Derailments," *Railway Mechanical Engineer* 100 (January 1926): 23–26; "Freight Car Derailments: A Discussion of Known and Unknown Wheel and Rail Reactions—Field for Research is Large," RA 83 (October 22, 1927): 763–764. See also "Axle Loads, the Track Structure and Rail Failures," REM 23 (June 1927): 293–294.

32. That business firms can be thought of as coalitions of interest groups that shape the direction of technological change is from W. Bernard Carlson, "The Coordination of Business Organization and Technological Innovation within the Firm: A Case Study of the Thomson Houston Electrical Company in the 1880s," in *Coordination and Information: Historical Perspectives on the Organization of Enterprise*, ed. Naomi Lamoreaux and Daniel Raff (Chicago: University of Chicago Press, 1995), 55–94.

33. E. Scott Geller, *The Psychology of Safety Handbook* (Boca Raton, FL: Lewis, 2001), Chapter 3.

34. Seely, *Building the American Highway System* (n. 27). On medicalization, see John Burnham, "Why Did the Infants and Toddlers Die? Shifts in Americans' Ideas of Responsibility for Accidents: From Blaming Mom to Engineering," *Journal of Social History* 29 (Summer 1996): 817–837.

35. For Dunn, see "Making Friends with the Public," RA 116 (February 12, 1944): 348–351, quotation on 350; "Pointing the Way to Future Progress," RA 116 (May 20, 1944): 950; "Planning Program Now on Home Stretch," RA 118 (January 4, 1945): 6; "Results of Research Will Guide Railroads," RA 119 (January 5, 1946): 6.

36. "The ICC Signal Order," RA 122 (June 28, 1947): 1295. Congress had clarified the ICC powers to inspect and mandate block signals as well as train control in the Signal Inspection Act of 1937.

37. For a broad assessment of the impact of the diesel, see Maury Klein, "Replacement Technology: The Diesel as a Case Study," *Railroad History* 162 (Spring 1990): 109–120. For more on the safety of diesels, see my "The Contribution of the Diesel Locomotive to Trainmen's Safety on US Railroads, 1927–1954," *Journal of Transport History* 25 (September 2004): 75–92.

38. "Train Communication on P.R.R.: Extensive Installation Authorized After Long and Painstaking Research and Experimentation," RA 117 (October 28, 1944): 652; ICC, *Ex Parte No. 176 Accident Near Jamaica, N.Y.* (Washington, 1950); "Tests Overdue on 17 of the Cars in L.I. Wreck," *New York Herald Tribune* (December 3, 1950); "Head of Pennsylvania Blames P.S.C. for Troubles of L.I.R.R.," *New York Herald Tribune* (January 7, 1951).

39. "Grade Crossings Again in the Limelight," RA 121 (July 27, 1946): 125; "G.N. Box Cars 'Light Up' at Crossings during Darkness," RA 123 (November 15, 1946): 246.

Chapter 2. Off the Tracks

1. NTSB, *Pennsylvania Railroad, Train PR11-A, Extra Train 2210 West and Train SW-6 Extra 2217 East, Derailment and Collision, Dunreith, Indiana, January 1, 1968* (RAR, Washington, 1968), quotation on 14; author's interviews with John C. Linegar, Joe Pitcher, and Michael Smith, February 2–3, 2016, Indiana State Police Museum, http://www.in.gov/isp/museum.htm.

2. The term "system accident" is from Charles Perrow, *Normal Accidents* (New York: Basic, 1984).

3. Accidents that did not meet the threshold with or without casualties were termed "train service accidents" by the ICC and later "train incidents" by the FRA.

4. See A. E. Schulman and C. E. Taylor, *Analysis of Nine Years of Railroad Accident Data, 1966–1974* (Washington: AAR, 1976), Appendix C-3.

5. For details of the calculations as well as calculations of corrected collisions, see Appendix 2.

6. ICC, *Accident Bulletin for 1948, No. 117* (Washington, 1949), 14; *Accident Bulletin for 1949, No. 118* (Washington, 1950), 13.

7. There was little awareness among contemporaries of just how badly changing prices and reporting thresholds distorted the train accident data. An exception to this generalization is the Railroad Retirement Board, *Reporting of Accidents*

and Casualties in the Railroad Industry (Washington, 1959). Congressional hearings reveal little understanding of this issue, while an AAR study, Schulman and Taylor, *Analysis of Nine Years* (n. 4), focused only on the period when accidents were overreported.

8. Railroad Retirement Board, *Reporting* (n. 7); "FRA Creates Storm on Safety," MR 19 (December 1974): 19–20; "Safety's Strangest Year," MR 30 (February 1975): 54–57.

9. For discussions of the demise of the Penn Central, see Joseph Daughen and Peter Binzen, *The Wreck of the Penn Central* (Boston: Little, Brown, 1971), and Stephen Salsbury, *No Way to Run a Railroad* (New York: McGraw Hill, 1982).

10. "The 'System' That Always Fails," RA 124 (January 10, 1948): 21.

11. Other contributing causes were the movement in economic activity away from the Northeast and the shift in GDP toward services and away from heavy industry. For a perceptive contemporary overview, see George Hilton, *The Northeast Railroad Problem* (Washington: American Enterprise Institute, 1975).

12. For an excellent contemporary assessment, see David P. Morgan, "Who Shot the Passenger Train?" *Trains* 14 (April 1959): 14–35, 45–49, passim. McGinnis is from US 85th Cong., 2d Sess., Senate Committee on Interstate and Foreign Commerce, Hearings, *Problems of the Railroads Part I, Testimony of the Railroads* (Washington, 1958), 305.

13. US 87th Cong., 1st Sess., Senate Committee on Interstate and Foreign Commerce, *Special Study Group on Transportation Policies in the United States* (Washington, 1961) (hereafter *Doyle Report*) provides a good review of railroad problems as they appeared about 1960.

14. For one of many similar articles urging better service, see "You Should Know What Shippers Think of Rail Service," MR 22 (January 1967): 60–64. Richard Saunders notes that the organization of rail labor by craft made it especially vulnerable to technological change. See his *Main Lines: Rebirth of the North American Railroads, 1970–2002* (DeKalb: Northern Illinois University Press, 2003). The Long Island example is from *Doyle Report* (n. 13), 495. "RRs Gain Ground in Crew Law Battle," RA 158 (February 22, 1965): 64.

15. The literature on ICC economic regulation is vast. A good review is William Burt, "Gabriel Kolko Revisited," *Railroad History* 214 (Spring–Summer 2016): 22–45. A good brief contemporary discussion of the ways that regulation shaped the carriers' technology and behavior is Hilton, *The Northeast Railroad Problem* (n. 11). The Florida East Coast is from "The FEC Story: Survival without Unions," RA 157 (July 27, 1964): 34. The cost of "featherbedding" is from "1959 Target 'Featherbedding,'" RA 146 (February 16, 1959): 9.

16. State over-taxation is from *Doyle Report* (n. 13), 487; "Taxes, Railroads are Special," MR 32 (July 1977): 69–71. The airport example is from ICC, *Railroad Passenger Train Deficit* (Docket 31954, Washington, 1959), 43. That the real money losers were branch lines with substantial traffic is from Hilton, *Northeast Railroad Problem* (n. 11). The reader interested in more detail on the causes of the railroads' decline should consult Robert Gallamore and John Meyer, *American Railroads: Decline and Renaissance in the Twentieth Century* (Cambridge, MA: Harvard University Press, 2014).

17. The store-door story is from Albert Churella, "Saving the Railroad Industry to Death: The Interstate Commerce Commission, the Pennsylvania Railroad, and the Unfulfilled Promise of Rail-Truck Cooperation," *Business and Economic History* (2006), http://www.thebhc.org/. See also his "Delivery to the Customer's Door: Efficiency, Regulatory Policy, and Integrated Rail-Truck Operations, 1900–1938," *Enterprise and Society* 10 (March 2009): 98–136. US 85th Cong., 2d Sess., Hearings, *Problems* (n. 12), contains data on states' decisions on service discontinuances (221–222) and Perlman's claim (231).

18. Symes's calculation is from US 85th Cong., 2d Sess., Hearings, *Problems* (n. 12), 110. The 2.2 billion is from US 94th Cong., 1st Sess., House Committee on Interstate and Foreign Commerce, Committee Print 17, *Materials Concerning the Effects of Government Regulation on Railroads and an Economic Profile of Railroads in the United States* (Washington, 1975), Table 1. The Big John story is told briefly in Saunders, *Main Lines* (n. 14), 24–25. Paul MacAvoy and James Sloss, *Regulation of Transport Innovation: The ICC and Unit Coal Trains to the East Coast* (New York: Random House, 1967). For a general review of the deadening effect of regulation on railroad innovation, see Richard Barber, "Technological Change in American Transportation: The Role of Government Action," *Virginia Law Review* 50 (June 1964): 824–895.

19. The "Avoidable" column in Table 2.1 includes the net costs that were solely related to passenger service and would therefore be saved if the service were ended. Total costs contain some that were joint

with freight operations. In fact, some of these could probably have been avoided too. See John Meyer et al., "Avoidable Costs of Passenger Train Service," in US 85th Cong., 2d Sess., Hearings, *Problems* (n. 12), 235–284.

20. For traffic density, see "Now or Never," MR 31 (November 1976): 48–51. Return on investment is revenues from all sources minus operating expenses, taxes, and rents, divided by total railway investment. It is essentially the average return on railroad capital available to stock and bondholders. "What Price Money?" MR 30 (February 1975): 51–53, quotation on 52.

21. Jan Heier and A. Lee Gurley, "The End of Betterment Accounting: A Study of the Economic, Professional and Regulatory Factors That Fostered Standards Convergence in the Railroad Industry, 1955–1983," *Accounting Historians Journal* 34 (June 2007): 25–55.

22. "Postwar History is Repeating Itself," RA 127 (October 8, 1949): 45. The adjustment calculation is as follows. If I is the rate of return, P the rate of inflation, and R the rate of return adjusted for inflation, then $R = I - P$.

23. The annual proceedings of the AREA for the 1960s reveal many concerns that the carriers were unable to attract top engineering talent. See A. Scheffer Lang, "Railroad Engineering from a New Vantage Point," AREA *Proceedings* 69 (1968): 746–750. Hamilton is quoted in Art Detman, "A Hell of a Way to Run a Railroad," *Sales Management* 96 (March 4, 1966): 27–30, quotation on 30. Roger Grant, *Erie Lackawanna: Death of an American Railroad, 1938–1992* (Stanford, CA: Stanford University Press, 1994). That managers were mediocrities is my judgment based on his work. Roger Grant, *Visionary Railroader: Jervis Langdon Jr. and the Transportation Revolution* (Bloomington: Indiana University Press), 79. See too his "The New B&O," *Trains* 69 (August 2009): 24–29. Don Hofsommer, *Grand Trunk Corporation: Canadian National Railways in the United States, 1971–1992* (East Lansing: Michigan State University Press, 1995). Many writers treat bad management as a cause of railroad collapse. Yet all industries have some bad managers, and the carriers' rapid turnaround after 1980 suggests that overall railroad management was not that bad when it had freedom to manage.

24. "Standards of Maintenance . . . Have Not been Observed," RA 151 (July 17, 1961): 16–18; "The Great Railroad Robbery," MR 29 (September 1974): 67–72; "Deferred Maintenance Reaches $950 Million," RA 130 (March 26, 1951): 46. The $5.7 billion figure is from "A Problem That Only Management Can Solve," RTS 70 (September 1974): 6. Accounting changes make maintenance data for 1978–1982 not comparable to either earlier or later figures.

25. The AREA data are from its *Proceedings*, various years.

26. Loomis is from "Needed: Money for Improvement," RA 146 (March 16, 1959): 22. The B&O is from "Standards of Maintenance" (n. 24). "How little" is from "Time for a Change in Basic Attitude," RTS 60 (February 1964): 8. "M/W Penny Pinching is Returned with Interest," MR 20 (September 1965): 91; "The Crisis in Roadway and Track," MR 28 (March 1973): 50; "The Cancer of Maintenance Deferral," MR 29 (March 1974): 50–52; "Time Bomb in the Track Department," RTS 70 (April 1974): 8; "Can a Crisis in M/W be Avoided?" RTS 70 (October 1974): 6. Symes is from US 87th Cong., 2d Sess., Hearings, *Problems* (n. 12), 125.

27. Above I have argued that unprofitability retarded the introduction of new technology. The big cars, like diesel locomotives, were an exception because external finance was available for equipment but not fixed plant. For a good history of containerization, see Marc Levinson, *The Box: How the Shipping Container Made the World Smaller and the World Economy Bigger* (Princeton, NJ: Princeton University Press, 2006); "Big Cars for Auto Parts," RA 154 (June 3, 1963): 9.

28. "New GATC Tank Car Scores with Shippers," RA 151 (September 4, 1961): 18; "PRR Will Appraise Giant Tank Cars," RLC 140 (February 1966): 38–39; "L&N Tests 100 Ton Hoppers," RA 152 (March 26, 1962): 18; "100 New L&N Cars to Replace 600 Old Cars," RA 152 (April 16, 1962): 29; "ACL Tests 'Whopper Hopper,'" RA 158 (April 26, 1965): 44; "The Big Cars Become Regulars," RA 159 (November 1, 1965): 11; "New Trends in Car Design Bring Promises, Problems," RA 151 (July 31, 1961): 49.

29. "History Repeats Itself," RLC 142 (June 1968): 24. The recommendations are contained in AAR Mechanical Division *Proceedings* 29 (1959): 499.

30. For a good history of concerns with larger cars, see George H. Way, "Heavy Cars—What are the Issues?" AREA *Proceedings* 76 (1975): 616–621; John Fishwick, "Luncheon Address," AREA *Proceedings* 72 (1971): 487–489; J. W. Read, "Luncheon Address," AREA *Proceedings* 82 (1981): 431–434.

31. "What Today's Heavier Wheel Loads are Doing to Your Rails," RA 150 (March 6, 1961): 16; "Heavy Cars Cut Rail, Wheel Life," RA 155 (December 2, 1963): 11; "Wide Gage Research Leads to a New Standard on UP," RTS 76 (March 1978): 14–17; "Abex: High Carbon Wheels Not Faulty," RA 175 (April 24, 1978): 13; Robert Sewell, Interview 185, August 11, 1975, Youngstown State University Oral History Project, quotation on 23; James Larson, "Ralph Rotten and the Chicago Great Western," *Railroad History* 214 (Spring–Summer 2016): 10–21.

32. "Causes of Shelly Spots and Head Cracks in Rail," AREA *Proceedings* 65 (1964): 576–606; "Heavier Cars Put New Pressure on Track Upkeep," RA 164 (March 18, 1968): 28; "Derailment Analysis: What Part Do Rail Failures Really Play?" RA 166 (March 10, 1969): 21; R. M. Owen, "The Rail Defect Picture and Advances in Detector Cars," RTS 67 (April 1971): 26–27; "Defective Rail," RTS 74 (December 1978): 26–27; "Rail Corrugations: Search for the Causes and the Cure," RTS 72 (February 1976): 24–27; J. Kalousek and R. Klein, "Investigation into Causes of Rail Corrugations," and F. E. King and J. Kalousek, "Rail Wear and Corrugation Studies," both in AREA *Proceedings* 77 (1976): 429–448 and 601–620, respectively.

33. L. F. Koci, "Derailments: Can More of Them be Explained?" RLC 146 (February 1972): 13–14.

34. "Report of the Committee on Freight and Passenger Car Construction," AAR Mechanical Division *Proceedings* 29 (1959): 482.

35. "Effects of the Bad One," MR 31 (June 1976): 59–63.

36. ICC, *Chicago, Milwaukee, St. Paul and Pacific Railroad Company, In Re Accident at Mauston, Wisconsin on April 2, 1957* (Report 3743, Washington, 1957). For an insightful and readable discussion of the engineering aspects of train accidents, see George Bibel, *Train Wreck: The Forensics of Rail Disasters* (Baltimore: Johns Hopkins University Press, 2012).

37. Schulman and Taylor, *Analysis of Nine Years* (n. 4), also show that train accident rates were higher among the bankrupt roads. A statistical analysis for the years 1959–1965, using a negative binomial formulation, yields the equation below. ROA is the decimal return on assets; car miles are in thousands; figures in parentheses are z-scores; there are 511 observations. The large standard error of ROA results because railroad profits are poorly measured, while derailments result from cumulative unprofitability as well as other factors such as track-train dynamics. An anonymous reviewer also points out that the ratio of branch lines to main track likely increased derailments.

$$\text{Ln (Derailments)} = \underset{(16.06)}{2.54} - \underset{(1.16)}{1.36}\,\text{ROA} + \underset{(6.91)}{6.87 * 10^{-7}}\ \text{Car Miles}$$

Average ROA for the sample was 2.3 percent. Doubling profitability would therefore have reduced derailments by $\exp^{(-1.36 * 0.023)} - 1 = -3.1$ percent.

38. ICC, *Union Pacific Railroad Company, Wamego, Kansas, December 21, 1967* (Report 4142, Washington, 1967). A reviewer points out that the shift from steam to diesel locomotives might have worsened risks from drowsy cab crew as it is virtually impossible to fall asleep while operating a steam locomotive.

39. FRA, *Illinois Central, Riverdale, Illinois, September 26, 1969* (Report 4163, Washington, 1969).

40. For concerns with loss and damage and the role that yard accidents were playing, see "L&D—Time for a New Look," RA 154 (March 25, 1963): 17–22, and "The L&D War: Bigger Role for the Shipper," RA 165 (September 30, 1968): 40–42.

41. "Safety Semantics," MR 30 (February 1975): 45. An analysis of company data for 1969–1978 similar to that in note 37 finds a statistically insignificant and weaker negative impact of profitability on collisions. NTSB, *Train Accidents Attributable to "Employee Negligence"* (RSS 72-1, Washington, 1972), found that the five most profitable carriers had a below-average accident rate for both human factor and all accidents.

42. Schulman and Taylor, *Analysis of Nine Years* (n. 4).

43. That accidents from track-train dynamics were more prevalent at lower speeds is from George Hilton, "Slack," *Trains* 36 (February 1976): 23–28; "Derailment Analysis: What Part Do Rail Failures Really Play?" RA 166 (March 10, 1969): 21–23. The newspaper headline is from NYT (September 18, 1968).

44. See, for example, "No Death from Rail Handling of Explosives in '51," RA 132 (June 23, 1952): 85.

45. David Vogel, *Fluctuating Fortunes* (New York: Basic, 1989), Chapter 4. Representative Culver is from US 91st Cong., 1st Sess., House Committee on Government Operations, Hearings, *Transportation of Hazardous Materials* (Washington, 1969), 6. For a discussion of hazmat yard accidents, see

NTSB, *Railroad Yard Safety—Hazardous Materials and Emergency Preparedness* (SIR 85/02, Washington, 1985).

46. Poten & Partners, "The Story of LPG" (2003), http://www.poten.com/document.aspx?id=4252&filename=The_Story_of_LPG.pdf. Ammonia production from Richard Sutch and Susan Carter, eds., *Historical Statistics of the United States Millennial Edition* (Cambridge: Cambridge University Press, 2000), Series Dd387. Chlorine is from J. S. Sconce, ed., *Chlorine: Its Manufacture, Properties and Uses* (Malabar, FL: Robert F. Krieger Co., 1972).

47. I searched *The New York Times*, the *Los Angeles Times*, the *Chicago Tribune*, and *The Boston Globe* for the period 1947–1974 for reports of hazmat accidents.

48. NTSB, *Railroad Accident Report, Southern Railway Company Train 154, Derailment with Fire and Explosion, Laurel, Mississippi, January 25, 1969* (Washington, 1969), quotation on 1. For good summaries of this and the following accidents, see FRA, *Major Railroad Accidents Involving Hazardous Materials Release: Composite Summaries, 1969–1978* (DOT-TSC-FRA 80-22, Washington, 1980).

49. NTSB, *Railroad Accident Report, Chicago Burlington and Quincy Railroad Company Train 64 and Train 824 Derailment and Collision with Tank Car Explosion, Crete, Nebraska, February 18, 1969* (RAR 71-2, Washington, 1971), quotation on 39.

50. NTSB, *Illinois Central Railroad Company Train Second 76 Derailment at Glendora, Mississippi, September 11, 1969* (RAR 70-2, Washington, 1970), block quotation on 10. For Eastland, see "30,000 Escape Menace of Gas after Derailment in Mississippi," NYT (September 13, 1969).

51. Callao is from FRA, *Major Railroad Accidents* (n. 48). See also Callao file, box 7, Railroad Accident Jackets 1968–1971, FRA Records, RG 399, NA.

52. For some of the literature on risk perception, see Chapter 1, n. 22. "Train Blast Rips Part of Town," CT (January 26, 1969); "Ammonia Gas from Leaking Tank Car Kills 8 in Nebraska," CT (February 19, 1969); "Blast Sets Town Ablaze," CT (June 22, 1970).

53. This and the previous paragraph are based on the following: NAS, *Science and Technology in the Railroad Industry, A Report to the Secretary of Commerce* (Washington, 1963); "Railroad Research Sets Record Pace," RA 160 (April 18, 1966): 18–23; "A New Research Effort for the Railroads," RA 166 (March 31, 1969): 30–33. Manufacturing is from William Leonard, "Research and Development in Economic Growth," *Journal of Political Economy* 79 (March–April 1971): 232–256. For Harris, see "'We Must Assign Substantial Priority to Research,'" RA 170 (January 11, 1971): 32–33, and "Dr. Harris to Retire from AAR," RA 186 (June 1, 1985): 81. For background, see "Contributions to Railway Research—No. 1: Casting for the Future," RA 141 (August 13, 1956): 38–42, which is the first of several articles in a series on research by railroad suppliers and briefly notes railroad research.

54. Gellman's views are in "Railroading Still Needs a Research Conception," MR 22 (October 1967): 96–98.

55. "Freight Brakes for the '80s," RA 182 (December 28, 1981): 38–43; "Braking Those Gold Stripers," RA 157 (September 14, 1964): 28–29; "Post Zephyr Braking," *Trains* 26 (January 1976): 40–46; "How the L&N Trains Engineers," RA 161 (December 19, 1966): 34–35; "Mid-Train Slaves Ease N&W Train Handling," MR 23 (September 1968): 130; "Freight Train Unbounded," *Trains* 70 (September 2010): 22–33; "Santa Fe's 'College' for Enginemen," RA 173 (November 27, 1972): 22–24; "The Faster Freight Brake Control Valve," RA 178 (March 28, 1977): 30.

56. "Today's biggest . . ." is from "Journal Box Debate Continues," RA 147 (July 20, 1959): 41–43. ICC, *The New York Central Railroad Company, Syracuse, NY, August 25, 1957* (RAR 3774, Washington, 1957).

57. "Norfolk & Western Finds Lubricator Pads Do Reduce Hot Boxes," RLC 131 (August 1957): 40–41. For Timken, see "Interchangeable Freight Car Bearing," RA 137 (July 12, 1954): 45. The cost-benefit study is in "If All Freight Cars Had Roller Bearings," RA 131 (December 31, 1952): 37–42. "Entirely charity" is from "Are We Really Serious about Hot Boxes?" RA 132 (February 11, 1952): 68. "Complete conversion . . ." is from "What's Ahead for Roller Bearings?" RLC (December 1970): 13–15. "Four hundred detectors" is from "Hot Box Detectors Prove Their Worth," RA 152 (February 19, 1962): 26–28; "Electronics Speeds Hot Box Data to Forestall Costly Train Delays," RA 144 (March 31, 1958): 22–23. Origins of detectors are from "How to Pinpoint Hot Boxes," RA 142 (April 1, 1957): 47. Fifteen hundred is from "Hot Box Sleuths Map New Strategy," RA 167 (September 11, 1969): 83–84.

58. "Hot Boxes Dropping to All-Time Low," RLC 136 (April 1962): 40; "Journal Problems Have Top Priority in AAR Research," RLC 137 (May 1963):

19–21; "Anti-Hot Box Campaign is Paying Off," RA 152 (June 25, 1962): 48–48; "Solving the Problem," RLC 143 (January 1969): 11.

59. "How Longer, Heavier Cars Affect Today's Traffic," RA 158 (March 22, 1965): 26–30; "High-Cant Tie Plates," RTS 65 (February 1969): 16–17; "Portec Rail and Flange Lubricators," RTS 65 (September 1969): 78; "What Detectors are Telling You," RA 163 (July 17, 1967): 14–18; "Rock and Roll: Is Reducing Wheel Lift Enough?" RLC 142 (June 1968): 22–23; "Progress is Goal of AAR Research," RLC 142 (August 1968): 13–15; "Construction Package Built with Panelized Turnouts," RTS 65 (April 1969): 25. For the L&N test track, see "AAR Mechanical Research Covers a Wide Range," RA 160 (June 27, 1966): 22–23.

60. Clearly, the men whose responsibility it was to maintain the carriers' track and equipment must have been well aware of the train accident problem. But for a long time such matters had been left for them to solve at their own pace and they had tended to ignore political matters. They were probably not aware that by the 1960s those days were over. George H. Way, "Remarks," AREA *Proceedings* 79 (1978): 409–412, quotation on 410; Rex Manion, "Up to Our Neck in Bugs," AAR Mechanical Division *Proceedings* 34 (1969): 169–175.

61. NTSB, *Railroad Accident Report Southern Railway* (n. 48); NTSB, *Safety Effectiveness Evaluation Analysis of Proceedings of the National Transportation Safety Board into Derailments and Hazardous Materials, April 4–6, 1978* (SEE 78-2, Washington 1978), 3.

62. This paragraph is based on NTSB, *Railroad Accident Report Southern Railway* (n. 48). For the response of the Mechanical Division, see Appendix 4 of that document. See also "Wheels, A Real Quandary," RLC 144 (March 1970): 13–17, which contains "far from satisfied."

63. "Harris to Retire from AAR and TTD," RA 186 (June 1, 1985): 81; "AAR Adopts Recommended Minimum Track Inspection Standards," RLC 144 (January 1970): 26; "New Role for the AAR's Engineering Division," RLC 144 (March 1970): 19; "Engineers Get the Score on Pending Safety Legislation," RLC 144 (April 1970): 22–26; "AAR Now Has Recommended Track Standards," RTS 66 (November 1970): 28–31.

Chapter 3. On the Right Track

1. US Railroad Retirement Board, *Safety in the Railroad Industry* (Washington, 1962), Chapter 4. In 1962, the commission published *Interstate Commerce Commission Activities, 1937–1962* (Washington, 1962), which provided a review of its work in all areas, including safety. Had the commissioners felt the need for new legislation this was the perfect spot to raise the issue but they did not. US 88th Cong., 2d Sess., House Committee on Government Operations, Hearings, *Interstate Commerce Commission Operations: Railroad Safety* (Washington, 1965). The commission's annual reports document efforts to broaden the power brake law and gain jurisdiction over train radio, but they reveal no awareness of broader safety problems until 1964. See ICC, *78th Annual Report for the Fiscal Year Ending June 30, 1964* (Washington, 1964), 73.

2. This point is made by Richard Barsness, "The Department of Transportation: Concept and Structure," *Western Political Quarterly* 23 (September 1970): 500–515. The best modern treatment of the origins of the DOT is Mark Rose et al., *The Best Transportation System in the World* (Columbus: Ohio State University Press, 2006), who also see bureaucratic centralization as the main motive for the DOT, but say nothing about safety. The NTSB became an independent, stand-alone agency in 1974.

3. David P. Morgan, "7:38 a.m., Oct. 30, 1972," *Trains* 33 (January 1973): 3–4, quotation on 3. NTSB concerns are publicized in "Train Accidents Up 71% in Six Years," NYT (April 11, 1968).

4. For reports noting the lack of federal authority governing accidents, see FRA, *Railroad Accident Investigation Report No. 4141, New York Central Railroad Company, Cold Springs, Ohio, January 8, 1968* (Washington, 1968), and its *Railroad Accident Investigation Report No. 4143, Louisville and Nashville Railroad Company, Casky, Ky., January 29, 1968* (Washington, 1968). On the changing political culture of the 1960s, see the essays in Brian Balough, ed., *Integrating the Sixties: The Origins, Structures and Legitimacy of Public Policy in a Turbulent Decade* (University Park: Penn State University Press, 1996). See also David Vogel, *Fluctuating Fortunes* (New York: Basic, 1989).

5. US 90th Cong., 2d Sess., House Committee on Interstate and Foreign Commerce, Hearings, *Federal Standards for Railroad Safety* (Washington, 1968); "financial officials" is on 91. Rose et al., *Best Transportation System* (n. 2), xxi, remark on the extent to which the robber baron image still shaped Washington thinking after World War II. Probably the AAR chose to ignore railroad finances because the story did not fit with its claim that there was not much wrong with railroad safety.

6. "Fiery Crash Kills Fourteen," LE 94 (March 11, 1960): 1, and "Photos Cry Out for Action to Stop Crossing Crashes," LE 95 (February 10, 1961): 2. US 90th Cong., 2d Sess., Hearings, *Federal Standards* (n. 5), quotations on 54 and 341–342.

7. "39 Hurt in Mississippi Butane Blast," NYT (January 25, 1969); "RR Wreck, Gas Cloud Kill Eight," *The Boston Globe* (February 19, 1969); "Perilous Cargos," WP (May 8, 1969); "DOT Order," February 28, 1969, and "Report of the Task Force on Railroad Safety, June 30, 1969," both in file 07.4, box 313, General Correspondence 1967–1972, DOT, Office of the Secretary, RG 398, NA (hereafter DOT files).

8. US 91st Cong., 1st Sess., Senate Committee on Commerce, Hearings, *Federal Railroad Safety Act of 1969* (Washington, 1969), quotation on 8.

9. The phosgene gas story is from "Congressmen Protest Rail Plan to Carry Poison Gas to Atlantic," NYT (May 8, 1969). US 91st Cong., 1st Sess., Hearings, *Federal Railroad Safety Act* (n. 8), quotations on 210 and 214.

10. "BLE Prepares for Participation in Hearings on Rail Safety Bill," LE 103 (May 9, 1969): 1.

11. For Glendora, see "30,000 Escape Menace of Gas," NYT (September 13, 1969). The wreck was specifically mentioned in the hearings. See also "Rail Accidents Stir Tempest," NYT (August 30, 1969); "Danger Grows in Shipping Deadly Cargo," LAT (September 28, 1969); and "Dangerous Rail Shipments Burden Innocent Bystander," WP (December 20, 1969). Volpe's discussion of his committee is from US 91st Cong., 1st Sess., Hearings, *Federal Railroad Safety* (n. 8), 336–339. For splits over the bills, see William Skutt [Brotherhood of Locomotive Engineers] to R. N. Whitman, September 19, 1969; Carl Lyon to Administration, November 21, 1969; and Benjamin Biaggini to John Volpe, December 4, 1969, all in file 07.4, box 107, DOT files.

12. US 91st Cong., 2d Sess., House Committee on Interstate and Foreign Commerce, Hearings, *Railroad Safety and Hazardous Materials Control* (Washington, 1970); P.L. 91-458 in 49 US Code, 20101-20144. On the act of 1893, see Steven Usselman, "Air Brakes for Freight Trains: Technological Innovation in the American Railroad Industry, 1869–1900," *Business History Review* 58 (Spring 1984): 30–50.

13. "A Bad Day for the Railroads," CT (June 23, 1970).

14. "AREA Members Hear FRA Plans 'More Rigid' Track Standards," RTS 67 (April 1971): 20–24; "Task Force Estimates Cost of FRA Track Standards," RTS (December 1971): 19.

15. The standards discussed here are from "FRA Track Standards," RTS 67 (November 1971): 30–35. Revisions are in "Modifications Issued of FRA Track Standards," RTS 68 (November 1972): 24–28. Leverage and respect are from Allen Boyd [Illinois Central; Secretary of Transportation, 1967–1969], "Luncheon Address," AREA *Proceedings* 72 (1971): 561–565. Barnes is from "What the FRA Track Standards will Mean for My Road," RTS 68 (January 1972): 24–27, quotation on 26. An FRA statistical study estimated that the standards resulted in a 1 percent increase in maintenance spending; see J. E. Tyworth et al., *Analysis of Railroad Track Maintenance Expenditures for Class I Railroads, 1962–1977* (FRA 81-20, Washington, 1982).

16. "FRA Makes Its Proposals for Freight Equipment Inspection Standards," RLC 146 (October 1972): 9–18; "Federal Rules Effective over Next Three Years," RLC 148 (January 1974): 14–15. The quotation is from "FRA Standards Get Quick Revision," RLC 148 (February 1974): 22–23.

17. NTSB, *Safety Effectiveness Evaluation of the Federal Railroad Administration's Hazardous Materials and Track Safety Programs* (SEE 79-2, Washington, 1979), 25.

18. GAO, *Railroad Safety: New Approach Needed for Effective FRA Safety Inspection Program* (Washington, 1990), quotation on 2, and *Railroad Safety Weaknesses Exist in FRA's Enforcement Program* (Washington, 1991). For the risk-based programs and other changes, see GAO, *Rail Transportation: Federal Railroad Administration's New Approach to Railroad Safety* (Washington, 1997); "Interim Waiver of FRA Track Standards for Penn Central," RTS 69 (November 1973): 17; "FRA Engineer Hits Railroads for Noncompliance with Safety Rules," RTS 70 (October 1974): 14–17.

19. The problems with the Sperry railer are described in "Are FRA Standards Restricting Rail Flaw Detection?" RA 176 (June 9, 1975): 42–43. Tyworth et al., *Analysis of Railroad Track Maintenance* (n. 15), also found that the regulations reduced speed but had no impact on accidents. NTSB, *Special Study of Proposed Track Safety Standards* (RSS 71-2, Washington, 1971), 3.

20. For the stepped-up enforcement plan, see US 94th Cong., 1st Sess., Senate Committee on Commerce, Hearings, *Implementation of the Federal Railroad Safety Act of 1970* (Washington, 1975), 92–101, and US 96th Cong., 2d Sess., House

Committee on Interstate and Foreign Commerce Hearings, *Railroad Safety* (Washington, 1980), 17–21. For a critique, see "FRA Creates Storm on Safety," MR 29 (December 1974): 19.

21. FRA claims about the L&N and the ICG are from US 97th Cong, 2d Sess., House Committee on Commerce, Science, and Transportation, Hearings, *FRA Railroad Safety Program* (Washington, 1982), 9–10. For a defense of the L&N, see "L&N Rebuilds for Growing Coal Tonnage," RA 181 (November 10, 1980): 36–41. Whatever the reasons, the L&N and the ICG together accounted for about 17 percent of the decline in derailments from 1980 to 1981.

22. This history and description is based largely on AAR Mechanical Division *Proceedings* 48 (1977): 27–37; Transportation Research Board, *Ensuring Tank Car Safety* (Washington, 1994). While DOT-105 cars were insulated, the insulation was not designed to provide high-temperature protection.

23. For AAR doings, see Appendix 4 of NTSB, *Southern Railway Company Train 154 Derailment with Fire and Explosion, Laurel, Mississippi, January 25, 1969* (RAR, Washington, 1969); AAR Mechanical Division *Proceedings* 40 (1969): 67–75, and 42 (1971): 35–44. The contract with the AAR is from US 94th Cong., 1st Sess., Senate Committee on Commerce, Hearings, *Implementation of the Federal Railroad Safety Act of 1970* (Washington, 1975), 19; tank car research funding is on 72. The tank car project is also discussed in US 95th Cong., 2d Sess., House Committee on Interstate and Foreign Commerce, Hearings, *Railroad Safety Authorization Fiscal Year 1979* (Washington, 1978), 198–199.

24. D. E. Adams et al., *Rail Hazardous Material Tank Car Design Study* (DOT-FR 20069, 1975); D. E. Adams, *Cost/Benefit Analysis of Thermal Shield Coatings Applied to 112A/114A Series Tank Cars* (FRA-OR&D 75-39, 1974); C. G. Interrante et al., *Analysis of Findings of Four Tank Car Accident Reports* (FRA-OR&D 75-50, 1975); W. A. Bullerdiek et al., *A Study to Reduce the Hazards of Tank Car Transportation* (FRA-RT 71-74, 1970). The rule is "Specifications for Tank Cars," *Federal Register* 42 (September 15, 1977): 46306–46315. For a discussion of the DOT-111 cars ca. 1990, see NTSB, *Safety Study: Transport of Hazardous Materials By Rail* (SS 91/01, Washington, 1991).

25. Reginald Whitman to John Volpe, May 19, 1970, and John Reed to John Volpe, July 7, 1970, both in hazmat file, box 3, Administrative Subject Correspondence 1967–1970, FRA Records, RG 399, NA; "133 Hurt in Decatur Tank Car Blast," CT (July 20, 1974); "Motorists Die in Chlorine Cloud," WP (February 27, 1978); "Gas in Derailed Tank Car Kills 8," NYT (February 27, 1978); "Tank Car Safety," WP (April 7, 1978). "Danger Rides the Rails," LAT (April 10, 1978), contains the quotation. The standards as of 1978 are from US 95th Cong., 2d Sess., *Railroad Safety Authorization* (n. 23), 96. The emergency order is available at www.fra.dot.gov.

26. "Our Unsafe Railroads," *Washington Star* (March 10, 1980); "When 20/20 Vision Failed," RA 181 (June 20, 1980): 46–47.

27. Author's interviews with John C. Linegar, Joe Pitcher, and Michael Smith, Indiana State Police Museum, http://www.in.gov/isp/museum.htm. Further reorganization resulted in hazmat regulations being moved to the Pipeline and Hazardous Materials Safety Administration. For the developments in this paragraph, see US 95th Cong., 2d Sess., *Railroad Safety Authorization, Fiscal Year 1979* (n. 23), 255–259, and US 96th Cong., 2d Sess., House Committee on Interstate and Foreign Commerce, Hearings, *Railroad Safety* (Washington, 1980), 108–128; "CSX Operation Respond," www.nysema.org/pdfs/ . . . /2015/CSX%20Operation%20Respond.pdf; "This Train Delivers Safety," *Trains* 75 (March 2015): 24, General OneFile web.

28. For example, AAR regulations were already phasing out high-carbon wheels when the FRA emergency order banned them.

29. NTSB, *The Accident Performance of Tank Car Safeguards* (Special Investigation Report HZM 80-1, Washington, 1980); Transportation Research Board, *Ensuring Tank Car Safety* (n. 22), 82. For discussion of the decision to use normalized steel, see NTSB, *Derailment of Canadian Pacific Railway Freight Train 292-16 and Subsequent Release of Anhydrous Ammonia near Minot, North Dakota, January 18, 2002* (RAR 04/01, Washington, 2004). Theodore Glickman, "Rerouting Railroad Shipments of Hazardous Materials to Avoid Populated Areas," *Accident Analysis and Prevention* 13 (May 1983): 329–335; Christopher Barkan et al., "Railroad Derailment Factors Affecting Hazardous Materials Transportation Risk," TRR 1825 (2003): 64–74; Robert Anderson and Christopher Barkan, "Railroad Accident Rates for Use in Transportation Risk Analysis," TRR 1864 (2004): 88–98; Morteza Bagheri et al., "Reducing the Threat of In-Transit Derailments Involving Dangerous Goods Through Effective Placement Along the Train Consist," *Accident Analysis and Prevention* 43 (May 2011): 613–620.

30. For the irrelevance of conspicuity devices on the end of freights, see the testimony of Harold Hall

of the Southern in US 94th Cong., 2d Sess., House Committee on Interstate and Foreign Commerce, Hearings, *Federal Railroad Safety Authorization Act of 1976* (Washington, 1976), 113–114.

31. Robert Gallamore and John Meyer, *American Railroads: Decline and Renaissance in the Twentieth Century* (Cambridge, MA: Harvard University Press, 2014), Chapter 11. John K. Brown has pointed out to me that abandonment of passenger traffic on a line might result in lower maintenance standards.

32. Dingell and Ingram are from US 93rd Cong., 2d Sess., House Committee on Interstate and Foreign Commerce, Hearings, *Railroad Safety* (Washington, 1974), 12, 68–69, 74. At virtually the same time as the safety hearings the carriers' financial woes were being aired in US 93rd Cong., 2d Sess., House Committee on Interstate and Foreign Commerce, Hearings, *Northeast Rail Transportation* (Washington, 1973).

33. Rooney and Hall are from US 94th Cong., 1st Sess., House Committee on Interstate and Foreign Commerce, Hearings, *Federal Railroad Safety Authorization Act of 1975* (Washington, 1975), 4, 8; for Amtrak, see 73–77. Deferred maintenance numbers are from Appendix F. Rooney's worry about management is from US 94th Cong., 2d Sess., House Committee on Interstate and Foreign Commerce, Hearings, *Federal Railroad Safety Authorization Act of 1976* (Washington, 1976), 2. See also "Chief Engineer's Team Takes Close Look at Penn Central," RTS 71 (February 1975): 34–37.

34. On Waverly, see, for example, "Homes Evacuated for Unloading Second Tanker at Blast Site," NYT (April 26, 1978); "Death Rides the Rails," *The Boston Globe* (April 9, 1978); "Gas in Derailed Car Kills 8," NYT (February 27, 1978); "Alabama Town Blocks Railroad After Three Derailments in Week," NYT (April 30, 1978).

35. For Staggers's remarks, see US 95th Cong., 2d Sess., House Committee on Interstate and Foreign Commerce, Hearings, *Railroad Safety Authorization Bill for Fiscal Year 1979* (Washington, 1978), 171–174.

36. US 95th Cong., 2d Sess., Hearings, *Railroad Safety Authorization Bill* (n. 35), 171–174; Peterson is from 77 and 81. Sullivan's testimony and the FRA's new position are from 87–113. Office of Technology Assessment, *An Evaluation of Railroad Safety* (Washington, 1978).

37. Florio is from US 96th Cong., 2d Sess., House Committee on Interstate and Foreign Commerce, Hearings, *Railroad Safety* (Washington, 1980), 1. Florio's role in economic deregulation is from Rose et al., *Best Transportation System* (n. 2), 207–208. Vogel, *Fluctuating Fortunes* (n. 4), discusses changes in business political power during these years but ignores railroad matters. For labor support, see "Rail Deregulation: One More Chance," Brotherhood of Maintenance of Way Employees *Journal* 89 (October 1980): 22–23. For a good review of the act, see "The Staggers Act: Back from the Brink," RA 191 (August 1990): 30–34. On the possibility of returning to economic regulation, see "Keep Railroads Free to Keep the Economy Moving," RA 214 (March 2013): 15–22, and "Rails Face Old Threats from New Congress," RA 215 (December 2014): 18.

38. The statistics are

$$\text{Ln(Derailment Rate)} = 0.72 - \frac{0.041}{6.41}\text{Trend};$$
$$R^2 = 0.33; N = 34, \text{ and}$$

$$\text{Ln(Collision Rate)} = 0.27 - \frac{0.046}{7.80}\text{Trend};$$
$$R^2 = 0.41; N = 35.$$

39. For an accessible discussion of the effects of economic deregulation, see "Free to Compete," *Trains* 70 (October 2010): 24–33.

40. "Speed and Consistency," MR 27 (July 1972): 39. See also "Do We Really Understand Marketing?" MR 27 (August 1972): 26–32; "As the Shipper Sees It," MR 28 (September 1973): 41–44; "A Look into Reliability," MR 28 (February 1973): 46–47. On the importance of accidents as a source of loss and damage, see "The L&D War: A Bigger Role for the Shipper?" RA 165 (September 30, 1968): 40–42. Richard Saunders also dates the carriers' turnaround from these years; see his *Main Lines: Rebirth of the North American Railroads, 1970–2002* (DeKalb: Northern Illinois University Press, 2003), Chapter 4. Non-railroaders also glimpsed the origins of the marketing turnaround; see Art Detman, "A Hell of a Way to Run a Railroad," *Sales Management* 96 (March 4, 1966): 27–60.

41. "How the C&NW Cuts Casualty Losses," RA 164 (March 25, 1968): 36–39. Chessie is from "L&D, The Fire is Under Control, But—," RA 175 (March 25, 1974): 17–19. "UP Sets '90s Priorities," RA 191 (February 1990): 27–30. "Rio Grande Converts All Hotshot Freights to Short Fast Concept," MR 22 (March 1967): 100–104; "Blending Service and Efficiency for Profit," RA 172 (February 28, 1972): 23–25, quotation on 25; "Catering to the Customer is Good Business," RA 172 (February 28, 1972): 26–28.

42. "Door to Door TOFC Pays Off," RA 182 (July 27, 1981): 16; "Conrail Intermodal: Improving a Winning Game," RA 190 (June 1989): 34–36. Chicago & Eastern Illinois is from "A Total Dedication to Intermodal Operations: TOFC/COFC," RA 177 (May 31, 1976): 19–20; "An Almost Fanatical Dedication to Service," RA 173 (July 31, 1972): 2. For more on the MoPac and TOPS, see Craig Miner, *The Rebirth of the Missouri Pacific, 1956–1983* (College Station: Texas A&M University Press, 1983). The SP is from "Running a Railroad with Dedication and Flair," MR 25 (July 1970): 90–93. "In MoPac Operations the Key Word . . . Change," RA 177 (March 31, 1976): 32–34. "Unit Train Operations," RA 179 (July 31, 1978): 25–31, reveals the range of commodities shipped in unit trains. For a good history of intermodal, see "Stacking the Deck," *Trains* 71 (November 2011): 32, General OneFile web.

43. "Contract Rates are Catching on," RA 181 (September 29, 1980): 32–32; "ICG Signs 20 Year Coal Rate Contract," RA 181 (June 28, 1980): 15. On the role of deregulation and contracting, see Marc Levinson, "Two Cheers for Discrimination: Deregulation and Efficiency in the Reform of U.S. Freight Transportation, 1976–1988," *Enterprise and Society* 10 (March 2009): 178–215. For a discussion of the importance of these kinds of long-term contracts, see Naomi Lamoreaux et al., "Beyond Markets and Hierarchies: Toward a New Synthesis of American Business History," *American Historical Review* 108 (April 2003): 404–433.

44. The potato example is from US 85th Cong., 2d Sess., Senate Committee on Interstate and Foreign Commerce, Hearings, *Problems of the Railroads Part I, Testimony of the Railroads* (Washington, 1958), 306. "Chessie/SCL Industries: A Merger of Equals," RA 182 (March 30, 1981): 26–29; "UP/MP/WP: A Whole Greater Than the Sum of the Parts," RA 182 (April 27, 1981): 20–26.

45. For the UP merger, see Saunders, *Main Lines* (n. 39), and Maury Klein, *Union Pacific: The Reconfiguration* (New York: Oxford University Press, 2011). Denis Breen suggests that some of the problems might have resulted because of the deteriorated state of the SP; see his "The Union Pacific–Southern Pacific Rail Merger: A Retrospective on Merger Benefits," *Review of Network Economics* 3 (September 2004): 283–322. For current commentary, see "Train Safety Inspections Follow Fatal Wrecks," NYT (August 31, 1997); "Study Faults Union Pacific Rail Safety," WP (September 11, 1997), which contains the quotation; and "The Wreck of the Union Pacific," *Fortune* 137 (September 30, 1998): 94–102. Most of Locomotive Engineer *News* 31 (September 1997) is devoted to the UP breakdown.

46. "High Productivity Railroading: Keep Rolling Stock Rolling," RA 187 (May 1, 1986): 32–33, quotations on 32.

47. "Derailments: The Problem Only Dollars Can Solve," RA 185 (September 9, 1984): 31–33. The productivity data are from P. E. Schoech and J. A. Swanson, "Patterns of Productivity Growth for U.S. Class I Railroads: An Examination of Pre- and Post-Deregulation Determinants," https://www.lrca.com. Over this same period, equipment maintenance and capital spending per owned freight car mile also increased.

48. The data for 1970–1975 are from GAO, *Information Available on Estimated Costs to Rehabilitate the Nation's Railroad Track and a Summary of Federal Assistance to the Industry* (Washington, 1976); for 1978–1984, see GAO, *Federal Assistance to Rehabilitate Railroads Should Be Reassessed* (Washington, 1980); for Conrail, 1976–1985, see CBO, *Economic Viability of Conrail* (Washington, 1986), and author's calculations. These figures exclude separate spending on the Northeast Corridor or any state aid. For examples of state spending, see "Iowa Points the Way . . . Blue Print for Track Rehabilitation," RTS 71 (September 1975): 14–17, and "Illinois: Away from Railroading, into Track Improvement," RA 180 (December 31, 1979): 31–32. In 1977, Wisconsin also began to provide aid to communities to help preserve freight service. In a few cases shippers also indirectly contributed funds by financing car repair; see "Shippers Finance Freight Car Repairs," RA 180 (April 30, 1979): 14.

49. "How Many Freight Cars are Too Many?" RA 180 (May 28, 1979): 32–33; Gallamore and Meyer, *American Railroads* (n. 31), Chapter 11. In the absence of productivity changes, real maintenance of way spending per track mile rose only about 7 percent between 1982 and 1990, but the rise of productivity increased actual purchasing power by about 80 percent.

50. "Conrail: Will a $5 Billion Investment in Roadway Rehabilitation Pay Off?" RA 177 (March 8, 1976): 20–23; "Conrail Rehabilitation: A Tremendous Job, Well Begun," RA 177 (November 8, 1976): 18–20; "Conrail: A Triumph of Men, Money and Machines," RA 183 (December 13, 1982): 22–25. The 20 percent figure is from GAO, *Railroad Revenues: Analysis of Alternative Methods to Measure Revenue Adequacy* (Washington, 1986), 57.

51. "ICG's Goal: Less Mileage, Better Maintained," RA 178 (March 14, 1977): 30–31, quotation on 31. Ahif is from "Matching M/W Spending Practice to Required Use of Track," RTS 75 (September 1979): 34–36, quotation on 36. "Why IC is Single Tracking," RA 191 (February 1990): 32–33; "How CN Stretches Its M/W Dollar," RA 181 (September 8, 1980): 65–67; "Grand Trunk: Good Track is Good Business," RA 183 (April 12, 1982): 15–18; Don Hofsommer, *Grand Trunk Corporation Canadian National Railways in the United States, 1971–1992* (East Lansing: Michigan State University Press, 1995), Chapter 3; "The Good-Track Roads: Seaboard System," RA 184 (May 1, 1983): 49–54, quotations on 49; "Milwaukee Road: Trimmed Down, Shaping Up," RA 184 (August 1, 1983): 54–55.

52. "M/W: A Change of Heart in the Executive Suite?" RA 182 (January 26, 1981): 50–52, quotation on 52.

53. Douglas Caves et al., "Productivity in U.S. Railroads, 1951–1974," *Bell Journal of Economics* 11 (Spring 1980): 166- 181, and Schoech and Swanson, "Patterns of Productivity Growth" (n. 47). A brief review that peers into possible future technological improvements is "6 High Tech Advances," *Trains* 68 (November 2008): 26–33.

54. "The Coming Explosion in Track Research," RA 174 (April 9, 1973): 35–37. FRA research funding for all purposes averaged $49 million a year from 1980 to 1984 and $27 million from 1985 to 1989. AAR budgets from "What's Needed in R&D," RA 192 (June 1991): 24–28, and "The Growing Role of R&D&T," RA 194 (May 1993): 61–64. "FRA, Reality through Research," MR 29 (January 1974): 62–65, contains a list of FRA research projects.

55. "The New Decision Makers?" RTS 74 (December 1978): 58. See also the editorial "The Spearhead of Technology," RTS 37 (June 1982): 69.

56. D. G. Ruegg, "Luncheon Address," AREA *Proceedings* 78 (1979): 454–458, quotation on 455.

57. "Penn Central Charts Economics of Hotbox Detection," MR 23 (September 1968): 76–78. For other examples of early use, see the following: "Improving Railroad Efficiency," RA 158 (January 25, 1965): 21–23; "Frisco's Industrial Engineers Have Big Computer Plans," RA 164 (June 24, 1968): 60–61; "Santa Fe: Better Performance through Industrial Engineering," RA 180 (February 12, 1979): 42–44; "Cost Benefit Analysis Establishes Project Priority," RTS 72 (November 1976): 36–37.

58. "U.S. Transit Agency Taking Over Center Operated by NASA," NYT (March 29, 1970). What role the FRA and the AAR played in choosing the research undertaken at Pueblo is not clear, but there may not have been much disagreement because most of the work was at least consistent with AAR priorities; see "Railroads Mount Widening R&D Attack; 12 Key Problem Areas," RA 175 (September 30, 1974): 20–25. For a modern evaluation of FRA research, see NAS Transportation Research Board, *Evaluation of the Federal Railroad Administration Research and Development Program* (Special Report 316, Washington, 2015).

59. "New M/W Technology Evolves at Pueblo's FAST Track," RTS 77 (March 9, 1981): 32–34. See also "FAST—Highlights of Test Results," RTS 78 (January 1982): 32–39. For a modern review of the TTC, see "Guiding the Way," *Trains* 69 (November 2009): 32–37.

60. "The Economic Loss Caused by 100-Ton Cars" and "The Implications of the 100-Ton Car," MR 37 (February 1980): 37 and 52–53, respectively. For HAL, see "Crucial to the Industry," RTS 86 (March 1990): 5. "Dragging the track" is from "Determining the Future of Heavy Haul," RTS 87 (September 1989): 11–17; "The Economics of Heavy Axle Loads," RA 192 (September 1991): 96–99. Two early votes in favor of the large cars are George Way, "Economics of Freight Car Size or Where's the Bottom of the Bathtub?" and John Slammon, "Preliminary Study of Rail Car Size," both in AREA *Proceedings* 80 (1979): 356–362 and 363–371, respectively. "Heavy Axle Loads Make Sense to BN," RTS 86 (April 1990): 31, 41, and R. R. Newman et al., "Burlington Northern's Assessment of the Economics of High Capacity Heavy Axle Load Cars," AREA *Proceedings* 91 (1990): 136–172. "Another 100 MGT for HAL," RTS 86 (December 1990): 11–13. "Level playing field" is from "Making History," RTS 86 (September 1990): 3. A modern discussion is Carl Martland, "Introduction of Heavy Axle Loads by the North American Rail Industry," *Journal of the Transportation Research Forum* 52 (Summer 2013): 103–125.

61. "Dr. Harris to Retire . . . ," RA 186 (June 1, 1985): 61; L. F. Koi, "Wheel and Rail Loadings from Diesel Locomotives," AREA *Proceedings* 72 (1971): 500–528. Individual carriers continue to support research and testing departments. For the BNSF, see "Poke, Scrutinize, Examine and Determine," *Trains* 70 (October 2010): 34–37.

62. "Dr. Harris to Retire . . ." (n. 61); C. W. Parker, "Track-Train Dynamics and the Overturned Rail," AREA *Proceedings* 75 (1974): 861–880. For early work by the SP and others, see "Individual Efforts Mark Truck Research," RLC 145 (June 1971):

14–16; "Putting the Computer into the Cab," RA 179 (September 25, 1978): 48–49, quotation on 49; "The Great Truck Race: Temporarily on Hold," RA 182 (April 27, 1981): 40–45; "High Performance High-Cube Covered Hopper Program," RA 184 (December 1, 1983): 83; "12 Years of TTD," RA 185 (June 1, 1984): 71; "TTD Bows Out at Year's End," RA 187 (September 1, 1986): 92; "A Brief Review of VTS," RA 188 (June 1, 1987): 45.

63. "Stop in the Name of Progress," RA 209 (April 2008): 48–49, quotation on 48. "Climbing the Pumpkin Vine," RA 208 (July 2007): 23, contains the fuel savings. "Wired for Progress," RA 209 (June 2008): 20–29; "New Braking Technology Taking Hold," RA 210 (October 2009): 20–24, quotation on 20.

64. "NDT Techniques are Constantly Refined," RLC 145 (April 1971): 24–25; "A Solution to Thermal Cracking," RA 183 (March 29, 1982): 43; "Perspective: Are Wheels Being Scrapped Unnecessarily?" RA 189 (April 1, 1988): 74; "Keeping Wheels in Shape," RA 186 (January 1, 1985): 39–43; "Wheel-Rail Impact Loading," RTS 82 (September 1986): 12–14; "Examining the Economics of High-Impact Wheel Loads," RTS 90 (March 1994): 29–32; "Economics of Wheel Impact Loading," RTS 91 (December 1995): 9–10. The one-sixteenth figure is based on data in "Keeping Wheels in Shape," RA 186 (January 1, 1985): 39–43.

65. Testimony of Edward Hamberger, in US 107th Cong., 2d Sess., House Committee on Transportation and Infrastructure, Hearings, *Recent Developments and Railroad Safety* (Washington, 2002), 41–43. See too "The Curious Case of Remote Control," RA 199 (January 1998): 54–56; "Locomotive Remote Control," RA 200 (February 1999): 24–25; and "The Power of Remote Control," RA 206 (January 2005): 25–32. "Making the Switch," RA 208 (March 2007): 23–25.

66. The formula is from Arnold Kerr, "Thermal Buckling of Straight Tracks: Fundamentals, Analysis, and Preventive Measures," AREA *Proceedings* 80 (1979): 16–47.

67. J. W. Farmer, "How Norfolk Southern Avoids Speed Restrictions on CWR Due to Heat," AREA *Proceedings* 94 (1993): 153–159; "Track Tester Takes Shape," RA 187 (March 1990): 33; "How UP Achieves Lateral Track Stability on CWR," RTS 87 (June 1991): 18–22; "Going Automatic with Tie Renewals," RTS 92 (December 1995): 12–14; Gary Carr and J. Kevin Kesler, "Gage Restraint Measurement—What is It Telling Us," AREMA *Proceedings* (1999): 324–332; "The Found Flaw," RTS 110 (January 2014): 27–28.

68. For flaw detector cars, see "New Refinements in Rail Detector Cars," RTS 62 (May 1966): 24–27; "Rail Flaw Detection, A Science That Works," RTS 86 (May 1990): 30–32; "Rail Flaw Detection Picks Up the Pace," RTS 91 (July 1995): 21–22. "Stopping Silent Killers," RA 207 (May 2006): 38–40; Joseph Palese and Thomas Wright, "Risk Based Ultrasonic Rail Test Scheduling on Burlington Northern Santa Fe," AREMA *Proceedings* (2000): 55–73. See too Allan Zarembski and Joseph Palese, "Characteristics of Broken Rail Risk for Freight and Passenger Railway Operations," AREMA *Proceedings* (2005): 515–525, which contains the 1 in 200 figure. NTSB, *Derailment of Norfolk Southern Railway Company Train 68QB119 with Release of Hazardous Materials and Fire, New Brighton, Pennsylvania, October 20, 2006* (RAR 08/02, Washington, 2008).

69. "Boston & Maine Expands Role of Mechanical Track Inspection," RTS 62 (April 1966): 26–29; "On the Southern: Track Inspection," RTS 67 (March 1971): 18–21; "Chessie's TGC-2—A Maintenance Management Tool," RTS 77 (November 1981): 26–28. For the FRA, see Arthur Clouse and Kevin Kesler, "Advancements in Track Inspection," AREMA *Proceedings* (2003): 24–32. "How Does Your Track Measure UP?" RA 205 (September 2004): 77–80; "Automating Track Inspection," RTS 106 (November 2010): 35–37; "Automated Ultrasonic Inspection Detects Cracks in Joint Bars," RTS 107 (April 2011): 16–18; "Can Drones Make Crude By Rail Safe?" *Trains* 75 (June 2015): 8, General OneFile web; "Wayside Data Advancing Fast," *Trains* 74 (March 2014): 20, General OneFile web.

70. "Testing of Subgrade Stabilization Fabrics," RTS 72 (July 1976): 20–21; "Pumping Track on FEC: Answer Found in Stabilization Fabric," RTS 74 (January 1978): 34–35; T. B. Hutchinson and R. F. Breese, "Field Evaluation of a Ballast-Subgrade Radar System," AREA *Proceedings* 81 (1980): 430–446; "Special Service Stretches Tie Life," RTS 83 (January 1987): 16–28; "Tiescan: Measuring Wood-Tie Condition with Sonic Waves," RTS 88 (March 1992): 22–24; "X-Rays See Through Tie Problems," *Trains* 76 (March 2016): 18, General OneFile web.

71. "Rail Technology—Where Do We Stand?" RTS 87 (November 1991): 15–16. See also "Clean Steel Manufacturing and Improved Rail Fatigue Life," RTS 89 (January 1993): 13–14. The defect rate is from Alan Zarembski et al., "The Effect of Improved Rail Manufacturing Process on Rail Fatigue Life," AREA *Proceedings* 92 (1991): 382–391.

72. The $100,000 figure is from "Extending Rail Life," RA 193 (March 1996): 49–53; "Scheduling Rail Testing," RTS 86 (November 1990): 9. "More Rail Grinding Economics," RTS 83 (February 1987): 12; John Stanford et al., "Burlington Northern Santa Fe Preventive Gradual Grinding Initiative," AREMA *Proceedings* (1999): 748–776; "Rail Replacement Criteria," RTS 89 (November 1993): 10–11; "Economics of Wheel Impact Loading," RTS 91 (December 1995): 9; "Effective Rail Inspection," RTS 83 (January 1987): 12; "Pushing the Planning Envelope on CP Rail," RTS 91 (February 1995): 14–15. For the CP, see M. D. Roney, "CP Rail's Track Maintenance Advisory System," AREA *Proceedings* 93 (1992): 315–330; Robert deVries et al., "Preventive Grinding Moves into the 21st Century on Canadian Pacific Railway," AREMA *Proceedings* (2001): 105–118; "Lightening Lateral Loading," RA 201 (April 2004): 43–45.

73. John Parola and Conrad Ruppert, "Systems of Track Infrastructure Safety at Amtrak," AREMA *Proceedings* (2002): 643–654.

74. "A Smarter Breed of Defect Spotters," RA 181 (November 10, 1980): 25–30; "Defect Detectors Get Smarter," RA 194 (March 1993): 47–48. For the smart bolt, see "What's Ahead in Bearing R&D?" RA 198 (November 1997): 41–42; Ronald Newman et al., "Hot Bearing Detection with the 'Smart Bolt,'" ASME/IEEE Joint Rail Conference *Proceedings* (April 17–19, 1990): 105–110; "10,000 Cars to Get On-Board Devices," RA 191 (May 1990): 18–19.

75. For hazmat sniffers, see "A Watchful Eye," RA 209 (August 2008): 33–36. The Remote Information Service is described by the AAR's Edward Hamberger in US 107th Cong., 1st Sess., House Committee on Transportation and Infrastructure, Hearings, *Recent Derailments and Railroad Safety* (Washington, 2002), 41, 73. See also "Detecting Failures Before They Occur," RA 201 (February 2000): 56–57, and "You Can't Manage What You Can't Measure," RA 201 (October 2000): 45–47. "Connective Intelligence," RA 208 (June 2007): 21–22; "Take a Look at This!" RA 211 (February 2010): 26–28. For "smart cars," see "Instrumented Rolling Stock," *Trains* 75 (July 2013): 20, General OneFile web. "Automating Track Inspection," RTS 106 (November 2010): 35–37; "The New View From Below," RA 213 (September 2012): 62–64; "Wayside Monitoring: Show Me the Data," RA 215 (July 2015): 32–34.

76. For a discussion and application of counterfactual analysis in history, see John K. Brown, "Not the Eads Bridge: An Exploration of Counterfactual History of Technology," *Technology and Culture* 55 (July 2014): 521–559. See also Niall Ferguson, *Virtual History: Alternatives and Counterfactuals* (New York: Macmillan, 1998), who argues that only considered alternatives should be treated as counterfactuals. Rose et al., *Best Transportation System* (n. 2), demonstrate that deregulation was considered at least as early as the Truman administration.

77. For Molitoris's views, see her testimony in US 103d Cong., House Committee on Energy and Commerce, Hearings, *Railroad Safety* (Washington, 1994), 3–25, 141–191, and US 104th Cong., 2d Sess., Senate Committee on Science, Commerce, and Transportation, Hearings, *Railroad Safety* (Washington, 1996), 39–54. "A Century of Great Railroaders," RA 200 (December 1999): 51.

78. For early criticism, see Railroad Retirement Board, *Safety in the Railroad Industry* (n. 1), Chapter 4, which cites a management evaluation of the ICC done in 1953. For Riley's testimony, see US 99th Cong., 1st Sess., Senate Committee on Commerce, Science, and Transportation, Hearings, *Railroad Safety* (Washington, 1985), 18, and US 100th Cong., 1st Sess., House Committee on Energy and Commerce, Hearings, *Railroad Safety* (Washington, 1988), quotation on 41. GAO, *Railroad Safety: New Approach Needed* (n. 18), and its *Railroad Safety Weaknesses* (n. 18).

79. For a statement of the new approach, see FRA, *Enhancing Rail Safety Now and into the 21st Century: The Federal Railroad Administration's Safety Programs and Initiatives, A Report to Congress* (Washington, 1996). The cooperative approach advocated by Molitoris seems to have remained after she left office in 1999, notwithstanding an article entitled "FRA's 'Aggressive' New Approach to Safety," RA 206 (June 2005): 13.

80. For the new approach, see the testimony of FRA Administrator Jolene Molitoris in US 103d Cong., House Committee on Energy and Commerce, Hearings, *Railroad Safety* (Washington, 1994), 145–191, and GAO, *Rail Transportation: Federal Railroad Administration's New Approach to Railroad Safety* (Washington, 1997). "Only since . . ." is from "Roundtable on Hazardous Materials," November 8, 1994, Administrator's Roundtable Discussions, 1993–1997, p. 90, FRA Records, RG 399, NA. "I'm Jolene Molitoris . . ." is from "FRA/Industry Dialogue on the State of the Industry," December 8, 1993, Administrator's Roundtable Discussions, 1993–1997, p. 3, FRA Records, RG 399, NA. "Number one . . ." is from US 105th Cong.. 2d Sess., House

Committee on Transportation and Infrastructure, Hearings, *Reauthorization of Federal Railroad Administration* (Washington, 1998), 16.

81. NTSB, *Head-On Collision of Consolidated Rail Corporation Freight Trains UBT-506 and TV-61 Near Thompsontown, Pennsylvania, January 14, 1988* (RAR 89-02, Washington, 1988); GAO, *Railroad Safety: Engineer Work Shift Length and Schedule Variability* (RCED 92-133, Washington, 1992); GAO, *Railroad Safety: Human Factor Accidents and Issues Affecting Engineer Work Schedules* (RCED 93-160BR, Washington, 1993). For a broader treatment of early interest in the relation between fatigue and accidents, see John Burnham, *Accident Prone: A History of Technology, Psychology and Misfits of the Machine Age* (Chicago: University of Chicago Press, 2010).

82. For the work review task force, see US 105th Cong., 2d Sess., House Committee on Transportation and Infrastructure, Hearings, *Reauthorization of Federal Railroad Administration* (Washington, 1998), 376. That fatigue can nullify an alerter is on 523. These hearings also contain the first AAR survey: Patrick Sherry, *Current Status of Fatigue Countermeasures in the Railroad Industry* (1998). They also include a survey of Canadian lines' efforts: Martin More-Ede et al., *Canalert '95 Alertness Assurance in the Canadian Railways, Phase II Report* (1996). The second AAR report with the same author and title is in US 109th Cong., 2d Sess., House Committee on Transportation and Infrastructure, Hearings, *Human Factor Issues in Rail Safety* (Washington, 2007). "Did Fatigue Cause This Wreck?" *Trains* 71 (July 2011): 13, General OneFile web.

83. Jordan Multer et al., *Developing an Effective Corrective Action Process: Lessons Learned from Operating a Confidential Close Call Reporting System* (DOT/FRA/ORD 13/12, Washington, 2013), and Joyce Ranney and Thomas Raslear, *Derailments Decrease at C3RS Site at Midterm* (FRA RR 12-04, Washington, 2012). "The 'Gotcha-Free' Zone," *Trains* 72 (March 2012): 18, General OneFile web. Vaccine is from James Reason, *Managing the Risk of Organizational Accidents* (Burlington, VT: Ashgate, 1997), Chapter 6.

84. FRA, *The FRA Risk Reduction Program: A New Approach for Managing Railroad Safety* (Washington, 2008), www.dot.fra.gov; "Cabooseless Trains: A Sign of the Times," RA 183 (November 29, 1982): 21–22; "The Facts are in, Cabooseless Trains are Safer," RA 188 (January 1, 1987): 20; "Monitoring Cabooseless Trains," RA 188 (March 1, 1987): 47–48; "Canada Goes Cabooseless," RA 191 (July 1990): 75–77. For FRA policy, see the testimony of James Hall of the NTSB in US 104th Cong., 2d Sess., Senate Committee on Commerce, Science, and Transportation, Hearings, *Railroad Safety* (Washington 1998), 28–39.

85. The NTSB representative is quoted in "Safety Plan; Calls for Rail Changes," *Riverside [CA] Press Enterprise* (November 30, 2005); "Dangerous Trains Pass Through Region," *Great Falls [VA] Connection* (January 20, 2015); "Keeping the Public Safe from Railroads' Cargoes," *Austin [TX] American Statesman* (January 18, 2005); "Deadly Leak Underscores Concerns about Rail Safety," NYT (January 9, 2005). "Deadly Trains," NYT (February 7, 2005) contains the quotations. "Trouble on the Tracks," NYT (October 17, 2005); "America's Vulnerable Railways," *The Boston Globe* (June 10, 2005).

86. See "Safe, Accountable, Flexible Efficient Transportation Equity Act: A Legacy for Users" (P.L. 109-59, section 9005, Washington, 2005). Pipeline and Hazardous Materials Safety Administration, "Hazardous Materials: Improving the Safety of Railroad Tank Car Transportation of Hazardous Materials," *Federal Register* 74 (January 13, 2009): 1770–1802.

87. For a brief report on the Lac-Mégantic disaster, see http://www.tsb.gc.ca/eng/rapports-reports/rail/2013/r13d0054/r13d0054-r-es.asp. For tank car problems, see "Tank Car Safety: What's the Right Mix?" *Trains* 74 (March 2014): 6, and "Tank Car Safety Takes Top Priority," *Trains* 74 (July 2014): 6, General OneFile web.

88. "Wrecks Hit Tougher Oil Railcars," WSJ (March 9, 2015), contains the rise in crude shipments and problems with the CPC-1232 cars. "Accidents Surge as Oil Industry Takes the Train," NYT (January 26, 2014); "DOT-117 Defined," RA 216 (June 2015): 17–18; "Freight Railroads Join U.S. Transportation Secretary Foxx in Announcing Industry Crude By Rail Safety Initiative," press release, February 24, 2014, https://www.aar.org. The new regulations are more complex than I have indicated. See "Summary: Enhanced Tank Car Standards and Operational Controls for High Hazard Flammable Trains," www.fra.dot.gov. "Data-Phobic FRA's Book of Mormon," RA 215 (June 2014): 13; "Crew Size Debate Heats Up," *Trains* 76 (June 2016): 7, General OneFile web.

Chapter 4. "Our Goal is Zero Accidents"

1. The ICC changed reporting requirements in 1957 that reduced reported fatalities, and under congressional pressure it reversed field in 1961. Thus,

the rise in the fatality rate after 1957 is, if anything, an undercount. Injury data before 1975 are not presented because the definitions employed by the ICC and later the FRA change so often as to make the figures largely worthless for discerning trends.

2. Earl Currie, "Keynote Address," ARBBA *Proceedings* 101 (1996): 63–69, quotation on 65.

3. J. M. Symes to Ethelbert W. Smith, January 8, 1948, box 342, PRRC, HML; "BRT Wins Action on Safety Hazards," *Trainman News* 3 (September 3, 1949): 2. The survey is from "Where Do We Stand on M of W?" MR 17 (March 1962): 71. Complaints about maintenance are voiced in US 91st Cong., 1st Sess., Senate Committee on Commerce, Hearings, *Federal Railroad Safety Act of 1969* (Washington, 1969). The quotation is from US 95th Cong., 2d Sess., House Committee on Interstate and Foreign Commerce, Hearings, *Railroad Safety Authorization for Fiscal Year 1979* (Washington, 1978), 352.

4. C. D. Young to Martin W. Clement, April 21, 1947, box 342, PRRC, HML. Ralph Strickland, Interview H-0180, Documenting the American South, Southern Oral History Program Collection, #4007; http://docsouth.unc.edu/sohp/H-0180/menu.html. On the development of car retarders, see "The Evolution of Control Systems in Retarder Yards," AREA *Proceedings* 74 (1973): 490–503. Edwin Mansfield, *Industrial Research and Technological Innovation, An Econometric Analysis* (New York: W. W. Norton, 1968), Chapters 7–9.

5. Employing Stata, I fitted a fixed effect, negative binomial equation to panel data for the period 1959–1965 to estimate the effects of profitability (net income) on casualties (C) with controls for assets and employee hours (EH). There were 511 observations on 75 usable groups (companies). Income, assets, and employee hours are all in thousands; figures in parentheses are z-scores. The results are:

$$\mathrm{Ln(C)} = 1.80 - \frac{5.19 * 10^{-6}}{2.11}\,\text{Net Income} - \frac{3.17 * 10^{-7}}{1.27}\,\text{Assets} + \frac{2.0 * 10^{-5}}{5.99}\,\text{EH}$$

The equation also includes a set of time dummies. The mean and standard deviations of net income (in thousands) are, respectively $9,817 and $16,819; a 1 standard deviation increase implies a decrease in casualties of $\exp^{(-0.00000519 * 16,189)} - 1$, or −8.8 percent.

6. For the safety engineer, see "Report of the Committee on Direction," AAR Safety Section *Proceedings* 28 (1948): 150–151, and 29 (1949): 26. For a critique of railroad safety programs that contains both surveys reported in the text, see US Railroad Retirement Board, *Safety in the Railroad Industry* (Washington, 1962), Chapter 2. Harry Brady, Interview 591, August 16, 1982, Youngstown State University Oral History Program, http://www.maag.ysu.edu/oralhistory/oral_hist.html), quotation on 18. James Larson, "Ralph Rotten and the Chicago Great Western," *Railroad History* 214 (Spring–Summer 2016): 10–21, quotations on 18. The survey is from Charles Bailey and Dan Petersen, "Using Perception Surveys to Assess Safety System Effectiveness," *Professional Safety* 34 (February 1989): 22–26.

7. For the role of the Southern in mechanization, see "How Bill Brosnan and the U.S. Cavalry Saved the Railroads," RTS 101 (April 2005): 25–35. Because the FRA stopped collecting employee hours by occupational group after 1983, the data in Table 4.1 are reported on a per-worker basis.

8. FRA, *Railroad Fatalities Investigated by the Federal Railroad Administration in 1975* (Washington, 1976). The first year of this program was 1973, but I have been unable to find that publication.

9. J. P. Hiltz, "Remarks," AAR Safety Section *Proceedings* 32 (1952): 169–172; E. L. Woods, "'PERT'—Program Evaluation and Review Technique," NSC Railroad Section *Proceedings* 54 (1966): 10–19. The Santa Fe, Soo, and Penn Central employees are from FRA, *Railroad Fatalities Investigated . . . 1975* (n. 8), nos. 59, 78, and 86.

10. The B&M employee is from FRA, *Railroad Fatalities Investigated . . . 1975* (n. 8), no. 57. Fatalities are from US 86th Cong., 2d Sess., Senate Report No. 1546, *Regulation of the Operation of Railroad Track Motorcars by the Interstate Commerce Commission* (Washington, 1960).

11. For the lack of setoffs, see W. E. Rader to R. H. Burkett, October 11, 1939, file 3, box 2, William Doble Papers, Collection 5182, KCCU. An accident that resulted because the track car would not trip a crossing signal is reported in L. R. Bloss to F. M. Brown, May 23, 1950, file 13, box 2, William Doble Papers, Collection 5182, KCCU. Refusal to provide a lineup is from R. M. Smith to W. E. Green, January 17, 1945, file 7, box 2, William Doble Papers, Collection 5182, KCCU.

12. The L&N accident is from "More Blood Has Been Shed," Brotherhood of Maintenance of Way Employees *Journal* 62 (May 1952): 5. "AAR Studying Track Car Safety," RA 134 (August 30, 1954): 9; Railroad Retirement Board, *Reporting of Accidents*

and Casualties in the Railroad Industry (Washington, 1959), Table 4-4B; US 85th Cong., 2d Sess., Senate Committee on Interstate and Foreign Commerce, Hearings, *Safety Regulation of Railroad Track Motorcars* (Washington, 1958).

13. For similar critiques of the ICC's safety work, see US Railroad Retirement Board, *Safety in the Railroad Industry* (n. 6), Chapter 4, and US 89th Cong., 2d Sess., House Committee on Government Operations, Hearings, *Interstate Commerce Commission Operations (Railroad Safety)* (Washington, 1966). Marver Bernstein, *Regulating Business by Independent Commission* (Princeton, NJ: Princeton University Press, 1955), quotation on 88. That the quotation applies to the ICC is my judgment.

14. The initial rise in injury rates in Figure 4.2 probably reflects improved reporting as companies adjusted to the new rules. Complaints about the blue flag law are from Charles Amos, "Remarks," AREA *Proceedings* 79 (1978): 403–408.

15. A. E. Shulman, *Analysis of Nine Years of Railroad Personnel Casualty Data 1966–1974* (Washington: AAR Research and Test Department, 1976). The first director of safety is from "Watching Washington," RA 166 (April 21, 1969): 36. The quotation is from John German, Vice President, MoPac, in US 94th Cong., 2d Sess., House Committee on Interstate and Foreign Commerce, Hearings, *Federal Railroad Safety Authorization Act of 1976* (Washington, 1976), 188–204, quotation on 189. The Safety Research Board is on 86.

16. US 101st Cong., 2d Sess., Senate Committee on Commerce, Science, and Transportation, Hearings, *FELA in Relation to Amtrak* (Washington, 1988), 43; "BN Pays Dearly for Unsafe Conditions That Cost Three Railroaders Their Lives," LE 119 (October 18, 1985), 2. Lost workdays are from US 110th Cong., 2d Sess., House Committee on Transportation and Infrastructure, Hearings, *The Impact of Railroad Injury, Accident, and Discipline Policies on the Safety of America's Railroads* (Washington, 2007), 246.

17. On the loss and damage campaign, see, for example, "Report Outlines L&D Attack Plan," RA 154 (June 17, 1963): 52, and "Quality Control is the Big Weapon," RA 161 (December 19, 1966): 26–28. "Miss Careful Handling" is from "The Girl from PRR," MR 21 (December 1966): 55.

18. For Richards's early work, see "The Campaign Against Accidents on the Chicago & North Western," RA 49 (September 2, 1910): 391–393, and my "Safety First Comes to the Railroads," *Railroad History* 162 (Spring 1992): 6–33. For more on morale problems and company responses, see "Railroads and Their Employees All Drawing Closer Together," RA 148 (July 8, 1974): 32–35.

19. "How the C & NW Cuts Casualty Losses," RA 164 (March 25, 1968): 36–39; "The Battle Continues: L&D," RA 166 (March 31, 1969): 33–34. See also Ben Heineman, "An Address," NSC Railroad Sessions *Proceedings* 55 (1967): 15–17; V. L. Preisser and K. L. Patrick, "Casualty Control—A New Approach," and "C & NW Train Accident Program," NSC Railroad Sessions *Proceedings* 56 (1968): 10–18 and 18–20, respectively.

20. The Soo is from "Starts New Safety Department," MR 21 (January 1966): 178; "Rio Grande Sets up a Brain Trust," RA 162 (February 13, 1967): 52; "Management Close-Up: Rio Grande, Blending Service and Efficiency for Profit," RA 172 (February 28, 1972): 23–25, quotation on 25; and "Why Are Some Railroads Safer Than Others?" RA 183 (January 11, 1982): 32–34. For the Frisco, see "The Rail Safety Record: How Can It be Improved?" RA 167 (November 10, 1969): 12; "From Chessie, A New Twist in Safety Training," RA 176 (February 10, 1975): 16–18; "Accident Prevention: No Accident at Chessie," MR 38 (March 1983): 50–52. The MoPac program is from the testimony of John German, US 94th Cong., 2d Sess., Federal Railroad Safety Authorization Act (n. 15).

21. "ATSF Safety Program Gets Big Results," RA 166 (June 9, 1969): 20–21; "Safety Trains Replace Traditional Trophies," RA 162 (June 9, 1967): 21. Reed is quoted in "The Operative Word is Quality," RA 177 (April 26, 1976): 16–21, quotation on 18. For the Santa Fe, see also "Why Are Some Railroads" (n. 20)

22. The MoPac program is from "Railroad of the Year: MoPac," MR 27 (June 1972): 41–45; "A Late Bloomer Maybe—But Look at MoPac Now," RA 177 (September 13, 1976): 24–29; and the testimony of John German, US 94th Cong., 2d Sess., Federal Railroad Safety Authorization Act (n. 15). See also Craig Miner, *The Rebirth of the Missouri Pacific 1956–1983* (College Station: Texas A&M University Press, 1983).

23. Interview with Sid Showalter, Oral History Program, Southern Arizona Transportation Museum, http://tucsonhistoricdepot.org/; "Locomotive Simulator Looks and Acts like the Real Thing," RA 167 (September 15, 1969): 36–37.

24. Among the many articles discussing the new view of training, see "CN Pushes a Huge Training Program," RA 159 (October 18, 1965): 20, 25; "The Training Scene: Busy and Getting Busier," RA 175

(February 25, 1974): 18–21, quotation on 20; "People Power on the Railroads," MR 44 (November 1989): 16–19; "Higher Capacity Railroads Need Higher Capacity People," MR 22 (October 1967): 100–104, quotation on 102. The NTSB conclusion is from the testimony of Patricia Goldman in US 97th Cong., 2d Sess., Senate Committee on Commerce, Science, and Transportation, Hearings, *FRA Railroad Safety Program* (Washington, 1982), 16–27, quotation on 19. Ed Butt, "Training," ARBBA *Proceedings* 99 (1994): 96–101.

25. Kenefick is from "Union Pacific Stresses Customer Awareness, Productivity, Service," MR 40 (May 1985): 28–38, quotation on 29; FRA, *Certain Railroad Employee Fatalities Investigated by the Federal Railroad Administration, Calendar Year 1978* (Washington, 1979). Ian Savage stresses the importance of unit trains; see his *Economics of Railroad Safety* (Boston: Kluwer, 1998), Chapter 2.

26. Buddliners were individual rail diesel cars produced by the Budd Company that the railroads used on lines with light traffic.

27. The L&N is from "You Ought to Know," RA 153 (August 27, 1962): 60; "PC Reports Rise in Theft, Vandalism," RA 170 (February 22, 1971): 33. "It seems to be a reflection . . ." is from "Vandalism: Time to Crack Down," RA 181 (December 29, 1980): 54–56, quotation on 55. The grocery cart is from "The Case of the Fast-Moving Target," *Trains* 31 (February 1971): 20–24. Amtrak accidents are from US 94th Cong., 2d Sess., House Committee on Interstate and Foreign Commerce, Hearings, *Federal Railroad Safety Authorization Act of 1976* (Washington, 1976), 179. The passenger death is from "On Vandalism," *Trains* 36 (July 1976): 66. The Buddliner is from "Congress Stalls, Vandals Strike," LE 112 (March 24, 1978): 1; "Santa Fe Operating Crews Spearhead the Fight Against Vandals," RA 180 (June 11, 1979): 18–19; "Vandals' Victim Awarded $180,000 in Lawsuit," LE 112 (July 21, 1978): 1. Grand Trunk is from "Vandalism is Rising and the Railroads Need Help," RA 180 (June 11, 1979): 16–18.

28. Florida East Coast is from "Vandalism is Rising" (n. 27); "Vandals Create 'Combat Zone' for Railroaders; a Federal Crime?" MR 34 (March 1979): 23–24; "Vandals Strike on Schedule: Fear Rides the Railroad," LAT (December 3, 1978).

29. "Helicopter to Fight Chicago Rail Crime," RA 168 (April 27, 1970): 41; "MTA Launches 'Copter Patrol," RA 170 (May 10, 1971): 13; "Copters Cut Commuter Line Vandalism . . . Protect Lives," MR 32 (February 1977): 77–78; "Railroads Fight Mounting Wave of Vandalism," RA 170 (March 29, 1971): 34–35.

30. "Long Island Railroad Sets an Example," LE 112 (December 1, 1978): 1; "Congress Stalls" (n. 27). The FRA rule is described in "Protective Glazing: Meeting FRA's New Standard," RA 181 (December 29, 1980): 56. The rule is contained in 49 CFR 223. "Railroads Fight Mounting Wave" (n. 29).

31. The FRA did not report casualties for contract workers separately until 1979 and it does not report their employee hours. From then on their fatalities amounted to about 10 percent of those of railroad employees on duty. The ratio of injuries to fatalities for contractors has been much less than for railroad workers, suggesting substantial underreporting for the former group. For accident investigation services, see "CSI: Train Derailments," *Trains* 71 (July 2011): 16, General OneFile web.

32. After 1996, the FRA no longer reported injuries and employee hours separately for Class I and other carriers. However, the employment shares of the current seven Class I carriers have been roughly constant since 2000.

33. "Why Are Some Railroads" (n. 20).

34. W. Edwards Deming worked for the Minneapolis & St. Louis as early as 1957. He was employed by the Chicago & North Western from 1966 to 1969 as a consultant dealing with railroad mergers. He did similar work for the KCS and the Illinois Central. Deming also worked for the Duluth, Missabe & Iron Range from 1971 to 1979 to help discover cost-effective sampling techniques for car weighing. See the W. Edwards Deming Papers, Library of Congress, boxes 72, 76, and 80.

35. There is no good history of the TQM Movement. I have relied on the following: W. Edwards Deming, *Out of the Crisis* (Cambridge, MA: MIT, 1986); Olice Embry, "Edwards Deming and the Early Contributions to the Quality Movement," *Essays in Economic and Business History* (1992–1993): 210–217; Richard Hackman and Ruth Wageman, "Total Quality Management: Empirical, Conceptual and Practical Issues," *Administrative Science Quarterly* 40 (June 1995): 309–342; Jeremy Main, *Quality Wars* (New York: Free Press, 1994); Robert Grant et al., "TQM's Challenge to Management Theory and Practice," *MIT Management Review* 35 (January 15, 1994): 25–35; and D. A. Garvin, *Managing Quality: The Strategic and Competitive Edge* (New York: Free Press, 1988).

36. A considerable literature began to appear in the late 1980s relating the TQM Movement to safety. See, for example, Roland Dumas, "Safety and

Quality: The Human Dimension," *Professional Safety* 32 (1987): 11–14; Kaoru Ishikawa, "TQM & Safety: New Buzz Words or Real Understanding?" *Professional Safety* 39 (June 1994): 31–36; Dan Petersen, *Analyzing Safety System Effectiveness*, 3d ed. (New York: Van Nostrand Reinhold, 1996); James Manzella, "Achieving Safety Performance Excellence through Total Quality Management," *Professional Safety* 42 (May 1997): 26–28; and Susana Herrero et al., "From the Traditional Concept of Safety Management to Safety Integrated with Quality," *Journal of Safety Research* 33 (Spring 2002): 1–20. Jolene Molitoris, "Address," ARBBA *Proceedings* 99 (1994): 91–96, quotation on 93.

37. Harold Roland and Brian Moriarty, *System Safety Engineering and Management* (New York: Wiley, 1990); Woods, "'PERT'" (n. 9). The unrocked boat is from James Reason, *Managing the Risks of Organizational Accidents* (Brookfield, VT: Ashgate, 1997), Chapter 1.

38. For the "Swiss cheese model," see Reason, *Managing the Risks* (n. 37), Chapter 1. The accidents recounted in this and the following paragraph are from FRA, *Certain Railroad Employee Fatalities Investigated by the Federal Railroad Administration, Calendar Year 1984* (Washington, 1985), nos. 6 and 7.

39. For behaviorism, see E. Scott Geller, *The Psychology of Safety* (Boca Raton, FL: Lewis, 2001), which is the source of the "Three E's." See too Thomas Krause, *The Behavior Based Safety Process*, 2d ed. (New York: Van Nostrand Reinhold, 1997), and Aubrey Daniels, "What is Behavior Based Safety?" http://aubreydaniels.com. The AAR introduction to behaviorism is from Dan Petersen, "An Experiment in Positive Reinforcement," *Professional Safety* 28 (May 1984): 30–35, and Bailey and Petersen, "Using Perception Surveys" (n. 6).

40. "Accident Prevention No Accident at Chessie," MR 38 (March 1983): 50–52.

41. "UP Makes Safety Commitment," MR 39 (December 1984): 58–59, taxi quotation on 59. "Union Pacific Stresses Customer Awareness, Productivity, Safety," MR 40 (May 1985): 28–38; DuPont executive quotation on 37. A review of DuPont's safety program is "Safety and Occupational Health: A Commitment to Action," *National Safety and Health News* 131 (October 1985): 55–59; for the history of DuPont's safety ideas, see my *Safety First* (Baltimore: Johns Hopkins University Press, 1997), Chapter 4.

42. The best discussion is Maury Klein, *Union Pacific* (New York: Oxford University Press, 2011), Chapter 17. The quotations are on 230 and 235. See also Main, *Quality Wars* (n. 34), and "The Competitors: Union Pacific," RA 195 (March 1994): 36–53.

43. For the Florida East Coast, see "Panel Discussion, Safety," ARBBA *Proceedings* 99 (1994): 30–44.

44. Portec is from "Using a Club to Maintain Quality," RA 181 (October 27, 1980): 48–49. For the AAR, see "Testimony of William H. Dempsey," in US 102d Cong., 1st Sess., House Committee on Energy and Commerce, Hearings, *Railroad Safety Programs* (Washington, 1991), 176–209. See also "Railroads Move to Curb Component Failures," RA 185 (May 1, 1984): 26, and "Quality Assurance Monitoring Starts with Bearings," RA 186 (November 1, 1985): 19.

45. For Norfolk Southern, see "Statement of Arnold McKinnon," US 102d Cong., 1st Sess., House Committee on Energy and Commerce, Hearings, *Railroad Safety Programs* (Washington, 1991), 210–232; Donald Turvey, "President's Address," AREA *Proceedings* 92 (1991): 135–139; P. R. Ogden, " Norfolk Southern's Engineering Department Action Plan for Safety," AREA *Proceedings* 98 (1997): 285–295. "Good business" is from Craig Webb, "Engineering Department's Safety History on Norfolk Southern," AREMA *Proceedings* (2002): 1061–1062. "What—Another Harriman [Award for the Norfolk Southern]?" RA 207 (June 2006): 4–5.

46. "Conrail's 'Wise Owls,'" *Inside Track* 2 (February 1980): 8; "Working Together: The Team Approach Grows at Conrail," *Inside Track* 4 (Fall 1982): 5–6; "In Safety, Delmarva People Look Out for Number One and for Each Other," *Inside Track* 7 (Winter 1985): 12–13, quotation on 13. DuPont is from "Conrail Enhances Safety Strategy with Tips from a Good Coach," *Inside Track* 8 (Fall 1986): 14. "New Safety Program Focuses on Maximum Teamwork for Maximum Results," *Inside Track* 10 (Fall 1988): 18–20; "A Safer Conrail: A New Era Begins," *Inside Track* 11 (April 1989): 14–21, quotation on 14; "One Year Later," *Inside Track* 12 (January–February 1990): 14–16.

47. "Where Quality Begins," *Inside Track* 12 (May–June 1990): 5–7; "The Process of Continuous Quality Improvement," *Inside Track* 12 (July–August 1990): 3–4; "The Competitors: Conrail," RA 195 (December 1994): 31–45, quotation on 31; R. J. Rumsey and R. Shiloh, "Conrail's Infrastructure Reliability Optimization," AREA *Proceedings* 98 (1997): 347–352; "Conrail Cuts the Risks," RA 199 (April 1998): 75–78. Main, *Quality Wars* (n. 35), Chapter 9, also discusses Conrail. "Confronting the

Safety Dilemma: Implementing B-Safe at Conrail," *Performance Management Magazine* 14 (Fall 1996): 7–10.

48. Amtrak is from "STOP—In the Name of Safety," MR 44 (March 1989): 60–62; "People Power on the Railroad," MR 44 (November 1989): 16–19. For the North Western, see James Zito, "Keynote Address," ARBBA *Proceedings* 93 (1988): 64–70, and Roger Grant, *The North Western* (DeKalb: Northern Illinois University Press, 1996), Chapter 11. For the 80 percent claim, see the statement of Arnold McKinnon, US 102d Cong., 1st Sess., *Railroad Safety Programs* (n. 45), 210. "Safety at the Head of the Class," RTS 99 (August 2003): 22–24.

49. "BNSF, UTU Join in Safety Initiative," RA 202 (October 2001): 6; "BNSF, BLE, UTU Seek 'New Safety Culture,'" RA 203 (April 2002): 12. For the BNSF, see also "Safety at the Head of the Class," RTS 99 (August 2003): 22–24, quotation on 24; "Listening Posts Spark New Initiatives for Transportation," *Railway* 1 (Summer 2010): 9. For Norfolk Southern, see, for example, *BizNs* 1 (July–August 2009). "Safe to Safer Emphasizes Collaboration," *Amtrak Ink* 14 (September 2009): 4–7; "Amtrak's Safe-2-Safer is Not Too Safe," *Trains* 75 (June 2015): 10, General OneFile web.

50. For the UP, see "Safety at the Head of the Class," RTS 99 (August 2003): 22–24, quotation on 24, and "Machinery Wish List," RTS 99 (August 2003): 25–26, quotation on 25. FRA, *Canadian Pacific Railway Mechanical Services' 5-Alive Safety Program Shows Promise in Reducing Injuries* (RR 06-14, 2006; http://www.volpe.dot.gov). "Six sigma" quality is from "Industry News," PR 44 (June 2001): 8. For scheduling, see "How CN Runs a Scheduled Railroad," RA 201 (December 2000): 14; "Propagating Process Change," PR 44 (May 2001): 14–16. For Norfolk Southern, see "A New Safety Culture," PR 55, November 2012, www progressiverailroading .com, n.p.

51. E. B. Burwell, "Keynote Address," ARBBA *Proceedings* 91 (1986): 57–61, quotation on 60; Earl Currie [Wisconsin Central], "Keynote Address," ARBBA *Proceedings* 101 (1996): 63–69, quotation on 67.

52. Cost figures are from "Fitting Work to Workers," RA 192 (August 1991): 55–56. Richard Reynolds, "Safety in the Railroad Industry Today," ARBBA *Proceedings* 95 (1990): 119–121, quotation on 121.

53. "Humble Announces the Blivet," MR 26 (January 1971): 104. "The Car Master," RA 192 (September 14, 1991): 14, which could move heavy equipment under a freight car, stressed that it did so "without the danger of back injury." "Personal Injuries: Biomechanics to the Rescue," RA 188 (December 1, 1987): 48–49; "Fitting Work to Workers" (n. 52).

54. Windell Pyles, "Quality—is It for Everyone?" ARBBA *Proceedings* 98 (1993): 40–44, quotation on 43.

55. These examples are from "Fitting Work to Workers" (n. 52); Vincent Terrill, "Investing in New Track Technology," ARBBA *Proceedings* 97 (1992): 27–34; Roger Cross, "A User's Perspective on Equipment Specifications," ARBBA *Proceedings* 98 (1993): 55–57, which is the source of the quotation; Vincent Terrill, "Remsa Presentation," ARBBA *Proceedings* 99 (1994): 168–175; Carter Jones, "Ergonomics and Machine Design," AREA *Proceedings* 97 (1996): 738–740; "No Ordinary Hump Yard," RA 201 (May 2000): 52–53.

56. In nearly every year after 2000, articles in *Railway Track & Structures* reveal new equipment safety features. See, for example, "Safety Again at the Top of Chief Engineers' Wish List," RTS 105 (August 2009): 21–29. For personal equipment, see "Keeping Safe from Head to Toe," RTS 102 (October 2006): 35–36.

57. The 57 percent figure is from "The Power of Remote Control," RA 206 (February 2005): 25–27; "BNSF Takes the Hi-Rail Road," PR 46 (June 2003): 4. Labor concerns may be followed in the UTU *News* and the Locomotive Engineer (and Trainmen) *Newsletter*. See, for example, "Detroit Bans Remote Control Locomotives," Locomotive Engineer *Newsletter* 16 (December 2002): 1.

58. For Hi-Rail, see "A Technology Bent," PR 48 (January 2005): 22–30. Descriptions of the track worker protection devices may be found at company websites (www.protrantechnology.com and www .smartrailroad.com). See "Keeping Safe in High Traffic Environment," RTS 104 (October 2008): 38–41. See too "Sounding the Alarm for Safety," *Trains* 73 (January 2013): 18, General OneFile web.

59. Interview with David Surratt, Interview 454, August 6, 1975, Youngstown Oral History Project, http://www.maag.ysu.edu/oralhistory/oral_hist .html; Larson, "Ralph Rotten" (n. 6).

60. Interview with Harry Brady (n. 6), quotation on 19.

61. US 105th Cong., 2d Sess., House Committee on Transportation and Infrastructure, Hearings, *Reauthorization of the Federal Railroad Administration* (Washington, 1998); the Conrail example and quotation are on 897.

62. For one of many articles, see "A Warning to Norfolk Southern Managers," UTU *News* 35 (September 2003): 4. The CSX quotation is from US 105th Cong., 2d Sess., *Reauthorization* (n. 61), 1133–1134. US 110th Cong., 2d Sess., House Committee on Transportation and Infrastructure, Hearings, *The Impact of Railroad Injury, Accident, and Discipline Policies on the Safety of America's Railroads* (Washington, 2007); "FRA Issues Questionable Remote Control Safety Audit," Locomotive Engineers and Trainmen *Newsletter* 18 (June 2004): 1–2.

63. "Rail Labor Protests Railroads' Safety Hypocrisy," Brotherhood of Maintenance of Way Employees *Journal* 106 (July 1997): n.p.

64. US 105th Cong., 2d Sess., Hearings, *Reauthorization* (n. 61); Prendergast is on 934; Rose is from 1037–1045, quotation on 1045. "OSHA Fines BNSF for Harassing Injured UTU Member," UTU *News* 42 (December 2010): 1. Behaviorists opposed most safety incentive programs as likely to result in injury underreporting. See, for example, Geller, *Psychology of Safety* (n. 39), Chapter 11.

65. The UTU claim is from US 110th Cong., 2d Sess., House Committee on Transportation and Infrastructure, Hearings, *Reauthorization of the Federal Railroad Administration* (Washington, 2007), vii. "Was It Time for the Harrimans to Go?" RA 213 (January 2012): 4.

66. The claim that the decline in train accidents had little to do with the improvement in worker safety is less true for trainmen than other workers. At their peak, train accidents accounted for about one-third of all trainmen's fatalities in 1969 (versus one-fifth for all workers) and the fall in train accidents accounted for about one-third of the decline in fatalities to trainmen.

67. "Intimidation/Harassment Roundtable," October 21, 1997, 69, Administrator's Roundtable Discussions, 1993–1997, FRA Records, RG 399, NA.

68. For Red Block, see "The Union Approach," RA 185 (July 1, 1984): 20, and "Railroad to Start Drug Abuse Fight," NYT (May 21, 1984). T. A. Mannello and F. J. Seaman, *Prevalence, Costs and Handling of Drinking Problems on Seven Railroads* (Washington: DOT, 1978). See also "How Do We Get the Drunks off Trains?" MR 39 (January 1984): 30–32.

69. The 2010 data are from Amtrak Office of Inspector General, *Railroad Safety: Amtrak is Not Adequately Addressing Rising Drug and Alcohol Use by Employees in Safety-Sensitive Positions* (Report OIG-E-2012-023, Washington, 2012). "Rule G Gets Some Muscle," MR 41 (March 1986): 53–54.

70. TQM ideas in the FRA probably predated Molitoris's tenure because in 1990 the agency instituted a Quality Improvement Program to improve its inspections. "FRA's Molitoris Says Safety and Service are 'Inextricably Linked,'" RA 200 (February 1999): 56.

71. The quotation is from David Brickey to Rodney Slater, November 9, 1999, Correspondence File, Project Case Files Relating to SOFA, FRA Records, RG 399, NA. "Knowing, Sharing—and Following—the Rules," RA 201 (November 2000): 64, briefly discusses SOFA.

72. FRA, *Behavior Based Safety at Amtrak-Chicago Associated with Reduced Injuries and Costs* (RR 07-07, 2007; www.fra.dot.gov); Michael Zuschlag et al., "Evaluation of a Safety Culture Intervention for Union Pacific Shows Improved Safety and Safety Culture," *Safety Science* 83 (March 2016): 59–73. See also *Operating Rule Compliance and Derailment Rates Improve at Union Pacific Yards with STEEL Process—A Risk Reduction Approach to Safety* (RR 09-08, 2009), and *A Pilot Examination of a Joint Railroad Management-Labor Approach to Root Cause Analysis of Accidents, Incidents, and Close Calls in a Diesel and Car Repair Shop Environment* (DOT/FRA/ORD 06/24). "Introduction to the Fames Committee," all at www.fra.dot.gov.

73. Further evidence that the new rule made few changes in the status quo is the absence of any concern evidenced in either the ARBBA or AREA *Proceedings*, or in RTS.

74. NTSB, *Special Investigation Report on Railroad and Rail Transit Roadway Worker Protection* (SIR 14/03, Washington, 2014); DOT and FRA, "Roadway Worker Authority Limits, Safety Advisory 2014-02," *Federal Register* 79 (November 25, 2014): 70268–70270.

75. "Inspection Rules for Self-Propelled Cars," RA 136 (May 31, 1954): 13. For the clean cab design, see John Zolock and David Tyrell, "Locomotive Cab Occupant Protection," ASME International Mechanical Engineering Congress (Rail Transportation) *Proceedings* (November 15–21, 2003): 57–64.

76. For a good overview, see "Cab Safety Evolves," *Trains* 73 (December 2013): 18, General OneFile web. For background on the carriers' standard, see the testimony of Larry McFather, president of the BLE, in US 102d Cong., 1st Sess., House Committee on Energy and Commerce, Hearings, *Railroad Safety Programs* (Washington, 1991), 287–289. AAR S-560 as modified is from FRA, *Collision Analysis Working Group Final*

Report, August 2006, Appendix G, www.fra.dot.gov. For the federal standard, see 49 US Code, ch. II, part 229. The payoff to the locomotive modifications is from "Consensus Rulemaking at the Federal Railroad Administration: All Aboard for Railway Safety Measures," *TR News* 236 (January–February 2005): 8–14; "Amtrak Orders New Crash-Worthy Locomotives," UTU *News* 42 (November 2010): 10.

77. Eckardt Johanning et al., "Back Disorder and Ergonomic Survey among North American Railroad Engineers," TRR 1899 (2004): 145–155; Srinivasan Kasturi et al., "Injury Mitigation in Locomotive Crashworthiness," ASME/IEEE Joint Rail Conference *Proceedings* (March 16–18, 2005): 1–8; Mehdi Ahmadian, "The Noise and Vibration Benefits of Soft Mounted Locomotive Cabs," ASME/IEEE Joint Rail Conference *Proceedings* (April 17–19, 2001): 227–237. Donovan is from "Death By Train," *New Jersey Star Ledger* (June 18, 2009).

78. "Railroads Tweak Critical Incident Stress Plans to Meet FRA Rules," PR (April 2015): n.p.; FRA, *Proposed Key Elements of a Critical Incident Intervention Program for Reducing the Effects of Potentially Traumatic Exposure on Train Crews to Grade Crossing and Trespasser Incidents*, DOT/FRA/ORD 14/06, www.fra.dot.gov.

79. FRA, *2006 Railroad Employee Fatalities: Case Studies and Analysis, February, 2008*, www.fra.dot.gov.

80. The figures in the text are calculated as follows. I first subtracted injury or fatality rates in 2013–2015 from those in 1975–1977 to compute the declines in risk. These are multiplied by employee hours for 2013–2015 and the result divided by 3 to generate annual casualties avoided.

Chapter 5. Passenger Safety in Modern Times, 1955–2015

1. ICC, *Accident Bulletin 125* (Washington, 1956); *Accident Bulletin 126* (Washington, 1957).

2. "Study Group Assays Passenger Outlook," RA 121 (November 16, 1946): 841–845, quotation on 844; "Train Travel is SAFE Travel," RA 130 (May 21, 1951): 74–75; "Pullman's Vicious Ad," *Aviation Week* 52 (June 26, 1950): 54; "Reshaping Public Opinion," RA 141 (July 23, 1956): 7–8, quotation on 7.

3. The statistics for 1981–2015 are as follows where FR is the fatality rate per billion passenger miles.

$$\text{Ln(FR)} = -0.826 - \underset{(0.57)}{0.013}\,\text{Trend};\ R^2 = 0.01;\ N = 35.$$

The value in parentheses is a t-ratio.

4. The changing definition of injuries over long periods makes any such comparisons suspect.

5. "Some PRR Trains are Held Together by 'Rubber Bands,'" *Railroad Trainman* 84 (February 6, 1967): 3; E. R. English, "Track and Structures—the Present Crisis in the Railroad Industry," ARBBA *Proceedings* 79 (1974): 18–22.

6. FRA, *The Baltimore & Ohio Railroad Company, Toll Gate, W. Va., February 24, 1967* (RAR 4116, Washington, 1967).

7. FRA, *Louisville & Nashville Railroad Company, Casky, Ky., January 29, 1968* (RAR 4143, Washington, 1968).

8. NTSB, *Penn Central Company Train Second 115 (Silver Star) Derailment at Glenn Dale, Maryland, June 28, 1969* (RAR 70-1, Washington, 1970). On the role of maintenance in causing accidents, see James Reason, *Managing the Risks of Organizational Accidents* (Brookfield, VT: Ashgate, 1997), Chapter 5.

9. NTSB, *Derailment of Amtrak Train No. 6 on the Burlington Northern Railroad, Batavia, Iowa, April 23, 1990* (RAR 91-05, Washington, 1991); NTSB, *Derailment of Amtrak Train No. 58, City of New Orleans, Near Flora, Mississippi, April 6, 2004* (RAR 05-02, Washington, 2005).

10. Some of the decline in derailments between 1975 and 1978 reflected the sharp rise in the reporting threshold discussed in Chapter 2. The initial FRA track regulations covered commuter track that also carried freight traffic. The leading railroad periodicals—*Modern Railroads, Railway Age*, and *Railway Track & Structures*—never mentioned the extension of these standards to dedicated commuter track, suggesting that their impact must have been quite modest.

11. Allan Zarembski and G. M. Magee, "An Investigation of Railroad Maintenance Practice to Prevent Track Buckling," AREA *Proceedings* 83 (1982): 1–63; J. M. Sundberg, "Union Pacific Railroad Laying and Maintenance Policies for CWR," AREA *Proceedings* 89 (1988): 353–356; Allan Zarembski et al., "Track Buckling Risk Analysis Methodology," AREMA *Proceedings* (2004): 1218–1249.

12. "Redesigning the Corridor," RA 185 (April 1, 1984): 55–58; "NEC Track Program Nears Completion," RA 183 (May 10, 1982): 16; APTA, *Transit Fact Book, 1985*, Table 22.

13. "Chicago: A Comeback Story," RA 185 (February 1, 1984): 51–53; "Growing Pains at 150," RA 184 (August 1, 1983): 61–62; "LIRR's Billion Dollar Fix," RA 187 (February 1, 1986): 43–45;

GAO, *Commuter Railroad Safety Activities on Conrail's Lines in New York Should Be Improved* (Washington, 1978); "Metro-North Modernizes," RA 184 (September 1, 1983): 82–83. Capital spending and passenger miles are from APTA, *2015 Public Transportation Fact Book, Appendix A, Historical Tables* (Washington, 2015), Tables 3 and 62.

14. FRA, *1985 Track Survey National Railroad Passenger Corporation (Amtrak)* (Washington, 1985).

15. Neither the FRA nor the Surface Transportation Board publishes data on types of signals and train control in use. The figures and claims in the text are from "Signals Plus," MR 33 (September 1978): 60–63, and "A Growth Market for CTC," RA 184 (February 1, 1983): 25. The shift to Amtrak resulted in the abandonment of much marginal passenger traffic that was likely poorly signaled.

16. "How the L&N Trains Engineers," RA 161 (December 19, 1966): 34–35; "Santa Fe's 'College' For Enginemen," RA 173 (November 27, 1972): 22–23. Training in the late 1980s is from testimony of AAR President William H. Dempsey, in US 100th Cong., 1st Sess., House Committee on Energy and Commerce, Hearings, *Railroad Safety* (Washington, 1988), 216–219. The NTSB found engine drivers' training to be better than that for other railroad workers. See NTSB, *Results of a Survey on Occupational Training in the Railroad Industry* (SIR 79-1, Washington, 1979).

17. NTSB, *Penn Central Company Collision of Trains N-48 and N-49 at Darien, Connecticut, August 20, 1969* (RAR 70-3, Washington, 1970). Later the NTSB investigated accidents resulting from human error and found little evidence of inadequate training but many other contributing causes. See NTSB, *Train Accidents Attributable to "Negligence" of Employees* (RSS 72-1, Washington, 1972).

18. NTSB, *Collision of Illinois Central Gulf Commuter Trains, Chicago, Illinois, October 30, 1972* (RAR 73-5, Washington, 1973). The rules are in 49 CFR 221.

19. The story of the Southern is recounted in Charles Morgret, *Brosnan: The Railroads' Messiah* vol. 1 (New York: Vantage, 1997), 615; NTSB, *Rear End Collision of Two Southern Pacific Transportation Company Freight Trains, Indio, California, June 25, 1973* (RAR 74-1, Washington, 1974).

20. For the SP, see "A New Look at Rule G: A Stronger Approach to Curbing Alcohol and Drug Use," RA 184 (April 1, 1983): 42–43; "BLE Warns BN Call off Your Dogs," LE 118 (September 7, 1984), and "BLE Wins Injunction," LE 118 (October 19, 1984): 1. A list of companies with programs is in US 97th Cong., 2d Sess., Senate Committee on Commerce, Science, and Transportation, Hearings, *Federal Railroad Safety Programs* (Washington, 1982), 61. For the UP program, see "Lines on Labor," RA 185 (July 1, 1984): 20.

21. NTSB, *Fire Onboard Amtrak Train No. 11, Gibson, California, June 23, 1982* (RAR 83-03, Washington, 1983); NTSB, *Derailment of Illinois Central Gulf Railroad Freight Train . . . at Livingston, Louisiana, September 28, 1982* (RAR 83-05, Washington, 1985), quotations on 40–41. There was more to the Livingston wreck than I have let on; the NTSB found that bad track and equipment as well as improper train makeup were contributory causes.

22. NTSB, *Rear-End Collision of Seaboard System Railroad Freight Trains Extra 8051 North and Extra 1751 North, Sullivan, Indiana, September 14, 1983* (RAR 84-2, Washington, 1984), contains materials presenting labor's position. See also "BLE Condemns Alcohol and Drug Use," LE 117 (September 16, 1982): 1, and "Setting the Record Straight on Alcohol and Drugs," LE 118 (October 7, 1983): 1. US 99th Cong., 1st Sess., House Committee on Energy and Commerce, Hearings, *Railroad Safety Reauthorization* (Washington, 1985), contains the views of the FRA beginning on 195 and the NTSB beginning on 288.

23. For the proposed rule, see "FRA Moves Cautiously on Alcohol, Drugs," RA 184 (August 1, 1983): 32; "Alcohol/Drug Rule Suspended Again," RA 187 (February 1, 1986): 20; "Alcohol/Drugs: What the New Rule Means," RA 186 (September 1, 1985): 61–63. For the final decision, see "The Right to Privacy vs. the Right to Live," RA 189 (March 1, 1988): 19.

24. NTSB, *Rear End Collision of Amtrak Passenger Train 94, the Colonial and Consolidated Rail Corporation Freight Train ENS-121 on the Northeast Corridor, Chase, Maryland, January 4, 1987* (RAR 88-01, Washington, 1988).

25. The US Supreme Court case is *Skinner v. Railway Labor Executives' Association* (489 US 602, 1989). For the regulations, see 49 CFR Part 219. T. A. Manello and F. J. Seaman, *Prevalence, Costs and Handling of Drinking Problems on Seven Railroads* (Washington, 1979), is the source of the number of accidents and percent of problem drinkers. See also the testimony of Jolene Molitoris in US 103d Cong., House Committee on Energy and Commerce, Hearings, *Railroad Safety* (Washington, 1994), 168; "How Do We Get Drunks Off the Trains?" MR 39 (January 1984): 30–32, quotation on 31.

26. NTSB, *Collision of Metrolink Train 111 with Union Pacific Train LOF65-12 Chatsworth, California, September 12, 2008* (RAR 10-01, Washington, 2008); "Metrolink Plans Live Video Cameras to Monitor Train Engineers," LAT (December 25, 2008); "Amtrak to Install Cameras in Locomotives," CT (May 26, 2015); "Cameras in the Cab: Federal Judge Sides with KCS on Inward Facing Cameras," Locomotive Engineers and Trainmen *News* 27 (August–September, 2013); "Railroads: The Post–Cell Phone Era," *Trains* 71 (August 2011): 6, General OneFile web; "The Great Locomotive Camera Caper," *Trains* 74 (April 2014): 12, General OneFile web. A critic points out that union opposition to cameras reflects not only privacy concerns but also the prospect that the cameras may change behavior, leading to overuse of emergency braking and thereby raising derailment risks.

27. However, not all track miles on the Northeast Corridor were governed by automatic train control.

28. For tampering, see US 100th Cong., 1st Sess., House Committee on Energy and Commerce, Hearings, *Railroad Safety* (Washington, 1988), 18 and 70. The act was P.L. 100-342.

29. Riley's testimony on training is in US 100th Cong., 1st Sess., Senate Committee on Appropriations, *Oversight Hearing on January 4 Amtrak-Conrail Accident* (Washington, 1987), 49. NTSB, *Near Head-On Collision and Derailment of Two New Jersey Transit Commuter Trains Near Secaucus, New Jersey February 9, 1996* (RAR 97-01, Washington, 1997).

30. NTSB, *Derailment of Amtrak Train No. 2 on the CSXT Big Bayou Canot Bridge Near Mobile, Alabama, September 22, 1993* (RAR 94-01, Washington, 1994). The FRA did publish bridge guidelines (49 CFR 213 App C) and it has issued emergency orders for unsafe bridges from time to time. In 2010, it published regulations requiring the railroads to implement bridge management and inspection programs and to know the load capacity of bridges.

31. NTSB, *Derailment of Amtrak Train No. 2* (n. 31); the costs of bridge safety are on 41. The logic and evidence for the $15 million figure are from Robert Hahn et al., *Do Federal Regulations Reduce Mortality?* (Washington: American Enterprise Institute, 2000). While one might reasonably question the size of this effect, the logic seems plausible.

32. For the Amtrak safety assessment, see the testimony of John Riley, FDA Administrator, in US 99th Cong., 1st Sess., Senate Committee on Commerce, Science, and Transportation, Hearings, *Railroad Safety* (Washington, 1985), 15–22.

33. The FRA was not involved in the investigation, as it has no jurisdiction over stand-alone transit systems. NTSB, *Collision of Two Washington Metropolitan Area Transit Authority Metrorail Trains Near Fort Totten Station, Washington, D.C., June 22, 2009* (RAR 10/02, Washington, 2010). For review of previous collisions and derailments, see FRA Accident Docket, Accident number DCA09MR007, documents 100 and 135, http://dms.ntsb.gov/pubdms/; "Transit M/W: Placing Safety First," RTS 109 (January 2013): 23–24; Luke Mullins and Michael Gaynor, "The Infuriating Story of How Metro Got So Bad," *Washingtonian* 51 (December 9, 2015): 56–71.

34. FRA, *Report to Congress, Operation Deep Dive: Metro North Commuter Rail Safety Assessment* (Washington, 2014), quotation on 2; NTSB, *Organizational Factors in Metro North Railroad Accidents* (SIR 14/4, Washington, 2014), quotations on 73 and 31; "Federal Safety Review Prompts Action on Metro North, LIRR," RTS 210 (April 2014): 5; NTSB, *Washington Metropolitan Area Transit Authority L'Enfant Plaza Station Electrical Arcing and Smoke Accident, Washington, D.C. January 12, 2015* (RAR 16/01, Washington, 2016).

35. "ATCS: A Management Tool," RA 187 (October 1, 1986): 35–37, quotation on 35; "CP Rail Moves toward ATCS," RA 188 (August 1, 1987): 45–46.

36. The best descriptions of Amtrak's system I have found are James Hoelscher and Larry Light, "Full PTC Today with Off the Shelf Technology: Amtrak's ACSES Overlay on Expanded ATC," AREMA *Proceedings* (2001): 1108–1124, and "Ahead of the Rest: Positive Train Control in the Northeast Corridor," *Trains* 71 (October 2011): 82, General OneFile web. See too "Amtrak 188 One Year Later," *Trains* 76 (May 2016): 6, General OneFile web; NTSB, *Derailment of Amtrak Passenger Train 188, Philadelphia, Pennsylvania, May 12, 2015* (RAR 16/02, Washington, 2016).

37. For early FRA activities, see US 103d Cong., 2d Sess., Senate Committee on Commerce, Science, and Transportation, Hearings, *Oversight and Reauthorization of Rail Safety Programs and S 2132 The Federal Railroad Safety Authorization Act* (Washington, 1994), testimony of James Hall, 28–38; "FRA, IDOT Revive PTC Project," RA 199 (March 1998): 25; "The Curtain Rises on the Next Generation," RA 199 (July 1998): 41–42.

38. "New Tech Train Control Takes Off," RA 203 (May 2002): 41–46. For the transit systems, see "A Breakthrough Year?" RA 205 (May 2004): 23–25; "Alaska Moves Toward Safer Operations," RA 204 (June 2003): 20. The Amtrak installation is from testimony of FRA Administrator Alan Rutter in US 107th Cong., 2d Sess., House Committee on Transportation and Infrastructure, Hearings, *Recent Derailments and Railroad Safety* (Washington, 2002), 116–150, which also contains a good review of FRA efforts to promote PTC. For costs and benefits, see testimony of Marion Blakely and Edward Hamburger in US 107th Cong., 2d Sess., Senate Committee on Commerce, Science, and Transportation, Hearings, *Railroad Safety* (Washington, 2005).

39. The quotation is from Mark Rosenker, Chairman, NTSB, in US 110th Cong., 1st Sess., House Committee on Transportation and Infrastructure, Hearings, *Rail Safety Legislation* (Washington, 2007), 35.

40. "Is Everyone On Board?" RA 211 (May 2010): 29–36; "Positive Train Control: Oh What a Mess," *Trains* 74 (March 2014): 16, General OneFile web. "PTC is about" is from "The Business of PTC is Safety, Period," RA 211 (May 2010): 4.

41. Failures in the block system (false clears) are rare but by no means unheard of. See, for example, "Amtrak Exceeds Yard Limits," *Trains* 73 (January 2013): 12, and "There is No Substitute for an Experienced, Cautious Engineer," *Trains* 74 (May 2014): 8, General OneFile web. Costs and benefits are from "Is Everyone On Board?" (n. 41). Costs are from FRA, *Department of Transportation Federal Railroad Administration 49 CFR Parts 229, 234, 235, and 236 [Docket No. FRA-2006-0132, Notice No. 1] RIN 2130-AC03 Positive Train Control Systems Regulatory Impact Analysis* (Washington, 2009), Tables 1–3.

42. Hahn et al., *Do Federal Regulations Reduce Mortality?* (n. 32); "PTC: A Question of Safety?" RTS 108 (October 2012): 3.

43. The $800 million estimate is based on FRA, *Department of Transportation . . . Regulatory Impact Analysis* (n. 41). The FRA estimated maintenance at 15 percent of initial costs.

44. "The Safest Seat," *American Railroad Journal* 27 (July 2, 1853): 426; "Vacationing by Train: How Safe Are the Passenger Cars?" WP (March 3, 1974).

45. For the AAR standards and that they were routinely exceeded, see the following: "Passenger Cars—1946 Models," RA 120 (January 1, 1946): 5–7; "Deluxe Service on a 117 Mile Route," RA 123 (August 16, 1947): 46–51; and "They Protect the Passengers," RA 129 (October 14, 1950): 16, which contains the quotation. For Budd, see "Design Considerations for Railway Passenger Cars," REM 120 (March 1, 1946): 111–113.

46. NTSB, *Boston and Maine Corporation Single Diesel Powered Passenger Car 563 Collision with Oxbow Transportation Company Tank Truck at Second Street Railroad Highway Grade Crossing, Everett, Massachusetts, December 28, 1966* (Washington, 1966).

47. NTSB, *Penn Central Company Train Second 115* (n. 9). Collinsville is from "U.S. Unit Tells Amtrak to Fix Safety Defects," WP (August 1, 1972). NTSB, *Derailment of Amtrak Train No. 1 While Operating on the Illinois Central Railroad Near Salem, Illinois, June 10, 1971* (RAR 72-5, Washington, 1972); NTSB, *Rear End Collision of Conrail Commuter Train No. 400 and Amtrak Passenger Train No 60, Seabrook, Maryland, June 9, 1978* (RAR 79-3, Washington, 1979); NTSB, *Fire Onboard Amtrak Passenger Train No. 11* (n. 22).

48. For Amtrak's actions, see GAO, *Amtrak Safety: Amtrak Should Implement Minimum Standards for Passenger Cars* (Washington, 1993), 28–30, which contains the quotation. See also testimony of James Hall, administrator of NTSB, in US 104th Cong., 2d Sess., Senate Committee on Science, Commerce, and Transportation, Hearings, *Railroad Safety* (Washington, 1996), 26–27. Early FRA research may be found in R. J. Cassidy and D. J. Romeo, *An Assessment of Crashworthiness of Existing Urban Rail Vehicles*, vol. 2 (Washington: DOT, 1975); M. J. Reilly, *Rail Safety/Equipment Crashworthiness*, vol. 4 (Washington: DOT, 1978); Edward Widmayer, *A Structural Survey of Classes of Vehicles for Crashworthiness* (DOT FRA/ORD 79-13, Washington, 1978). See also *1975 Annual Report by the President to Congress on the Administration of the Federal Railroad Safety Act of 1970* (Washington, 1976).

49. GAO, *Amtrak Safety* (n. 49). A review of the NTSB's unsuccessful prodding of the FRA is available at the NTSB website (www.ntsb.gov) under recommendation "R-83-076." A modern review of early auto safety work is in James Wetmore, "Delegating to the Automobile: Experimenting with Automobile Restraints in the 1970s," *Technology and Culture* 56 (April 2015): 440–463.

50. In defense of the railroads, it can be argued that the long dry spells between passenger car orders that drove most American builders from the market and discouraged entry might well have

retarded safety innovation. In addition, foreign producers, used to the much less rigorous standards abroad, sometimes have trouble designing cars for the US market ("Delays May Derail Funding for Amtrak Rail Cars," WSJ [April 10, 2016]). Fair enough, but this hardly excuses the federal government's tardy interest in researching safer rail passenger cars.

51. NTSB, *Collision between Northern Indiana Commuter Transportation District Eastbound Train 7 and Westbound Train 12, Near Gary, Indiana, on January 18, 1993* (RAR 93-03, Washington, 1993), quotation on 20.

52. "Administration Unveils $1.3 Billion High-Speed Rail Proposal," NYT (April 29, 1993). The law is the Swift Rail Development Act, 49 US Code 20133, section 215.

53. NTSB, *Collision and Derailment of Maryland Rail Commuter MARC Train 286 and National Railroad Passenger Corporation Amtrak Train 29 Near Silver Spring, Maryland, on February 16, 1996* (RAR 97-02, Washington, 1997), quotation from Appendix H, 2; "Emergency Order No. 20," www.fra.dot.gov.

54. Emergency preparedness regulations are in 49 CFR Parts 223 and 239; the equipment standards are at 49 CFR Parts 216, 223, 229, 231, 232, and 238. Both are also available at the FRA website (www.fra.dot.gov) under "Regulations and Rulemaking." The APTA standards are available at www.apta.com/resources/standards/press/Pages/default.aspx.

55. "The Silver Creek Catastrophe," *Railroad Gazette* 18 (September 17, 1886): 643.

56. George Bibel, *Train Wrecks: The Forensics of Rail Disasters* (Baltimore: Johns Hopkins University Press, 2012), Chapter 3.

57. William Haden et al., *Accident Research* (New York: Harper & Row, 1964), reprints some of De Haven's work in Chapter 9. Amy Gangloff, "Safety in Accidents," *Technology & Culture* 54 (January 2013): 40–61, recounts De Haven's impact on automobile safety. For early FRA studies, see n. 49. For British work, see "Crashworthiness Moves from Art to Science," *International Railway Journal* 151 (April 1995): 227–230. See also A. Sutton, "The Development of Rail Vehicle Crashworthiness," Institution of Mechanical Engineers Proceedings, Part F, *Journal of Rail and Rapid Transit* 216 (March 2002): 97–108; "All-New Cutlass," *Automotive Engineering* 104 (October 1996): 46–47.

58. Kristine Severson et al., "Analysis of Collision Safety Associated with Conventional and Crash Energy Management Cars Mixed Within a Consist ASME," *International Mechanical Engineering Congress & Exposition* (November 16–21, 2003): Paper IMECE2003-44122; Michelle Priante et al., "The Influence of Train Type, Car Weight, and Train Length on Passenger Train Crashworthiness," ASME/IEEE Joint Rail Conference *Proceedings* (March 16–18, 2005): 89–96.

59. Analyses of crashworthiness are available at www.volpe.dot.gov/infrastructure-systems-engineering/structures-and-dynamics/rail-equipment-crashworthiness.

60. The final rule is "Passenger Equipment Safety Standards," *Federal Register* 64 (May 12, 1999): 25540–25705. The rules for cab cars and MU passenger locomotives have subsequently been modified and more fully embrace the role of CEM; see "Passenger Equipment Safety Standards; Front End Strength of Cab Cars and Multiple-Unit Locomotives," *Federal Register* 75 (January 9, 2010): 1180–1232.

61. MU cars weigh at least 50 tons each. Hence, the kinetic energy of a 3-car train weighing 300,000 pounds moving at 80 mph (117 feet per second) is $E_k = 0.5[300{,}000/32 * (117)^2]$ = about 64 million foot-pounds.

62. For Glendale, see FRA, *Southern California Regional Rail Authority (SCRX), Glendale, California, January 26, 2005* (RAR HQ-2005-08). Metrolink specifications are from David Tyrell et al., "Overview of a Crash Energy Management Specification for Passenger Rail Equipment," ASME/IEEE, Joint Rail Conference, *Proceedings* (April 4–6, 2006): 131–140. Glendale and several subsequent accidents have also raised questions about the safety of push-pull operations. However, FRA claims push-pull is not relatively dangerous; see its *Report to the House and Senate Appropriations Subcommittees: The Safety of Push Pull and Multiple Unit Locomotive Passenger Rail Operations, June 2006* (Washington, 2006).

63. APTA, *Standard for the Design and Construction of Passenger Railroad Rolling Stock* (APTA PR-CS-S 034-99, October 1999). For Metro North, see "Crashworthiness Standards Pass the Test," RA 214 (June 2013): 2. For Oxnard, see "2015 Oxnard Train Derailment," on Wikipedia. See too the NTSB video briefing on www.youtube.com/watch?v=7hSlciKwelc.

64. I chose the years from 1958 on because after 1957 the ICC no longer treated travelers not on trains as passengers. For the whole period from

1947 to 2010, "little accident" fatalities are about two-thirds of those from train accidents.

65. I have contacted both Amtrak and the FRA and neither could explain the causes of the increase.

66. Trip length is from APTA, *Public Transit Fact Book, 2013* (Washington, 2013), Appendix A.

67. Of course, some smaller airports do not have jet ways and passengers enter and exit using stairs from the tarmac, thus experiencing some of the exciting aspects of train travel.

68. NTSB, *Long Island Rail Road Company Door Accident, Huntington Station, New York, December 1, 1974* (RAR 75-5, Washington, 1975). Other incidents are available at www.fra.dot.gov, Safety Data, Table 4.06.

69. "DOT Ready to Get Moving on $1.9 Billion Corridor Upgrading," RA 177 (October 11, 1976): 32–34; "Boston Rebuilds Its Regional Rail System," RA 198 (November 1997): G1–G13.

70. NTSB, *Long Island Railroad* (RAB 09-01, Washington, 2009); FRA, *FRA Approach to Managing Gap Safety, December 7, 2007*, www.fra.dot.gov.

71. APTA, *Developing a Gap Safety Management Program* (APTA AC-GSM-RP 001-10, March, 2013); "Making Amtrak-Served Stations Accessible" and "Ann Arbor Station Expands Accessible Boarding," *Amtrak Ink* 20 (July–September 2015): 10–12 and 14, respectively. For the dangers of various types of boarding, the payoff to improvements, and the Virginia Rail Express, see the following: Edward Morlok et al., "Boarding and Alighting Injury Experience with Different Station Platform and Car Entranceway Designs on U.S. Commuter Railroads," *Accident Analysis and Prevention* 36 (March 2004): 261–271; Edward Morlok and Bradley Nitzberg, "Reducing Boarding and Alighting Accident Rates on Mixed High and Low Platform Railroad Lines through Car and Station Design," *Journal of Transportation Engineering* 131 (May 1, 2005): 382–391. The latter article presents statistics showing that lines with mixed high and low platforms are the most dangerous but in the absence of a persuasive explanation, that may reflect inadequate controls.

Chapter 6. Look Out for the Train

1. "Rail Wreck! 17 Killed in Flaming Crash: 68 Injured," LAT (March 2, 1960); "17 Are Killed in Train Crash: 'Chief' Hits Oil Truck," WP (March 2, 1960); "Train Hits Truck: 17 Die in Wreck 55 Hurt," NYT (March 2, 1960); "Wreck Yields Last of 14 Bodies," WP (March 3, 1960).

2. "Speeding Train Hits School Bus; 20 Dead," CT (December 15, 1961); NTSB, *Waterloo, Nebraska, Public School Bus–Union Pacific Railroad Company Freight Train Accident, Waterloo, Nebraska, October 2, 1967* (Washington, 1968); "5 Boys Killed in Train Auto Collision in Ill.," WP (February 18, 1973); "School Cheer Leaders Killed in Desert Crash," LAT (January 27, 1964); Michael Smith, Interview, February 3, 2016, Indiana State Police Museum, http://www.in.gov/isp/museum.htm.

3. Costs for the 1970s are from "The Approach," MR 33 (August 1978): 40–43. For costs in the 1980s, see FHA, *Rail Highway Crossing Study* (FHWA/SA 89/001, Washington, 1989), Chapter 7. Marsh is from "Train Tank Truck Crash Kills Two," RA 148 (March 14, 1960): 61.

4. "Highway Tanker Curbs Urged," RA 168 (March 7, 1970): 36; ICC, *Prevention of Rail-Highway Grade-Crossing Accidents Involving Railway Trains and Motor Vehicles* (Report 33440, Washington, 1964), 1–97; "Div. 261 Member Wonders When Tanker-Truck 'Demon' Will Strike Again," LE 99 (March 3, 1965): 3; "Highway-Grade Crossing CARNAGE Continues Unabated," LE 90 (February 1956): 80–82; "Fiery Crash Kills 14," LE 94 (March 11, 1960): 1. "Ten years later" is from "Why We Blow Whistles," Locomotive Engineer *Newsletter* 14 (April 2000): 8.

5. Barbara Welke, *Recasting American Liberty: Gender, Race, Law and the Railroad Revolution, 1865–1920* (New York: Cambridge University Press, 2001), Chapter 3.

6. The statistics on grade crossings, accidents, and casualties contain a number of inconsistencies. Before 1975, grade crossings come from the ICC *Transport Statistics*, and appear to be only public crossings on Class I railroads. Accidents and casualties are from the *Accident/Incident Bulletin* and are for all railroads reporting to ICC/FRA at all crossings. It is unlikely that these discrepancies matter much, however. Beginning in 1975, the FRA began to publish data on both public and private crossings. In addition, changes in reporting to include all accidents irrespective of property damage or casualties ballooned the accident data. Thus, the data on public crossings and casualties are likely comparable throughout the entire period, whereas there is a discontinuity in 1975 for the data on total crossings and accidents. Finally, all of these data exclude accidents, crossings, and casualties to carriers that do not report to FRA (e.g., subways and stand-alone light rail systems), which amounted to

about 2 percent of the FRA totals in the period 2003–2010.

7. "North Western Installs Automatic Gates at Nine Crossings," RSC 52 (August 1, 1959): 18–21; "Crossings are Safer After Ten Years," RSC 53 (January 1, 1960): 21–23. Installation of flashing lights at crossings where the watchman had been on duty less than 24 hours a day probably improved safety.

8. For a review of some of the mistakes people make in estimating probabilities, see Colin Camerer and Howard Kunreuther, "Decision Processes for Low Probability Events: Policy Implications," *Journal of Policy Analysis and Management* 8 (Fall 1989): 565–592. For the origin of these careful crossing campaigns, see my *Death Rode the Rails: American Railroad Accidents and Safety, 1828–1965* (Baltimore: Johns Hopkins University Press, 2006), Chapter 9. "A.C.L. Works to Eliminate Grade Crossing Accidents," RA 131 (August 27, 1951): 41–43; "Educating Drivers to Reduce Crossing Accidents," RA 133 (September 8, 1952): 54. The Illinois Central is from H. F. Davenport to A. Scheffer Lang, January 25, 1968, grade crossing file, box 1, Administrative Correspondence Files, 1967–1970, FRA Records, RG 399, NA.

9. For state efforts, see "Crossing Safety Improves, Part I," RSC 62 (March 1, 1969): 20–26; "The Crossing Problem . . . Getting More Help from a Town," RA 140 (April 23, 1956): 31–32; "Grade Crossing Deaths Cut 97.4%: Record of Three Illinois Crossing Protection Plans," RA 136 (March 22, 1954): 88–89. The Southern Pacific commuter experience is from "Crossing Safety Improves, Part II," RSC 62 (April 1, 1969): 20–23.

10. Expenditures are from DOT, *Report to Congress, Railroad Highway Safety, Part I: A Comprehensive Statement of the Problem* (Washington, 1971), Tables 7–11. The MKT accident is from "Train Tank Truck Crash Kills Two," RA 148 (March 14, 1960): 61; "Highway-Grade Crossing CARNAGE" (n. 4). Heavy trucks were a threat to job security as well as safety and so the rail unions favored restrictions on truck carriage of explosives and flammables as well as better crossing laws and guards. See "Protest Hauling of Dangerous Explosives over Nation's Highways," LE 86 (February 1952): 78–79, and "Launches Drive to Ban Tank Trucks from Highway," LE 91 (January 1, 1957): 1.

11. ICC, *Prevention of Rail-Highway* (n. 4), 3.

12. NTSB, *Waterloo, Nebraska, Public School Bus* (n. 2); "High Speed Trains Spur Safety Drive," WP (June 5, 1966). For the DOT initiative, see Lowell Bridewell to the Secretary [Alan Boyd], November 3, 1967, grade crossing file, box 1, Administrative Subject Correspondence, 1967–1970, FRA Records, RG 399, NA, and DOT, *Report to Congress* (n. 10), 43. For a general review of these developments, see "Crisis at the Crossing," RA 164 (January 15, 1968): 19–20; "NTSB Wants a New Safety Unit," RA 166 (June 2, 1969): 32.

13. US 91st Cong., 1st Sess., Senate Committee on Commerce, Hearings, *Federal Railroad Safety Act of 1969* (Washington, 1969), quotation on 245; "BLE Demands Passage of Rail Safety Bill," LE 103 (June 6, 1969): 1.

14. DOT, *Report to Congress* (n. 10), and *Part II: Recommendations for Resolving the Problem* (Washington: DOT, 1972). For labor support, see "BLE and Signalmen Join in Supporting Grade Crossing Plea," LE 106 (May 26, 1972): 1. For brief discussions of the acts, see "The Big Breakthrough in Grade Crossing Improvement," RTS (June 1974): 15–16; "Safety at the Crossroads," MR 29 (August 1974): 35–39; and "$760 Million More for Grade Crossing Safety," RTS (May 1979): 28–32.

15. GAO, *Report to Congress by the Comptroller General of the United States: Rail Crossing Safety—At What Price?* (CED 78-83, Washington, 1978).

16. I estimated a negative binomial time series/cross-section analysis of crossing deaths and accidents using data from California, Florida, Michigan, Illinois, Indiana, North Carolina, Ohio, Pennsylvania, and Texas for 1947–1974. I employed state gasoline consumption as a proxy for automobile miles and state shares of train accidents as proxies for train miles. The equations also contain time dummies (not shown); results for crossing deaths (with z-scores in parentheses) are

$$\begin{aligned}\text{Ln(Deaths)} = -19.95 &+ \frac{1.976}{(9.67)}\text{Ln(Crossings)}\\ &+ \frac{0.334}{(6.91)}\text{Ln}\left(\frac{\text{Train Miles}}{\text{Crossings}}\right)\\ &+ \frac{1.203}{(7.88)}\text{Ln}\left(\frac{\text{Auto Miles}}{\text{Crossings}}\right) - \frac{1.359}{(2.95)}\end{aligned}$$

Percent Protected Crossings
$N = 336$

The costs of protection are from DOT, *Report to Congress* (n. 10), Table 8.

17. The statistics are from

$$\mathrm{Ln}\left(\frac{\text{Fatalities}}{\text{Train Mile}}\right) = 0.637 - \frac{0.039}{(7.10)}\text{Trend};$$

$R^2 = 0.49; N = 35.$

The figures in parentheses are t-ratios.

18. "In Deaths at Rail Crossings, Missing Evidence and Silence," NYT (July 11, 2004); DOT Inspector General, *Reauthorization of the Federal Railroad Safety Program* (Washington, 2007). Out of more than 17,000 crossing accidents between 1999 and 2004, the inspector general found that 139 were reported late. Ian Savage and Shannon Mok, "Why Has Safety Improved at Rail-Highway Grade Crossings?" *Risk Analysis* 25 (August 2005): 867–881.

19. Savage and Mok, "Why Has Safety Improved?" (n. 18), Table 1. These researchers also find the elasticity of fatalities with respect to the number of crossings was about 0.11. That is, a 10 percent reduction in crossings would reduce fatalities by 1.1 percent. Of course, while on average the guarding of 1,000 crossings may have a low payoff that does not preclude a high payoff to individual projects.

20. John Redden, "Train Horn Noise Mitigation," AREMA *Proceedings* (2003): 847–869.

21. FRA, *Florida's Train Whistle Ban* (Washington, 1995); FRA, *Nationwide Study of Train Whistle Bans* (Washington, 1995).

22. The railroads' support of preemption is from Edwin Harper [President, AAR], "Statement," in US 103d Cong., House Committee on Energy and Commerce, Hearings, *Railroad Safety* (Washington, 1994), 267–300, quotation on 293. "BLE Position on Use of Locomotive Horns," Locomotive Engineer *Newsletter* 14 (May 2000): 2.

23. FRA, "Train Horn Rule—History and Timeline," www.fra.dot.gov; "Whistle Law Steams Suburbs," *Chicago Sun-Times* (September 25, 1995). For reviews of these issues, see Cheryl Greene, "An End to Quiet Neighborhoods or Improved Public Safety?" *Journal of Legislation* 22 (1996): 223–239, and Mark Gruenes, "The Swift Rail Act: Will Sleepless Citizens be Able to Quiet Train Whistles and at What Cost?" *Northern Illinois University Law Review* 19 (1999): 567–587.

24. The Chicago resident is quoted in US 106th Cong., 2d Sess., House Committee on Transportation and Infrastructure, Hearings, *Implementation of the Federal Railroad Administration Grade-Crossing Whistle Ban Law* (Washington, 2000), 10–11. "Train Whistles Plugged with Court Order," CT (August 26, 1988), contains the other quotations. See also "Good Idea Whistles Up Rail Storm," CT (August 27, 1988); "Exemption on Whistle Blowing Urged for Most Rail Crossings," CT (August 31, 1988); "Regulators Silence Train Whistles," CT (October 13, 1988).

25. The quotation is from FRA, "Train Horn Rule" (n. 23). The notice of proposed rulemaking is "Use of Locomotive Horns at Highway Rail Grade Crossings," *Federal Register* 65 (January 13, 2000): 2230–2270.

26. US 106th Cong., 2d Sess., House Committee on Transportation and Infrastructure, Hearings, *Implementation of the Federal Railroad Administration Grade-Crossing Whistle Ban Law* (Washington, 2000). LaHood is quoted on 4.

27. US 106th Cong., 2d Sess., *Implementation* (n. 26), quotations on 145–146.

28. The final rule is "Use of Locomotive Horns at Highway Rail Grade Crossings," *Federal Register* 70 (April 27, 2005): 21844–21920. Train horns emit sound at up to 114 dB; wayside horns sound at about 90 dB, but because it is concentrated at the crossing it is less disruptive.

29. For details on Operation Lifesaver, see the statement of Thomas Simpson in US 102d Cong., 2d Sess., Senate Committee on Commerce, Science, and Transportation, Hearings, *Railroad/Highway Grade Crossing Safety and S, 2644, The Increased Railroad Locomotive Visibility Act* (Washington, 1993), 40–44. See too "Preventing Near Misses Like This," RA 207 (July 2006): 30–34. On the value of Operation Lifesaver in court, see "Safety Group Closely Echoes Rail Industry," NYT (November 14, 2004). "Where BLE Stands on Grade Crossing Safety," LE 114 (June 27, 1980): 1. Ohio is from GAO, *Railroad Safety: Status of Efforts to Improve Railroad Crossing Safety* (GAO/RCED 95-191, Washington, 1995).

30. "Video Protection Pays Off," RA 192 (August 1991): 48–49, and "More Than an Ounce of Prevention," PR 49 (January 2006): 37–38. Cameras on UP locomotives are from "Railroads Rely on Education, Engineering to Reduce Grade Crossing Accidents/Incidents," PR 52 (January 2009), n.p., www.ProgressiveRailroading.com. The staged crash is from "Locomotive Demolishes Auto: Staged Train-Car Crash Makes a Safety Point," LAT (May 22, 1985).

31. "Lessons from Fox River Grove," RA 198 (March 1997): 47–49. For Los Angeles, see "Enforcement Up, Collisions Down," RA 195 (April 1994): 71–72.

32. Details on Idaho and Nebraska are from "Grade Crossing Safety: E-E-E," MR 33 (August 1978): 46–47. The statistical study is Ian Savage, "Does Public Education Improve Rail-Highway Crossing Safety?" *Accident Analysis and Prevention* 38 (March 2006): 310–316.

33. FRA, "Rail Highway Crossing Safety Action Plan Support Proposals, June 1994," http://ntl.bts.gov/DOCS/APSP.html, and FRA, *Accidents That Shouldn't Happen* (Washington, 1996). See also "U.S. DOT Unveils Grade Crossing Safety Program," RA 195 (July 1994): 16.

34. NTSB, *Safety Study—Passenger/Commuter Train and Motor Vehicle Collisions at Grade Crossings* (SS 86-04, Washington, 1986). Where train speeds are 110–125 mph, the FRA requires an "impermeable barrier" at the crossing (www.fra.dot.gov/page/P0103). For North Carolina, see "The Lessons from Fox River Grove," RA 198 (March 1997): 47–50; "The 'Sealed Corridor' Search for Safer Crossings," RA 198 (July 1997): 47–50; and FRA, *North Carolina 'Sealed Corridor' Phase I U.S. DOT Assessment Report, 2003*, www.fra.dot.gov.

35. For a $15 million award, see *Burlington Northern v. Whitt*, 575 So. 2d 1011 (1990); a $4 million award is from *Sheets v. Norfolk S. Corp.*, 109 Ohio App. 3d 278 (1996); an $11 million judgment is from *Gollihue v. Consol. Rail Corp.*, 120 Ohio App. 3d 378 (1997). The corridors are from "Preventing Near Misses" (n. 29) and "9th Annual Grade Crossing Update," PR 51 (January 2008), n.p., www.progressiverailroading.com.

36. FRA, *Private Highway-Rail Grade Crossing Safety Research and Inquiry*, 2 vols. (DOT/FRA/ORD 10-02, Washington, 2010).

37. For the BNSF's efforts, see *Grade Crossing Safety 2014*, www.BNSF.com.

38. Herschel Leibowitz, "Grade Crossing Accidents and Human Factors Engineering," *American Scientist* 73 (1985): 558–562; Theodore Cohn and Tieuvi Nguyen, "Sensory Cause of Railroad Grade Crossing Collisions," TRR 1843 (2003): 24–30.

39. For locomotive pillows, see "Impact Absorber Urged for Trains," WP (May 20, 1969), and Charles Frisk to James MacAnanny, July 7, 1969, grade crossing file, box 1, Administrative Subject Correspondence, 1967–1969, FRA Records, RG 399, NA. For the carriers' experimentation with locomotive visibility, see US 102d Cong., 2d Sess., Senate Committee on Commerce, Science, and Transportation, Hearings, *Railroad/Highway Grade Crossing Safety and S 2644, The Increased Railroad Locomotive Visibility Act* (Washington, 1993); FRA, *Use of Auxiliary External Alerting Devices to Improve Locomotive Conspicuity* (DOT/FRA/ORD 95/13, 1995); "G.N. Box Cars 'Light Up' at Crossings during Darkness," RA 123 (November 15, 1946), 246; FRA, *Freight Car Reflectorization* (DOT/FRA/ORD 98-11, 1998); "Benefit Cost Analysis of Freight Car Reflectorization August 13, 1999," correspondence file, Project Case Files, SOFA, FRA Records, RG 399, NA.

40. NTSB, *Southern Pacific Railroad Company Fruitridge Road Grade Crossing, Sacramento, California, February 22, 1967* (Washington, 1968). For the "lulling effect," see W. Kip Viscusi, *Fatal Tradeoffs: Public and Private Responsibilities for Risk* (New York: Oxford University Press, 1992), Chapter 13. NTSB, *Safety Study—Passenger* (n. 34); NTSB, *Safety at Passive Grade Crossings, Volume 1: Analysis* (SS 98/02, Washington, 1998); "8 Hurt in Md. When Train Hits Truck," WP (October 9, 1996).

41. NTSB, *Safety at Passive Grade Crossings* (n. 40).

42. Research on the effectiveness of stop signs versus crossbucks is mixed. See Charles Schartung et al., "Crossing Paths: Trend Analysis and Policy Review of Highway Rail Grade Crossing Safety," *Journal of Homeland Security and Emergency Management* 8 (January 2011): 1–17. The KCS is from "Railroads Rely on a Mix of Time-Tested and New Equipment to Prevent Grade Crossing Accidents," PR 50 (January 2007), www.ProgressiveRailroading.com. The BNSF is from "Railroads Rely on Education" (n. 30).

43. The origin of constant warning crossing guards is from "Signaling Technology Sets Pace," RA 155 (October 21, 1963): 16.

44. NTSB, *Collision of National Railroad Passenger Corporation Amtrak Train 59 With a Loaded Truck-Semitrailer Combination at a Highway/Rail Grade Crossing in Bourbonnais, Illinois, March 15, 1999* (RAR 02/01, Washington, 2002); FRA, *High-Speed Passenger Rail Safety Strategy*, www.fra.dot.gov. The motorists are from NTSB, *Safety Study—Passenger* (n. 34); "Derailing Illusions that Kill," LAT (November 20, 2003). Welty is from "Safety: What about Grade Crossings?" RTS (April 1996): 44.

45. The UP installation is from FRA, *Illinois High-Speed Rail Four-Quadrant Gate Reliability Assessment* (DOT/FRA/ORD 09/19, Washington, 2009); FRA, *A Radar Vehicle Detection System for Four-Quadrant Gate Warning Systems and Blocked Crossing Detection* (DOT/FRA/ORD 12/24, Washington, 2012). For reviews of crossing technol-

ogy from the 1990s on, see the following: "High Tech at the Crossing," RA 199 (October 1998): 57–61; "Working Toward Safer Grade Crossings," RA 202 (May 2001): 37–38. The effectiveness figures are from FRA, *North Carolina "Sealed Corridor" Phase I, II, and III Assessment* (DOT/FRA/ORD 09/17, Washington, 2009). Offsetting behavior is from Byungkon Ko et al., "Evaluation of Flexible Traffic Separators at Highway-Related Grade Crossings," *Journal of Transportation Engineering* 133 (July 1, 2007): 397–405.

46. "Making the Grade," RA 203 (August 2002): 17–21; "Preventing Near Misses like This" (n. 29); "Railroads Rely on a Mix" (n. 42); "More Than an Ounce of Prevention," PR 49 (January 2006): 37–38.

47. Individuals killed at a crossing with active warning devices (lights or gates) are defined as trespassers and have been discussed above. They are excluded from trespassers discussed in this section.

48. For a discussion of trespassing during these early years, see my *Death Rode the Rails*. The 5 percent figure is from FRA, *Rail Trespasser Fatalities; Developing Demographic Profiles, 2008*, www.fra.dot.gov.

49. Railroads have not entirely disappeared from city streets. See "Railroads in the Streets," *Trains* 68 (April 2008): 22–31. The statistics are

$$\text{Ln}\left(\frac{\text{Trespasser Fatalities}}{\text{Train Miles}}\right) = 0.133 - * \frac{0.015}{(3.81)}\text{Trend;}$$

$R^2 = 0.14; N = 25$. Another quantitative study of trespassing finds that declining train miles and rising per capita GDP have been the main causes of declining fatalities. See Ian Savage, "Trespassing on the Railroad," *Research in Transportation Economics* 20 (2007): 199–224. The declines in traveling farmworkers and of individuals walking the tracks are examples of the ways that structural changes and rising well-being reduced casualties.

50. While the FRA does not include suicides in its official data it does count them and from 2012 to 2015 including them as trespassers would have increased measured trespassing deaths about 61 percent. For the statistical treatment of suicides, see FRA, *Countermeasures to Mitigate Intentional Deaths on Railroad Rights-of-Way: Lessons Learned and Next Steps* (DOT/FRA/ORD 14/36, Washington, 2014).

51. FRA, *Rail Trespasser Fatalities* (n. 48), and *Rail Trespasser Fatalities: Demographic and Behavioral Profiles, 2013*, www.fra.dot.gov. See also FRA, *Demographic Profile of Intentional Fatalities on Railroad Rights-of-Way in the United States* (RR 13-36, Washington, 2013); Ian Savage, "Analysis of Fatal Train-Pedestrian Collisions in Metropolitan Chicago, 2004–2012," *Accident Analysis and Prevention* 86 (January 2016): 217–228.

52. "Better Brothel Access: Corridor Puts Trains on Fast Track in Elko," LAT (November 8, 1982). The quotations are from "Latest Death Puts Focus on Rail Dangers," *The Boston Globe* (December 24, 1994); "2 Women Die After Being Hit By Train," LAT (December 6, 1990); "They Won't Forget," LAT (December 9, 1991).

53. "Colorado College Student Crushed by Moving Freight Train While Trying to Hop Aboard, Loses Both Legs," *New York Daily News* (September 6, 2011); "Male Model Dies After Catching Fire while Train Surfing Metro-North Car in Connecticut," *New York Daily News* (November 11, 2014). The two BU students are from "Without Warning," *The Boston Globe* (February 5, 2006).

54. The boy with electrical burns is from "The Rail Yard's Dangerous Lure," WP (June 9, 1982). "Jury Awards 24.2 Million to Boys Injured on Amtrak Tracks," http://www.prnewswire.com/news-releases/jury-awards-242-million-to-boys-injured-on-amtrak-tracks-56625212.html; "Weekend Hobos Try to Recapture a Romantic Past," WSJ (January 28, 1992); "Cold Case: Keeping Snowmobilers off the Straight and Narrow," WSJ (February 22, 2007). A trestle as an attractive nuisance is from "Death on the Tracks at Lake Accotink: Search for Solution Continues," WP (July 6, 1983); "'Bungee War': Getting the Drop on Illegal Jumpers," WP (September 14, 1993). For a group that obtained permission to use the tracks, see "Putt-Putting along the Rails," NYT (August 8, 2008); "Taking Photos on Train Tracks: A Mistake You Can't Undo," WP (October 2, 2015).

55. "Rash of Train-Track Deaths Stirs Plan for Safety Awareness Program," LAT (October 23, 1990); "Latest Death Puts Focus on Rail Dangers," *The Boston Globe* (December 24, 1994).

56. That a high probability of detection and punishment may help deter trespassing is from Brenda Lobb, "Trespassing on Tracks: A Review of Railway Pedestrian Safety Research," *Journal of Safety Research* 37 (Winter 2006): 359–365; FRA, *Railroad Infrastructure Trespassing Detection Systems Research in Pittsford, New York* (DOT/FRA/ORD 06-03, 2006); FRA, *State of the Art Technologies for Intrusion and Obstacle Detection for Railroad Operations* (DOT/FRA/ORD 07/06, 2007), www.fra.dot.gov.

Essay on Sources

This book is based mostly on primary printed sources available as trade publications and from professional associations, labor unions, and government agencies. I have also found a few archival sources valuable. There is a small secondary literature on railroad safety, but considerable informative writing on a variety of contributory topics. This essay describes the primary sources I consulted as well as the most important secondary works.

Archival Sources

The records of the Federal Railroad Administration (FRA) (RG 399) at the National Archives are disappointing. They contain little about policy and almost nothing modern from the office of the administrator. The Administrative Subject Correspondence, 1967–1970, contains modest insights into early work on grade crossing safety and hazmat regulations. The Administrator's Roundtable Discussions, 1993–1997, highlight Jolene Molitoris's skills. The Project Case Files Relating to the Switching Operations Fatality Analysis, 1999–2002, document the birth of that program. I also found some relevant materials on the Confidential Close Call program and there are two groups of "Railroad Accident Investigation Jackets" for 1968–1971 and 1969–1971 that are not jackets at all but case files; they contain some National Transportation Safety Board (NTSB) investigations as well. These are not listed online. Records of the Department of Transportation (DOT) (RG 398) include the Records of the Office of the Secretary of Transportation, 1967–1972, and contain useful information on the maneuvering that led to the Federal Railroad Safety Act of 1970.

The NTSB apparently has no archives—which ought to be a scandal. Here is what a 2014 National Archives report had to say: "NTSB has not submitted schedules in more than 15 years; NTSB indicates on the RMSA [Records Management Self-Assessment] that they do not have any permanent records; it is unclear what NTSB has been doing with their records in terms of disposition" (National Archives and Records Administration, *Records Management Oversight Inspection Report 2014, National Transportation Safety Board Records Management Program* [Washington, 2014]).

Private-sector archives for the modern period are similarly thin. Few contain much on safety and fewer still extend much beyond the 1960s. The Pennsylvania Railroad Archives at the Hagley Museum and Library contains tidbits on the 1960s, while the records of various labor unions maintained at the Kheel Center in the Catherwood Library at Cornell also contain small amounts of useful modern materials. The W. Edwards Deming collection at the Library of Congress contains extensive materials on his work with railroads, but most of it relates to statistical quality control rather than Total Quality Management (TQM).

I found three small but valuable oral history collections available online. These document the perspective of railroad workers from the 1940s well up into modern times. They are the University of North Carolina, Library, "Documenting the American South" collection, http://docsouth.unc.edu/sohp/index.html, and the Southern Arizona Transportation Museum's oral history program, http://tucsonhistoricdepot.org/?page_id=255. The Youngstown State University Oral History Program, http://www.maag.ysu.edu/oralhistory/oral_hist.html#M, also contains a number of insightful interviews. I have also done several interviews with first-responders to the Dunreith, Indiana, accident and the transcripts are at the Indiana State Police Museum, http://www.in.gov/isp/museum.htm.

Primary Printed Materials

The FRA publishes a wealth of information on railroad safety (as did its predecessor, the Interstate Commerce Commission [ICC]). The basic data on accidents and injuries are available in the ICC/FRA, *Accident Bulletin* (now *Safety Statistics*), and the variously named *Rail-Highway Accident Incident Bulletin*. From 1975, some of these data are also available on the FRA website, www.fra.dot.gov. ICC accident reports from 1911 to 1994 are available at http://dotlibrary.specialcollection.net/Home, which also includes reports from the FRA and the NTSB. ICC, *Prevention of Rail-Highway Grade-Crossing Accidents Involving Railway Trains and Motor Vehicles* (Report 33440, Washington, 1964), and DOT, *Report to Congress Railroad Highway Safety, Part I: A Comprehensive Statement of the Problem, and Part II: Recommendations for Resolving the Problem* (Washington: DOT, 1972), provide a wealth of information on grade crossing problems as they appeared at the time. Recent (January 2005 on) individual accident reports are also on the FRA's website.

The FRA e-library, https://www.fra.dot.gov/eLib/Find, also provides access to its various dockets, research reports, testimony, safety advisories, proposed and final rules, and much else. Unfortunately, the agency also announces that if you cannot find something it may be because the "document is historical and has been removed from the Library to avoid confusion with current publications." There appears to be no way to know what these documents may be and where they have gone. As a result, some of the documents from the FRA website referenced here may no longer be available. An increasing number of these works are available online through a Google search, and at Internet Archive, https://archive.org, or HathiTrust, https://www.hathitrust.org.

The NTSB's Railroad Accident Reports (and Hazardous Materials Accident Reports) are available online at the agency's website, www.ntsb.gov, back to 1996. That site also contains its safety recommendations. As noted, earlier NTSB accident reports are also sometimes available with the ICC and the FRA historical accident reports in the DOT's Special Collections cited above. Some of the NTSB's Safety Studies and Special Reports going back to the 1970s are also available at its website, as is a link to its accident dockets. Although I have made little use of Federal Transit Administration (FTA) materials, its *Commuter Rail Safety* (Washington, 2006) along with the American Public Transit Administration's (APTA) *Public Transit Fact Book* (Washington, various years) provide valuable information on the commuter rail network. The Federal Highway Administration's (FHA) *Highway Statistics* (Washington, various years) performs similar functions.

Congressional documents provide a rich source on general railroad matters along with many aspects of railroad safety from the late 1950s on. The various Senate and House hearings on railroad safety, an almost annual event from 1970 on, document congressional concerns and include discussions and critiques of policy by the FRA and the NTSB in addition to testimony from the AAR, various railroads, railroad supply companies, labor groups, and others. Other government agencies that provide useful information include the Railroad Retirement Board, which published *Report on Accidents and Casualties in the Railroad Industry* (Washington, 1959) and *Safety in the Railroad Industry* (Washington, 1962). The General Accounting Office (GAO), http://www.gao.gov/, has done a number of valuable evaluations of FRA programs, as have the inspectors general for Amtrak, https://www.amtrakoig.gov/, and the DOT, https://www.oig.dot.gov. The John A. Volpe National Transportation Systems Center (or Volpe), https://www.volpe.dot.gov/, is also a largely government-funded operation that provides research on a host of transport-related issues. I found its work on rail vehicle crashworthiness invaluable.

In the private sector, the railroad journals provide a mine of information and opinion on virtually all aspects of railroading. But as the carriers have declined in number, so has the range of trade publications. Current publications include *Railway Age*, *Railway Track & Structures*, and *Progressive Railroading*. *Trains* has recently begun to increase the number of articles that bear on safety. Older journals include *Railway Locomotives and Cars*, which ended publication in 1974, and *Modern Railroads*, which *Railway Age* absorbed in 1991. *Railway Signaling and Communication* (under various titles) has also occasionally proved useful. The National

Safety Council's (NSC) *Transactions* available up through 1978 include a railroad section that contains occasional valuable material. Industry technical publications have become comparatively scarce in modern times. The *Proceedings* of the American Railway Engineering Association (AREA) and those of the American Railway Bridge and Building Association (ARRBA) to 1997 are online at Internet Archive, https://archive.org. In 1997, these two organizations and others merged to form the American Railway Engineering and Maintenance of Way Association (AREMA) and their proceedings and other publications have become hard to find. The AAR Mechanical Division *Proceedings* usefully document that group's concerns but are also hard to find and end in 1989. Two valuable AAR publications for the 1970s are A. E. Shulman's *Analysis of Nine Years of Personal Casualty Data 1966–1974* and *Analysis of Nine Years of Railroad Personnel Casualty Data 1966–1974* (Washington: AAR Research and Test Department, 1976). The annual issues of *Railroad Facts* provide economic and sometimes safety data on Class I railroads. The AAR Safety Section largely petered out in the late 1950s. The AAR also publishes a great variety of other material on railroad safety, but only a small amount is publicly available; I have made little use of it because it is hard to find and expensive to purchase.

Railroad employee magazines help document the carriers' changing approach to safety. Some older publications are now online, including those of the Erie-Lackawanna and preceding lines, http://elmags.railfan.net/magindex.cgi?ERLA. The more recent *Amtrak Ink*, https://www.amtrak.com/servlet/ContentServer?c=Page&pagename=am%2FLayout&cid=1241267284550, is modestly useful. Conrail's *Inside Track* helps document that company's embrace of TQM and its evolving safety culture in the 1990s. The BNSF's *Railway*, http://www.bnsf.com/employees/communications/railway-magazine/, also provides hard-to-find detail. The Norfolk Southern's *BizNS*, http://www.nscorp.com/content/nscorp/en/bizns.html, is, as the title suggests, more about business than safety. The Union Pacific's *Community Ties*, http://www.up.com/aboutup/community/community_ties/community_ties_031813.htm, contains company safety doings from time to time. I found no company magazines for CSX, the Canadian National, the Canadian Pacific, or the Kansas City Southern, although each carrier's website contains useful safety information.

Railroad labor journals provide the unions' perspective on safety developments and include much detail on working conditions. Just as the railroad journals have shrunk with the declining number of carriers and workers, so the union publications have merged and changed names. The variously titled *Locomotive Engineer* is a gold mine of union concerns from the 1950s on. Also valuable are *Trainman News*, and the also variously named *Locomotive Engineers and Trainmen News*, http://www.ble-t.org/pr/newsletter/, as well as the Brotherhood of Maintenance of Way Employees *Journal* and UTU *News*, http://utu.org/utu-news-online/.

For popular concerns with railroads and rail safety I have relied on a number of major newspapers that are available and searchable through Proquest. These include *The Boston Globe*, the *Chicago Tribune*, the *Los Angeles Times*, *The New York Times*, *The Washington Post*, and *The Wall Street Journal*. In addition, I have employed Newsbank and www.newspaperarchive.com to search a large number of other newspapers for writing on specific topics such as hazmat concerns.

Secondary Sources

Among the vast numbers of works on railroad history only a comparative few come up to modern times. While most of them have little that bears directly on safety, they are invaluable for context. I found the following especially useful. Eric Beshers, *Conrail: Government Creation & Privatization of an American Railroad* (Washington: World Bank, 1989); Seth Bramsen, *Speedway to Sunshine* (Boston: Boston Mills, 1984). Among H. Roger Grant's many works, I found valuable material in *Erie Lackawanna: Death of an American Railroad, 1938–1992* (Stanford, CA: Stanford University Press, 1994). See, too, his *The North Western: A History of the Chicago & North Western Railway System* (DeKalb: Northern Illinois Press, 1996) and *Follow the Flag: A History of the Wabash Railroad Company* (DeKalb: Northern Illinois University Press, 2004). See Don Hofsommer, *The Southern Pacific, 1901–1985* (College Station: Texas A&M University Press, 1986), and his *Grand Trunk Corporation: Canadian National Railways in the United States, 1971–1992* (East Lansing: Michigan State University Press, 1995); Maury Klein, *Union Pacific: The Reconfiguration: America's Greatest Railroad, from 1969 to the Present* (New York: Oxford University Press, 2011); and Craig Miner, *The Rebirth of the Missouri Pacific 1956–1983* (College Station: Texas A&M University Press, 1983).

For overviews of various aspects of railroads and railroading, see Robert Gallamore and John Meyer, *American Railroads, Decline and Renaissance in the Twentieth Century* (Cambridge, MA: Harvard University Press, 2014). For these matters as they appeared before deregulation, see James Nelson, *Railroad Transportation and Public Policy* (Washington: Brookings Institute, 1959). Maury Klein, *Unfinished Business: The Railroad in American Life* (Hanover, NH: University Press of New England, 1994), and H. Roger Grant, *Railroads and the American People* (Bloomington: Indiana University Press, 2012), discuss railroads and American culture. See, too, H. Roger Grant, "The Writing of Railroad History," *Railroad History* 148 (Spring 1983): 9–12. A fascinating study of containerization is Marc Levinson, *The Box: How the Shipping Container Made the World Smaller and the World Economy Bigger* (Princeton, NJ: Princeton University Press, 2006).

A number of works focus on specific aspects of railroading. Two valuable biographies are H. Roger Grant, *Visionary Railroader: Jervis Langdon Jr. and the Transportation Revolution* (Bloomington: Indiana University Press, 2008), and Charles Morgret, *Brosnan: The Railroads' Messiah*, 2 vols. (New York: Vantage, 1996). On modern rail mergers the bible is Richard Saunders, *Main Lines: Rebirth of the North American Railroads 1970–2002* (DeKalb: Northern Illinois University Press, 2003). See, too, Denis Breen, "The Union Pacific Southern Pacific Rail Merger: A Retrospective on Merger Benefits," *Review of Network Economics* 3 (September 2004): 283–322. Joseph Daughen and Peter Binzen, *The Wreck of the Penn Central* (Boston: Little, Brown, 1979), and Stephen Salsbury, *No Way to Run a Railroad* (New York: McGraw Hill, 1982), detail that important debacle. George Hilton, *The Northeast Railroad Problem* (Washington: American Enterprise Institute, 1975), provides a contemporary overview of railroad problems. ICC constraints on railroads are explored in Albert Churella, "Saving the Railroad Industry to Death: The Interstate Commerce Commission, the Pennsylvania Railroad, and the Unfulfilled Promise of Rail-Truck Cooperation," Business and Economic History Online, 2006, and in his "Delivery to the Customer's Door: Efficiency, Regulatory Policy, and Integrated Rail-Truck Operations, 1900–1938," *Enterprise and Society* 10 (March 2009): 98–136. Similarly, see Paul MacAvoy and James Sloss, *Regulation of Transport Innovation: The ICC and Unit Coal Trains to the East Coast* (New York: Random House, 1967). Richard Barber, "Technological Change in American Transportation: The Role of Government Action," *Virginia Law Review* 50 (June 1964): 824–895, provides a general review of the deadening effect of regulation on railroad innovation. Post-regulatory contracting is discussed in Marc Levinson, "Two Cheers for Discrimination: Deregulation and Efficiency in the Reform of U.S. Freight Transportation, 1976–1988," *Enterprise and Society* 10 (March 2009): 178–215.

The following works discuss other aspects of railroading. Jan Heier and A. Lee Gurley, "The End of Betterment Accounting: A Study of the Economic, Professional and Regulatory Factors That Fostered Standards Convergence in the Railroad Industry, 1955–1883," *Accounting Historians Journal* 34 (June 2007): 25–55; Douglas Caves et al., "Productivity in U.S. Railroads, 1951–1974," *Bell Journal of Economics* 11 (Spring 1980): 166–181; and P. E. Schoech and J. A. Swanson, "Patterns of Productivity Growth: An Examination of Pre- and Post-Deregulation Determinants," https://www.lrca.com. Valuable discussions of research are NAS, *Science and Technology in the Railroad Industry, A Report to the Secretary of Commerce* (Washington, 1963), and NAS Transportation Research Board, *Evaluation of the Federal Railroad Administration Research and Development Program* (Special Report 316, Washington, 2015). Carl Martland, "Introduction of Heavy Axle Loads by the North American Rail Industry," *Journal of the Transportation Research Forum* 52 (Summer 2013): 103–125; Byungkon Ko et al., "Evaluation of Flexible Traffic Separators at Highway-Related Grade Crossings," *Journal of Transportation Engineering* 133 (July 1, 2007): 397–405.

I also found a host of useful technical articles, too numerous to cite individually in the following. ASME/IEEE, *Joint Rail Conferences*, http://ieeexplore.ieee.org/xpl/conhome.jsp?punumber=1000613, and the (British) Institute of Mechanical Engineers, *Journal of Rail and Rapid Transit.*

Works on railroad or other safety that I found valuable are Ian Savage, *Economics of Railroad Safety* (Boston: Kluwer, 1998); Ian Savage and Shannon Mok, "Why Has Safety Improved at Rail-Highway Grade Crossings?" *Risk Analysis* 25 (August 2005): 867–881; Ian Savage, "Trespassing on the Railroad," *Research in Transportation Economics* 20 (2007): 199–224; his "Does Public Education Improve Rail–Highway Crossing Safety?" *Accident Analysis and Prevention* 38 (March 2006): 310–316; and his "Analysis of Fatal Train-Pedestrian Collisions in Metropolitan Chicago 2004–2012," *Accident Analysis and Prevention* 86 (January 2016): 217–228. George Bibel, *Train Wrecks: The Forensics of Rail Disasters* (Baltimore: Johns Hopkins University Press, 2012), provides valuable insights into the engineering aspects of disasters. For whistle bans, see Cheryl Greene, "An End to Quiet Neighborhoods or Improved Public Safety?" *Journal of Legislation* 22 (1996): 223–239, and Mark Gruenes, "The Swift Rail Act: Will Sleepless Citizens be Able to Quiet Train Whistles and at What Cost?" *Northern Illinois University Law Review* 19 (1999): 567–587. T. A. Manello and F. J. Seaman, *Prevalence, Costs and Handling of Drinking Problems on Seven Railroads* (Washington, 1979), provides a window into that problem. Christopher Barkan, "Improving the Design of Higher Capacity Tank Cars for Hazardous Materials Transport: Optimizing the Trade-Off between Weight and Safety," *Journal of Hazardous Materials* 160 (December 2008): 122–134, provides insight into the complexities of engineering-economic trade-offs. Treatments of passenger station injuries are comparatively rare. See Edward Morlock et al., "Boarding and Alighting Injury Experience with Different Station Platform and Car Entranceway Designs on US Commuter Railroads," *Accident Analysis and Prevention* 36 (March 2004): 261–271. See also Edward Morlock and Bradley Nitzberg, "Reducing Boarding and Alighting Accident Rates on Mixed High and Low Platform Railroad Lines Through Car and Station Design," *Journal of Transportation Engineering* 131 (May 2005): 382–391. A good review of trespassing is Brenda Lobb, "Trespassing on Tracks: A Review of Railway Pedestrian Safety Research," *Journal of Safety Research* 37 (Winter 2006): 359–365.

For the behavioral approach to safety, see E. Scott Geller, *The Psychology of Safety* (Radnor, PA: Chilton Book Company, 1996); Dan Petersen, *Analyzing Safety System Effectiveness*, 3d ed. (New York: Van Nostrand

Reinhold, 1996); and Thomas Krause, *The Behavior Based Safety Process*, 2d ed. (New York: Van Nostrand Reinhold, 1997). Also useful are James Reason, *Managing the Risks of Organizational Accidents* (Brookfield, VT: Ashgate, 1997), and Harold Roland and Brian Moriarty, *System Safety Engineering and Management* (New York: John Wiley, 1990). Many articles on behaviorism and its links to railroads and the TQM movement are in the *Journal of Safety Research*, *Professional Safety*, *Performance Management Magazine*, and *Safety Science*. See also Herschel Leibowitz, "Grade Crossing Accidents and Human Factors Engineering," *American Scientist* 73 (November–December 1985): 558–562.

For insight into regulation and deregulation, from the vast literature available I consulted the following: Marver Bernstein, *Regulating Business by Independent Commission* (Princeton, NJ: Princeton University Press, 1955); Thomas McCraw, "Regulation in America," *California Management Review* 27 (Fall 1984): 116–124; and Robert Higgs, *Crisis and Leviathan* (New York: Oxford University Press, 1987). On the politics of transport regulation and the origins of the DOT, see Mark Rose et al., *The Best Transportation System in the World* (Columbus: Ohio State University Press, 2006), and Richard Barsness, "The Department of Transportation: Concept and Structure," *Western Political Quarterly* 23 (September 1970): 500–515. For the political climate that contributed to policy changes, see Brian Balough, ed., *Integrating the Sixties: The Origins, Structures and Legitimacy of Public Policy in a Turbulent Decade* (University Park: Penn State University Press, 1996). See also David Vogel, *Fluctuating Fortunes* (New York: Basic, 1989); James Q. Wilson and John J. Dilulio Jr., *American Government: The Essentials: Institutions and Policies*, 12th ed. (Boston: Wadsworth, 2011). The case for economic deregulation as it appeared at the time is accessible in Paul MacAvoy and John Snow, *Railroad Revitalization and Regulatory Reform* (Washington: American Enterprise Institute, 1977). Robert Hahn et al., *Do Federal Regulations Reduce Mortality?* (Washington: American Enterprise Institute, 2000), gives the economists' view of the opportunity costs of health and safety regulations. David Vogel, *National Styles of Regulation: Environmental Policy in Great Britain and the United States* (Ithaca, NY: Cornell University Press, 1986), provides an international perspective as does Stephen Kelman, *Regulating America, Regulating Sweden: A Comparative Study of Occupational Health Policy* (Cambridge, MA: MIT, 1981).

I have especially benefited from the following modern writings that attempt to make economics relevant to business history and to move beyond the work of Alfred Chandler: Naomi Lamoreaux et al., "New Approaches to the Study of Business History," *Business and Economic History* 26 (Fall 1997): 57–79, and their "Beyond Markets and Hierarchies: Toward a New Synthesis of American Business History," *American Historical Review* 108 (April 2003): 404- 433. See too Naomi Lamoreaux, "Reframing the

Past: Thought about Business Leadership and Decision Making under Uncertainty," *Enterprise and Society* 2 (December 2001): 632–659; Richard Langlois, "The Vanishing Hand: The Changing Dynamics of Industrial Capitalism," *Industrial and Corporate Change* 12 (April 2003): 351–385. That regulatory agencies can make markets work better, I take from Mancur Olson, *Power and Prosperity* (New York: Basic, 2000). See also Omar Azfar, *Market Augmenting Government: The Institutional Foundations for Prosperity* (Ann Arbor: University of Michigan Press, 2003).

There is no good history of the TQM Movement. I consulted the following: W. Edwards Deming, *Out of the Crisis* (Cambridge, MA: MIT, 1986); Olice Embry, "Edwards Deming and the Early Contributions to the Quality Movement," *Essays in Economic and Business History* (1992–1993): 210–217; Richard Hackman and Ruth Wageman, "Total Quality Management: Empirical, Conceptual and Practical Issues," *Administrative Science Quarterly* 40 (June 1995): 309–342; Jeremy Main, *Quality Wars* (New York: Free Press, 1994); Robert Grant et al., "TQM's Challenge to Management Theory and Practice," *MIT Management Review* 35 (January 15, 1994): 25–35; and D. A. Garvin, *Managing Quality: The Strategic and Competitive Edge* (New York: Free Press, 1988). In addition, numerous works in the *Journal of Safety Research*, *Professional Safety*, and *Performance Management Magazine* discuss the relationships between quality and safety.

A number of broader historical works also provide insights into the forces shaping railroad safety and the following have helped structure my thinking: Aaron Wildavsky, "Richer is Safer," *Public Interest* 60 (Summer 1980): 23–39, and his "Why Health and Safety are Products of Competitive Institutions," in *Market Liberalism: A Paradigm for the 21st Century*, ed. David Boaz (Washington: Cato Institute, 1993), 379–387; Barbara Welke, *Recasting American Liberty: Gender, Race, Law and the Railroad Revolution, 1865–1920* (New York: Cambridge University Press, 2001); Joanne Yates, *Control Through Communication: The Rise of System in American Management* (Baltimore: Johns Hopkins University Press, 1989); Constance Nathanson, *Disease Prevention as Social Change* (New York: Russell Sage Foundation, 2007); and John Burnham, "Why Did the Infants and Toddlers Die? Shifts in Americans' Ideas of Responsibility for Accidents—From Blaming Mom to Engineering," *Journal of Social History* 29 (Summer 1996): 817–837. Bruce Seely, *Building the American Highway System: Engineers as Policy Makers* (Philadelphia: Temple University Press, 1987), yields insights into the increasing importance of engineers in molding social policy. Although I make little use of counterfactuals I have benefited from Niall Ferguson, *Virtual History: Alternatives and Counterfactuals* (New York: Macmillan, 1998), and John K. Brown, "Not the Eads Bridge: An Exploration of Counterfactual History of Technology," *Technology and Culture* 55 (July 2014): 521–559.

Writing on risk perception is vast. The classic work is Amos Tversky and Daniel Kahneman, "Judgment under Uncertainty: Heuristics and Biases," *Science* 185, no. 4157 (September 27, 1974): 1124–1131. Useful assessments are Daniel Kahneman et al., eds., *Judgment under Uncertainty: Heuristics and Biases* (New York: Cambridge University Press, 1982); James Flynn et al., *Risk Media and Stigma* (London: Earthscan, 2001); Paul Slovic, ed., *The Perception of Risk* (London: Earthscan, 2000); and Nick Pidgeon et al., eds., *The Social Amplification of Risk* (New York: Cambridge University Press, 2003). I have also learned from Colin Camerer and Howard Kunreuther, "Decision Processes for Low Probability Events: Policy Implications," *Journal of Policy Analysis and Management* 8 (Fall 1989): 565–592; W. Kip Viscusi, *Fatal Tradeoffs: Public and Private Responsibilities for Risk* (New York: Oxford University Press, 1992); Timur Kuran and Cass Sunstein, "Availability Cascades and Risk Regulation," *Stanford Law Review* 51 (February 1999): 683–768, and Cass Sunstein and Richard Zeckhauser, "Overreaction to Fearsome Risks," *Environmental and Resource Economics* 48 (March 2011): 435–449.

Index